# ENERGY AND FEEDSTOCKS IN THE CHEMICAL INDUSTRY

# ENERGY AND FEEDSTOCKS IN THE CHEMICAL INDUSTRY

*Editor:*

**ANDREW STRATTON**
International Consulting Engineer and Applied Systems Analyst
formerly Imperial Chemical Industries
Corporate Research and Technology Department

Published for the
SOCIETY OF CHEMICAL INDUSTRY, London

by ELLIS HORWOOD LIMITED
Publishers · Chichester

First published in 1983 by
**ELLIS HORWOOD LIMITED**
Market Cross House, Cooper Street, Chichester, West Sussex, PO19 1EB, England

*The publisher's colophon is reproduced from James Gillison's drawing of the ancient Market Cross, Chichester.*

**Distributors:**

*Australia, New Zealand, South-east Asia:*
Jacaranda-Wiley Ltd., Jacaranda Press,
JOHN WILEY & SONS INC.,
G.P.O. Box 859, Brisbane, Queensland 40001, Australia

*Canada:*
JOHN WILEY & SONS CANADA LIMITED
22 Worcester Road, Rexdale, Ontario, Canada.

*Europe, Africa:*
JOHN WILEY & SONS LIMITED
Baffins Lane, Chichester, West Sussex, England.

*North and South America and the rest of the world:*
Halsted Press: a division of
JOHN WILEY & SONS
605 Third Avenue, New York, N.Y. 10016, U.S.A.

© 1983 Society of Chemical Industry/Ellis Horwood Ltd.

**British Library Cataloguing in Publication Data**
Energy and feedstocks in the chemical industries.
1. Chemical industries — Energy conservation
I. Stratton, Andrew   II. Society of Chemical Industry
660     TJ163.5C54

ISBN 0-85312-492-2 (Ellis Horwood Ltd., Publishers)
ISBN 0-470-27396-8 (Halsted Press)

Typeset in Press Roman by Ellis Horwood Ltd.
Printed in Great Britain by Unwin Brothers of Woking.

# Table of Contents

# Editor's Preface

This book records the proceedings of an International Conference held in Brussels 15-17 March 1982 by the Society of Chemical Industry in association with the Institution of Chemical Engineers. The objective of the conference was to examine the longer term strategy for energy and feedstocks for the European Chemical Industry. It covered the following aspects,

- the European energy situation, set in a global context;
- the feedstock futures for the chemical industry;
- technological developments.

This introduction surveys the main findings of the papers and suggests areas for future research, development and economic assessment.

## GLOBAL ENERGY FUTURES

Some of the results of a major study by the International Institute for Applied Systems Analysis (IIASA) are illustrated in Chapter 7. The scenarios in the full report should be viewed as a range of possible energy futures. They allow for the higher growth of energy usage in the developing countries and vigorous conservation, and are based on a doubling of world population during 1975-2030. The scenario selected for illustration has the following characteristics.

**Oil** (excluding centrally planned economies)
There is likely to be a continuation and even an increase in oil from the Middle East and North Africa (33 Mbd in 2030).

Unconventional oil (tar sands, oil shales, heavy crudes, enhanced recovery) will start to contribute around 1985 and rise to 45 Mbd in 2030.

Coal liquefaction may start around 2000 and rise to 34 Mbd in 2030.

**Coal**
A fivefold increase in coal is likely from 1981 to 15 000 Mtce/yr (195 Mbd) in 2030, half of which would be liquefied.

The technologies for coal liquefaction are reviewed in Chapters 9 and 10. Currently only the Sasol plants in South Africa are operating on a commercial scale. Four other types of plants with inputs of 100 t/d of coal, or more, are in operation under construction or in preparation. The earliest date for commercial application outside South Africa is considered to be the 1990s but in 2000–2020 coal liquefaction could be a well-established technology.

### Nuclear electricity

It is estimated that an installed capacity of 5000 Gw(e) will be available in 2030, compared with 160 GW(e) in 1981; a primary energy equivalent of 113 Mbd.

With a combination of conventional and Fast Breeder Reactors, the energy available from the world's uranium reserves would be equivalent to 3.5 times the total world fossil fuel reserves (Chapter 5).

### Hydro, solar and other energy

There could be a combined total by 2030 equivalent to 39 Mbd. The rate of development, in particular of large-scale solar power, is limited by the high capital cost (Chapter 7).

## FUTURE PROSPECTS FOR WESTERN EUROPEAN ENERGY

Within the long-term global context, the more immediate prospects for Western European energy are outlined below.

For the EEC over the period 1981–1990, the total energy demand is expected to rise by no more than 25%. Of this increase of 4.6 Mbd, 93% is expected to come from sources other than oil and 58% of this increase from imports, bringing the total EEC dependence on imports to 50% in 1990 (Chapter 1).

### Oil (Chapter 2)

Whilst North Sea oil from all sectors is only 4% of world production, it supplies 30% of European demand and is thus significant in reducing European demand on OPEC, in giving security of supply and in helping to stabilise international supply.

Provided that economic conditions set by various governments are favourable to substantial exploration and development, availability should continue well into the next century.

Currently there is an international surplus of crude and prices are falling. The key to future stability lies in keeping demand on OPEC oil below the maximum that the OPEC members are willing and able to produce. The possibility of another oil crisis will always remain, the probability of occurrence, extent, and timing, being most difficult to forecast.

## Gas (Chapter 4)

The supply prospect from all sectors of the North Sea is a decline of about 20% by the year 2000, from that of 1985-1990 (3.3 Mbd). Whilst some augmentation (25-40% of the decrease in natural gas production) might come from gasifying 30-40 million tons of coal, the bulk of this compensation, together with an increase of 35% in the total Western European market for gas, would have to come from imports. This would raise the proportion of imported gas from 14% to 47% in 2000.

The Soviet Union could become the largest supplier (25% of Western European gas supply) with North and West Africa (19%) and a small contribution from the Middle East (3%).

The price of natural gas is expected to remain oriented to the highest value to the user, relative to the value of the alternative fuels. This is likely to lead to a shift towards the residential and commercial market and away from the use of natural gas in power generation and large industrial steam market.

## Coal (Chapters 1 and 3)

Although there are small reserves in Belgium and France, Western European coal reserves (nearly 13% of the world's total reserves) are concentrated in the United Kingdom and West Germany. Whilst much of the EEC coal-producing capacity is competitive with imported coal, some 15% has production costs more than double the delivered cost of imports.

The cost of supporting this uneconomic tail was $4300 million in 1981 and excludes any real prospect of substantial increase in Community coal production. The expectation is that EEC production will not increase substantially above the present 250 million tonnes a year (3.5 Mbd).

Coal demand in the EEC is forecast to increase by 30% in 1990 but could double by the year 2000. This increase in imports is likely to come mainly from the United States, Australia and South Africa.

The growing world demand for coal, coupled with increased mining and infrastructure costs in the exporting countries, will affect coal prices, which rose from about $38/t in 1978 to about $68/t in 1981 (imported steam coal cif Western Europe). The best insurance against rising import prices is investment in overseas mines, which is the policy of Japan (Chapter 14).

## Electricity

Development of nuclear electricity in Western Europe, as in other parts of the world, has lagged badly behind expectations; France however is an exception. According to current forecasts, EEC capacity should increase by 2.7 times to 110 GW(e) by 1990, and 4.4-6 times by the year 2000. Nuclear could thus be providing the equivalent of 5-6.6 Mbd by 2000, making it the single largest source of additional energy supply in the Community (Chapter 1).

The fuel cost for a nuclear power station is about 25% of the total cost of generating electricity, as compared with 60% for imported coal and 85% for oil. Uranium constitutes only one-third of the fuel cost. Control can thus be established over cost. Although there have been substantial increases in capital cost of plant, mainly to meet safety restrictions, the cost of nuclear power per kWh is about 65% that of coal generation (Chapter 5).

The selling price of electricity to the consumer however, is a complex function of the mix of power stations (nuclear, coal, oil, gas) and their load factors.

Chapter 18 discusses seasonal and daily load variations in demand and how energy management involving both supplier and consumer could be applied to minimise overall system costs.

### New and Renewable Sources

Renewable sources of energy, predominantly hydro or geothermal power, currently provide less than 2% (0.3 Mbd) of EEC total energy supply. The consensus of opinion is that a 7–8% contribution is about the upper limit that can be expected by the year 2000, and even this may be on the high side (Chapter 1).

The options available for North Western Europe are reviewed in Chapter 20. The main sources of renewable energy are likely to be the combustion of industrial and perhaps domestic waste, probable sources are through the supply of geothermal heat and the passive solar design of buildings, and possible sources are through electricity generation from winds and tides. Their use will help to limit rise in energy prices; increased prices however will have to occur for renewables to find widespread application in North Western Europe. Their contribution is unlikely to be big enough to affect any of the main energy markets substantially, before well into the next century.

### ENERGY FUTURES: CONCLUSIONS

Globally, oil and natural gas should be available well into the next century. The amounts could be significantly in excess of current production, albeit at higher cost due to the exploitation of unconventional sources.

The amount of oil and gas, however, will be insufficient to meet the needs of an increasing world population and the developing nations. The contribution expected from renewables is small and to fill this gap a many-fold increase in the production of coal and nuclear energy will be required.

In the more immediate time scale, Western Europe is faced with a decline in indigenous oil and gas. Even the modest forecast increase in energy consumption will have to be met from other sources, of which nuclear electricity is the only indigenous source capable of expansion on the required scale. The security of energy supply will inevitably decrease, and price uncertainty increase with the necessary substantial increase in gas and coal imports. Coal importation will have a higher security than natural gas.

Finally, whilst attention tends to focus on the environmental issues of nuclear power, the increasing use of coal also faces major environmental issues (Chapter 5).

With the development of environmental awareness as a political issue and the interdependence of Western Europe with the Third World (Chapter 6), the realisation of such energy features will face many problems.

## CHEMICAL FEEDSTOCK FUTURES

A very wide range of products are derived from a small range of organic chemical feedstock (Chapter 10). Figure P1 distinguishes between the primary feedstock, as it naturally occurs, and the secondary feedstock, the basic chemicals from which the down-stream chemicals and final products are derived. Synthesis gas, in addition to being the feedstock for methanol, is also the feedstock for

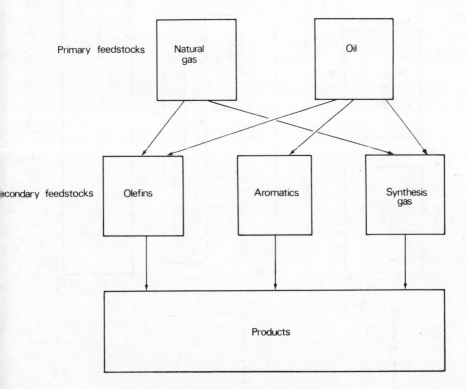

Fig. P1 – Chemical feedstock: The present.

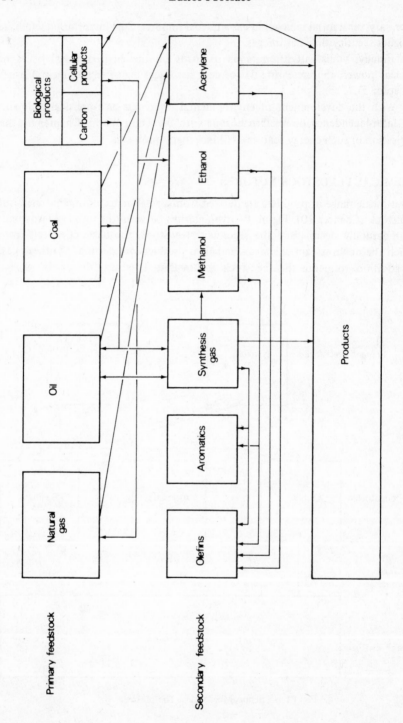

Fig. P2 – Chemical feedstock: Potential future options.

ammonia, in which carbon serves as a chemical reactant for the production of hydrogen (Chapter 8). Currently in Europe, natural gas is the major primary feedstock for synthesis gas, and oil products, in particular naphtha, the major primary feedstock for olefins and aromatics.

Figure P2 shows the potential new options for the future, namely coal and biological products as primary feedstock, and methanol, synthesis gas, ethanol and acetylene as secondary feedstock (for clarity only the additional links are shown).

## Coal

Coal has a complex aromatic chemical structure (Chapter 10). It is deficient in hydrogen compared with liquid fuels and most chemical feedstocks (Chapter 8). The conversion of coal to liquid products involves the breakdown of these complex molecular structures and either the addition of hydrogen or the rejection of carbon (Chapters 10 and 13). Methods adding hydrogen are of two types: direct liquefaction by solvent extraction that leaves most of the aromatic structure intact; and indirect liquefaction via gasification in which the coal is reduced to carbon monoxide and hydrogen (synthesis gas) before catalytic conversion to liquid products.

The direct liquefaction routes (Chapters 9 and 10) differ in procedure rather than principle; they produce an oil which is substantially different from natural crude oil and requires the further addition of hydrogen and refining (Chapter 13).

There are two routes from synthesis gas: direct synthesis by the Fischer-Tropsch process (as used by Sasol); and indirectly through the dehydration of methanol by zeolite catalysis (the Mobil process) (Chapter 10). The Mobil process has a much higher selectivity and produces a higher grade gasoline with a minimum of refining of the synthesis products. The first commercial application will be in New Zealand using natural gas as the feedstock. The Fischer-Tropsch process, however, is more appropriate when a requirement for diesel fuel obtains.

Chapter 13 describes a pyrolysis process in which carbon is rejected and substitutes for coal in a power station linked directly to the conversion plant. The ratio of hydrogen to carbon in the raw oil is much nearer to that of crude oil than the extracted coal liquids (Chapter 13); this will reduce the subsequent refining costs. The full calculation of plant economics is given. If applied to some 98 large coal-fired power plants in Europe, some 400 000–600 000 bbl/day of products would be produced for an additional coal consumption of 55–80 million tonnes of coal.

Whatever the process, however, the products of coal liquefaction have to sell in the international oil market in competition with those of natural or unconventional crude oil. Similarly as chemical feedstock their opportunity cost will be similar to feedstock from other sources of oil. Such differences as exist will be in the detailed economics of conversion; in particular, high aromatic yields can be obtained (Chapter 10).

## BIOLOGICAL PRODUCTS

The production processes for biological products are agriculture, hydrolysis and fermentation, and anaerobic digestion (Chapters 14 and 19). Two process routes then obtain (Fig. P2): the use of the product as a source of carbon for energy/ feedstock (for example, wood or straw); and the extraction of the cellular products as a primary feedstock (methane), a secondary feedstock (for example, ethanol) or an end chemical product.

### Carbon source

The technological routes for the conversion of biological products (biomass) to liquid/gaseous fuels or chemical feedstock parallel those for coal (Chapters 14 and 20), with gasification being the most likely route.

The distributed source, the low energy density (Chapters 14 and 21) and the consequent interaction with transport costs, mean that only small-scale conversion plants can be supported, which would have to compete with world-scale plants operating on natural gas, or coal that have 10–30 times the capacity (Chapters 20 and 21).

The increase in capital cost of the plant, with reduction in size, thus becomes a dominant factor in the economics. If the conventional two-thirds power rule is applied, then, even with zero feedstock cost, ammonia from straw would be more than twice the cost of current commercial prices, delivered to the farm and applied (Chapter 21).

Chapter 20, however, gives the economics of a wood to methanol conversion plant in which the processing cost of $6/GJ (at 5% DCF) is only 50% greater than for a large-scale coal to methanol plant at the same DCF (Chapter 8). With the energy price of wood about twice that of imported coal, this comparison supports the conclusion in Chapter 20 that the most likely use for wood is directly in the heat market, substituting for coal, gas and liquid fuels.

### Methane

The generation of methane by anaerobic digestion of biomass waste is a slow and unstable process (Chapter 19) and the density of the feed to the plant is low; it is of little interest commercially, other than for waste disposal (Chapter 14).

### Ethanol

The dehydration of ethanol to ethylene is a simple low capital process. If ethanol were available at the opportunity cost of gasoline substitution, it would be a highly economic ethylene feedstock, compared with cracking of any hydro-carbon feedstock, including ethane. The key factor in the economics is then the cost and price of ethanol by fermentation. Sugarcane to ethanol at high growth rates, with cheap labour and concentration of the sugar and processing done close to the fields using bagasse as a fuel, is an optimum combination (Chapter 14) and is used in Brazil to produce ethanol as a gasoline substitute. In Western

Europe with about half the agricultural productivity per hectare of the tropics (Chapter 14) and high labour costs and competition with food production, the unsubsidised economics are most unlikely to be favourable.

**Chemical products**
Chapter 21 reviews the unfavourable factors in the economics of a large-scale commercial fermenting plant, in comparison with chemical plants using hydrocarbon feedstock. These are:

- the low concentration of the product stream (3-5%), which requires high capital cost and energy usage to remove the water;
- the low reaction temperature (around 40°C), which means that the exothermic heat of reaction has no practical value and must be dissipated in cooling plant.

Chapter 14 shows how the generalised cost of production of a wide range of products by fermentation increases with decreasing volume production rate of the reaction.

The conclusion is that the future of biotechnology lies in the production of complex high value products and not in meeting future needs for energy and feedstocks.

## SYNTHESIS GAS AND METHANOL

In South Africa (Sasol) the Fischer-Tropsch synthesis of liquid fuels produces appreciable quantities of ethylene as a by-product. Dehydration of methanol by zeolite catalysis can also produce high yields of olefins and aromatics (Chapters 10 and 14). The additional possibility obtains for direct conversion of synthesis gas by zeolite catalysis to give a similar product spectrum to the methanol route (Chapter 10).

The cost of synthesis gas and methanol as feedstocks for olefins and aromatics will be a key factor in the process economics of olefins and aromatics from synthesis gas.

Chapter 12 gives an extensive survey of the economics of the production of synthesis gas from carbon fuels, covering the steam reforming of methane, partial oxidation of methane and residual oil, and coal gasification. The effect of the synthesis gas composition on energy efficiency and economics are shown. The choice of feedstock and process is very much dependent on future relative movement of feedstock costs. Based on anticipated price movements, the conclusion is that in Western Europe partial oxidation of residual oil is the most likely process to be selected, provided that methanol from low cost natural gas sources such as Saudi Arabia does not supply the entire European market. In the United States coal is likely to become the preferred feedstock.

Western Europe is well placed in the technology of coal gasification and Chapter 15 describes development and tests of a version of the Lurgi gasifier, with a higher throughput and ability to handle all sizes of coal.

## Underground gasification of coal

Chapter 17 describes experimental trials that are being prepared on the generation of synthesis gas by underground gasification of coal. The method is designed to operate at great depths with pressures of 20–30 bar, thereby increasing the rate of gasification. The gasification medium is oxygen, $CO_2$ and water. A large surface area of workings is required and the method combines conventional mining to prepare the faces, with boreholes to feed the gasification media and extract the gas. The mining force is estimated at 15–20% of conventional coal mining and the first assessments show a significant advantage even over imported coal.

## The role of hydrogen

Gasification of coal typically incurs capital costs 2.5–3 times those of steam reforming of methane. Whilst some of this cost lies in the extra cost of solid handling, disposal of ash, treatment of effluent and gas cleaning, a comparison of the thermodynamics of the two processes (Chapter 8) shows that the extra capital cost (and lower efficiency) are inherent in the upgrading of coal to a liquid or gaseous fuel through the addition of hydrogen and are determined by the cost and efficiency of power generation from coal.

Production of hydrogen by the carbon–water reaction also adds to the $CO_2$ in the atmosphere which could in the long term be a problem.

One possibility is to use electrolytic hydrogen generation by nuclear (or other non-fossil fuel) electricity (Chapter 8). Another is to use hydrogen from natural gas to gasify coal, with process heat supplied by a high temperature nuclear reactor (Chapter 7).

## ACETYLENE

Before oil became the main feedstock of the chemical industry, acetylene, produced from calcium carbide, was the basis of many chemical processes. After a brief period during which it was produced from natural gas and naphtha, acetylene lost its competitive position to naptha-derived olefins. Chapter 11 reviews the technology for acetylene production and the economics in competition with oil-based ethylene.

A major factor in the economics is the low energy efficiency (30% for carbide); new methods for coal offer the prospect of energy efficiencies of 60–65% and for natural gas and oil feedstocks 70% or more. Lower capacity compared with olefin plants also makes investment and running costs about twice as high.

The advantages of acetylene are two-fold:

- it can be produced from practically any carbon feedstock directly in one process step (including from natural gas);
- in comparison with ethylene, it can be converted to a greater number of chemicals and intermediates with greater ease and with high selectivity.

As an example, at current prices, natural gas–acetylene–VCM has a substantial cost margin (at 25% ROI) over naphtha–ethylene–VCM.

Optimum integration into a chemical complex (for example, to use the large volumes of synthesis gas that some processes generate) and improvements in processes, could make acetylene chemistry, based first on natural gas and in the longer term on coal, economically attractive compared with olefin chemistry.


## THE ROLE OF ELECTRICITY

Chapter 16 discusses the relationship between the electricity and chemical industries and finds that they have much in common. Increased supplies of electricity will displace oil and gas and thus ease demand and benefit the chemical industry. It is necessary however that the technology exists for using electricity effectively and some of the technology that has been developed for other industries is illustrated.

Given nuclear electricity at a cost almost wholly determined by the expected return on capital investment, there is a potential for replacing thermal chemistry by electrochemistry and increasing the use of electricity in chemical processes (Chapter 8).


## FEEDSTOCK FUTURES: CONCLUSIONS

Referring to Fig. P2, the liquefaction of coal (and the development of unconventional oil sources) will be a major factor in increasing world oils supplies and eventually, as the marginal source, stabilising oil prices, albeit at a much higher price than today. For the chemical industry, however, it offers only marginal differences in the cost of producing olefins and aromatics from oil and the likelihood of much closer integration of chemical feedstock production with oil refining.

The use of biological products, particularly in Western Europe, is most unlikely to contribute to the supply of economic chemical feedstock and will be limited to high value speciality products, where there is considerable future potential.

The greatest potential lies in the use of synthesis gas/methanol and acetylene, based on whatever carbon source gives the cheapest final product.

A key factor in reducing the dependence on feedstock cost would be the exploitation of nuclear electricity and high-temperature nuclear process heat to substitute for the use of carbon in the production of hydrogen and process energy. Such developments would be responsive to a perceived need in the future to reduce $CO_2$ emission into the atmosphere.

The development of alternative routes to petrochemicals via synthesis gas and acetylene could extend the threat from Middle East gas to Western Europe, United States and Japan. Much depends, however, on how the price of methanol is determined. If it becomes a major contribution to future liquid fuel supplies, either as a direct substitution for gasoline or diesel, or indirectly by conversion to gasoline through the Mobil process, then the opportunity cost of methanol will be fixed relative to oil products at a level substantially below that of chemical methanol today, and the opportunity cost of synthesis gas will likewise be fixed. The economics of the conversion of synthesis gas or methanol to olefins are susceptible to cost reduction by the development of high-yield catalysts and the product could be significantly cheaper to produce thereby than from cracking oil feedstock [8.2].

Those countries with supplies of coal that are cheap to mine and have low opportunity cost for export (high-ash bituminous and sub-bituminous coal) should, however, be able to produce methanol at a competitive cost for the fuel market, albeit with less margins than the Middle East producer.

The advantage of cheap feedstock is reduced by processes in which capital investment forms a high proportion of the product cost (provided, of course, that the final product is economic). Nuclear electricity and nuclear process heat are particular examples and have the added advantage of high security of fuel supply. With the much higher construction costs that obtain in the Middle East, Western Europe, United States and Japan may then be able to withstand Middle East competition by moving towards high capital intensity processes.

For Western Europe the potential of underground gasification of coal, if realised, could materially improve the competitive position *vis-a-vis* the United States and Japan with its investment in Australian coal; again a matter of investment by Western Europe.

## CONCLUSIONS

Without positive action and response to the challenge, it is unlikely that the heavy chemical industry in Western Europe can survive (Chapter 14). Realisation of the potential for survival will require a concerted effort by the industry, within the framework of long-term policies by governments on energy and large-scale investment (Chapter 8).

The conference has established the constraints and the framework of a sustainable future for the Western European chemical industry. The papers and the findings, however, are applicable to the chemical industry world-wide. Particular areas to which research, development and economic assessment should be directed are:

- synthesis gas and acetylene as chemical feedstocks;
- the reduction in carbon usage in chemical processes;
- the substitution of electro-chemistry for thermal chemistry;
- the efficient use of electricity and nuclear process heat in chemical plant;
- improvements in the economics of biological feedstock, in particular reduced capital cost of small scale plant and improved thermodynamic efficiency.

July, 1982                                          ANDREW STRATTON

# Part 1

# ENERGY: THE GLOBAL SITUATION

*Chairman:* **C. J. Audland,** Director General for Energy,
Commission of the European Communities

# 1

# The energy situation today

C. J. Audland, Director-General for Energy, Commission of the EEC

---

This is an opportune time to interpret the significance of recent developments in the oil markets and to discuss energy business within the framework of the EEC. The following discussion is set in a general context by looking at the overall energy situation in which the EEC finds itself and the main preoccupations for the future to which the Commission is directing the attention of the Council. It is hardly surprising that many of those preoccupations find their echoes in other chapters of this volume.

## THE FALL IN ENERGY AND OIL DEMAND

As we contemplate the present situation it is instructive to look back over all that has happened in the last ten years on the energy scene. In so doing, one is struck by an uncomfortable paradox. Perceptions of the short- and medium-term outlook for the oil markets can change almost overnight, but structural change to diversify energy supply and use takes years to accomplish. This paradox has bedevilled policy-making on energy ever since the first 'oil crisis' of 1973-74.

In the mid- to late-seventies the markets softened, there was growing optimism in some quarters about the prospects for oil supplies and prices and growing complacency about the need for the industrialised countries to reduce their dependence on oil. The in 1979 there was a panic reaction by some consumers to the consequences of the Iranian revolution, prices doubled and prognoses of future oil supplies and prices became gloomy. Now, oil markets are soft again; prices are falling; OPEC is unable to sell as much as it would like; and already there is a growing belief that the risk of a gap between world oil supply and demand is receding into a distant future.

One lesson to be learnt from the past is to be extremely cautious about this growing optimism. It is easy to be mesmerised by the immediate situation and to extrapolate from it. It is easy to ignore or to play down all the uncertainties about the future which may have been increased rather than reduced by recent developments.

How should we then assess the significance of recent events? Firstly, some basic facts.

No one can deny that, in the Community, the fall in energy and oil demand in the past two years has been impressive. Energy demand fell by over 4% in 1980 and by a similar percentage in 1981. Oil demand fell by 8% in 1980 and by about the same in 1981. Both are still falling. Thanks to falling demand for oil and increased production from the North Sea, out net oil imports fell to just over 7 Mbd last year, nearly 20% below the 1980 figure. The comparable figure in 1973 was 12 Mbd.

There has been a similar trend throughout the industrialised world, where oil demand has fallen more quickly than energy demand as a whole.

### The short-term benefits
The softening of the markets and the fall in oil import levels are already bringing benefits to the Community balance-of-payments after the adverse effects of the rise of about 25% in the $US against the ECU last year. That is good news for inflation and for growth in the year or two ahead.

### The longer term worries
But we have not removed the risk of longer term energy constraints on economic growth simply because of a remarkable turnaround on the oil markets or because OPEC appears to be on the defensive. There are four main reasons for caution.

In the first place we can only guess at levels of total world oil supply over the decade ahead. Chapter 2 will clarify whether we should be optimistic or pessimistic about OPEC and non-OPEC supply levels.

In the second, it is exceptionally difficult to forecast likely levels of total world oil demand. A major question mark hangs over the degree of pressure from the developing countries. There is something of a consensus that the developing world will be the single largest − perhaps the only − significant source of new oil demand given the needs of industrialisation and urbanisation. But how much oil will developing countries need? This depends on their prospects for economic growth, and also on their success in increasing domestic energy production and the efficiency of energy use. Major oil importers such as Brazil are already making significant efforts in this respect.

Thirdly, and to bring the question closer to home, no one in the Community can satisfactorily explain just why energy and oil demand have fallen so much more quickly in the recent past than anyone dared predict. Specifically, it is extremely difficult to be certain how far the fall is based on durable economies and durable changes in economic and industrial structures, and how much is due simply to the recession.

Quite clearly there has been something of a break over the past ten years in the link between economic growth and the growth in energy demand. Between 1973 and 1980 energy consumption in the Community was virtually static while GDP grew by around 17%. In 1980 the fall in oil and energy demand occurred while GDP grew by 1.4%. We have also seen a steady fall in the share of oil in total energy consumption (from 62% in 1973 to around 50% in 1981).

This process has been encouraged by the political commitments made within the Community framework, and in the framework of the Tokyo and Venice summits, in 1979 and 1980 respectively. But no one can be sure what will happen when growth picks up again.

Finally, and most importantly, the present market situation could well slow down the process of structural change. The fall in oil prices is already eroding the competitive edge of coal. It will make less attractive the economics of some new energy technologies, and therefore investment in energy saving.

More generally, the signals from the market are confusing to consumers, who may be tempted to 'wait and see' rather than to commit themselves now to modes of energy use which are more rational in a longer-term perspective.

## THE PROSPECTS FOR THE FUTURE

Assuming that economic recovery begins by late 1982 and that there are reasonable rates of economic growth for the rest of this decade (3% a year) — and we can argue about whether that is a sensible hypothesis — the EEC could need up to 15% more energy in 1990 than it consumes today. This figure assumes continuing progress in breaking the link between economic growth and growth in energy demand.

Over the same period we plan to reduce the share of oil in total energy consumption from 50% to around 40%. That means holding oil consumption in absolute terms around its present levels, with the whole of the increase in energy supply coming from non-oil sources. The latest figures on EEC energy balance for 1981 and 1990 are given in Table 1.1.

There are four main non-oil sources, and three of them (coal, gas and nuclear) are considered in more detail in later chapters. The fourth is the group of what are usually described as 'new' energy sources although some of them (hydro and geothermal power) are really rather old. This fourth group of energy sources is discussed later, but first a few words on each of the main non-oil fuels.

### Coal

Coal at present accounts for nearly one-quarter of out total energy demand in the Community, more or less the same share as in 1973, before the first 'oil shock'. Despite a widespread consensus that coal use must be expanded, absolute levels of coal consumption in the community today are almost exactly the same as they were ten years ago.

## Table 1.1

EEC energy balance

*Total consumption of primary energy*
(allowing for stock changes)†

| | 1981 | | 1990 | |
|---|---|---|---|---|
| | Mtoe | % | Mtoe | % |
| Total | 926 | 100 | 1164 | 100 |
| Oil | 459 | 50 | 493 | 42 |
| Solid fuels | 229 | 25 | 288 | 25 |
| Gas | 168 | 18 | 211 | 18 |
| Nuclear | 56 | 6 | 145.5 | 12.5 |
| Hydro, geothermal | 12.5 | 1.3 | 13 | 1.1 |
| New sources | 1.7 | 0.2 | 13.5 | 1.2 |

*Imports*

| | Mtoe | percentage of energy consumption | Mtoe | percentage of energy consumption |
|---|---|---|---|---|
| Oil | 358 | 38.7 | 388 | 33.3 |
| Solid fuels | 42 | 4.5 | 89 | 7.6 |
| Gas | 43 | 4.6 | 101 | 8.6 |
| Others | 2 | 0.2 | | |
| Total | 445 | | 578 | |

*Incremental energy requirements (1981/1990)*

| | Mtoe | % |
|---|---|---|
| Total | 238 | 100 |
| Oil | 34 | 14 |
| Solid fuels | 59 | 25 |
| Gas | 43 | 18 |
| Nuclear | 89.5 | 38 |
| Hydro, geothermal | 0.5 | 0.2 |
| New sources | 11.8 | 5 |

† The sum of domestic production plus net imports.
*Source:* COM(82)326 final, 'Review of Member States' Energy Policy Programmes'.

Only last year Member States were forecasting a 30% increase in coal consumption over this decade. That looks optimistic, particularly if the differential between oil and coal prices continues to be eroded. An increase of perhaps half that amount is now more reasonable, unless very major steps are taken within the Community to enhance coal use by helping to reduce the costs and risks of conversion of plants and the doubts about longer term price relativities.

Increased coal use on this scale would provide one-quarter of our additional energy needs in the rest of this decade.

Community coal production is currently running at about 250 million tonnes a year (177 Mtoe or 3.5 Mbd). Much of our coal-producing capacity is competitive with imported coal or has the potential to become so†. But some 15% of Community coal capacity is extremely uncompetitive, with production costs more than 100% above the delivered cost of imports.

The maintenance of this uneconomic capacity is becoming increasingly expensive. The cost of supporting it has tripled in nominal terms and doubled in real terms since 1975 (from 1350 MECUs $1510 million in 1975 to 3844 MECUs $4290 million in 1981). The weight of this financial burden is such as to exclude any real prospect of a substantial increase in Community coal production. So we shall have to look outside, with perhaps 30% of our supplies in 1990 coming from third countries, compared with around 20% today.

There is plenty of potential export capacity overseas. The bulk of it is the OECD area, and notably in the United States (perhaps exporting net around 85 Mtoe in 1990), Australia (potential exports of 78 Mtoe in 1990) and Canada (5-10 Mtoe)‡.

The essence of Community concern is to ensure that the numerous links in the coal chain between producer and consumer are closed and that the European consumer has therefore increasing assurance about security of overseas supplies. Without it, he will have a further disincentive to convert to coal. This means developing a stable framework for supply, together with our partners in the main exporting countries.

## Gas

Gas currently meets about 18% of our needs for primary energy, a considerably higher share than ten years ago. Gas has been the big growth sector of the past fifteen years. In 1965 Community gas consumption was around 20 billion† cubic metres. Five years later it had quadrupled to 80 billion cubic metres. By

---

† 50–60 million tonnes (20–25%) fully competitive
  140–150 million tonnes (60–65%) only marginally unprofitable
  40 million tonnes (15%) uneconomic.
‡ NB Community coal imports in 1981 were 41 Mtoe (70 million coal tons).
  Figures for United States, Australia and Canada from IEA/SLT (82) 24 of 23 Feb. 1982.
  *Conclusions of the coal review.*
† 1.3 billion cubic metres = 1 Mtoe.

1980, the figure was almost 220 billion cubic metres. This expansion followed the exploitation of the huge Groningen field in the Netherlands and of the 'Southern Basin' resources in the North Sea.

Gas has now reached something of a crossroads. Community gas production is forecast to decline slowly over the decade. But if gas is to maintain or to increase its share of total energy demand, and therefore to make a continuing contribution to our diversification away from oil, its consumption must continue to rise. Once again, the inescapable conclusion is that imports must grow.

Currently gas imports meet 27% of total gas demand. By 1990 they could be up to 50% of the total and they could then meet about 10% of the Community's primary energy needs. That would mean that gas met nearly one-fifth of our additional energy requirements over this decade. The main sources of supply for the moment are Norway, the Soviet Union and Algeria, in order of importance.

It is possible that by 1990 the Soviet Union could overtake Norway as the largest supplier, meeting perhaps one-fifth of total Community gas requirements (domestic and imported supplies together).

This underlines the importance of the approaches which we have suggested, within a Community framework, to maximise security of supply (interruptible contracts, additional storage facilities, increased gas interconnections, maintenance of spare production capacity, et. al.

## Nuclear

In the Community today 16% of electricity and 6% of our total energy is supplied by nuclear energy through a network of power stations with an available capacity of 41 gigawatts. According to the forecasts of Member States this nuclear capacity should increase by 2.7 times (to 110 GW) by 1990. Nuclear energy could therefore be providing the equivalent of up to 3 Mbd (150 million tonnes) of oil in 1990 compared with a little over 1 Mbd (56 million tonnes) in 1981.

These figures may be unduly optimistic and the past history of forecasting in this sector counsels caution. But even if the figure is estimated nearer 2 Mbd than 3 Mbd in 1990, nuclear would still be the single largest source of additional energy supply in the Community during this decade, meeting nearly 10% of our additional needs between now and 1990. So it is very important indeed; and it is very important that momentum should be revived.

The reasons for the generally slow progress so far (France is of course the big exception) are well known. It is partly to do with reduced electricity demand as a result of low economic growth. But it has more to do with opposition to nuclear power from a variety of sources and on a variety of grounds: understandable concerns about the safety of day-to-day operations; the risk of major accidents; terrorism; the risk of nuclear proliferation; and doubts about the macro-economic and employment benefits of such highly capital-intensive technology.

The strength of this opposition has varied and continues to vary in different

Member States. It would be rash to generalise about which way the wind is blowing. There are some signs at least of greater public and political awareness of the potential economic disadvantages to those countries who choose not to go down the nuclear route. But the softening of the oil markets could have the effect of encouraging a number of Community governments to continue to sit on the fence and to put off any clear commitment for a later stage.

The Commission views this with concern and will do all it can to allay public concern, to help resolve problems in the supply of nuclear materials, in the tackling of waste-disposal, and in promoting sensible informed discussion.

### New sources
New and renewable sources currently provide less than 2% of the Community's total energy supply, and the sources are predominantly hydro and geothermal power.

On Member States' forecasts the share could double by 1990, practically all the increase coming from sources such as solar and biomass. The French government is the most optimistic with regard to new and renewable sources. Their new plans for energy supply involve a target of 4-6% for 1990, the bulk of this being biomass (methanol-from-wood plants, biogas digesters, etc.). It remains to be seen whether this is realistic or not.

By the end of the century the contribution should be considerably higher. The consensus of opinion, however, seems to be that a 7-8% contribution is about the upper limit one can expect in the Community as a whole by the end of the century, and even that may be on the high side.

Whether this figure is reached, whether the consensus is unduly pessimistic or optimistic, will depend on the success of R & D, still more on commercial demonstration of technically feasible new processes, and on prices and availability of more conventional fuels as well. The Community's R & D programmes and its programmes of demonstration projects, are therefore very much directed towards developing these new sources, as well as the techniques of coal gasification and liquefaction discussed in later chapters.

New and renewable sources are going to make a small but significant contribution to our total additional energy needs over the next ten years, and increasingly so beyond the end of the decade. But it would be wrong to think that we can jump from a world dominated by so-called 'hard' energy to one of 'soft' energy, over a short or medium timescale.

### CONCLUSIONS
Given here is a very broad-brush picture as seen from the Community perspective, some forecasts and some policy issues which they raise. It is a matter of conjecture as to how correct those forecasts will turn out to be and it remains to be seen how governments and the private investor and consumer will respond to the

energy challenge as it is now posed, and to the specific issues which we have put to the Council. But there is one, final thought about the present situation which is of relevance to the Community's energy strategy as a whole.

Recent events seem to have underlined the mutual interest of both oil consumers and oil producers for greater predictability and stability in oil price movements. In the past two to three years both sides have experienced all the ills that follow violent swings in the pendulum of prices and supplies. With greater stability in prices, we in the industrialised world could ensure steady progress out of oil while avoiding sharp and severe shocks to our balances-of-payments. It would also make it easier for the producers to plan satisfactory depletion policies and the long-term financing of their programmes of economic development.

This is the moment to look carefully at what can be done to improve energy relations with the producers, and to help start a useful dialogue on energy with them and with the oil-importing developing countries, so as jointly to reduce the longer-term risks to the stability and growth of the world economy.

Finally, it may be pertinent to make some comments on petrochemical issues. In past years the petrochemicals industry in the Community has been preoccupied with the problem of high prices for naptha and gasoil feedstock — high both in relation to crude oil and in comparison with the prices paid by its international competitors, particularly in the United States, for oil and gas based materials.

At least on the first point, naphtha versus crude oil prices, I believe there is room for optimism at present. The availability of naphtha from Community refineries should be ample for foreseeable requirements, thanks to a large increase in conversion capacity and the reduction in crude runs. By 1985, we estimate that the ratio of conversion capacity to crude runs will have increased threefold since 1973. On the demand side the situation has also been improved by the increased flexibility of the petrochemicals industry to use a variety of feedstocks, including LPG and gasoil. Moreover we expect the availability of naptha in international trade to rise as a result of the fall in motor gasoline consumption in the United States and elsewhere.

One cannot unfortunately be optimistic about the relative costs of energy and feedstocks in Europe, North America, and the Middle East. Although oil prices have been deregulated, controls in the United States continue to keep the price of natural gas (and ethane in times of surplus) below free market levels; the United States has a natural advantage in cheap coal and exports to Europe will of course bear a substantial freight cost. In Canada all sources of energy will be below import parities for some years at least. The new petrochemicals plants in the Gulf must be expected to enjoy very low prices for energy and feedstock and Saudi Arabia has shown the way with its offer of natural gas at 50 US cents MBtu.

# 2

# Oil: World supplies and North Sea development

**J. M. Raisman**, Chairman and Chief Executive, Shell UK Limited

## INTRODUCTION

One aim of the papers presented in this volume is to provide some insight into the future. This is laudable and has been pursued over the centuries with varying degrees of success: so far as oil forecasting is concerned, the industry's record over the past two decades has not been outstanding. Two brief examples come to mind that have occurred within the last ten years. In 1972 there was virtually unanimous agreement amongst forecasters that oil demand in the United Kingdom would reach 140 million tonnes a year by 1980. In the event, actual demand for oil in that year was just over half that amount. An inevitable consequence of this is the serious problem of surplus refinery capacity we currently face in Britain, and it is no comfort to us that the same situation has been repeated throughout most of Western Europe. The second example relates to the North Sea. In 1974 it was estimated that annual production in the United Kingdom sector would reach 150 million tonnes by 1980. In practice the actual figure was 80 million tonnes – again, about half the forecast amount.

Failures such as these do not mean that attempts to look ahead should be abandoned. Business planning must continue, but on a different basis. In Shell we believe that this should take the form of looking not at a single forecast of the future, but at a range of possible futures against which we can test our projects. With these reservations in mind, I hope this chapter will be of use by looking at the oil scene and attempting to answer the general question as to whether there is likely to be adequate feedstock for the petrochemical industry over the forseeable future. The contribution of North Sea oil is a part of that outlook. Although small in the world context, it is of considerable importance to the countries bordering the North Sea, especially if one takes into account the availability of gas and gas condensates. Gas is a separate, but closely related, subject which is covered fully in Chapter 4 and is here only mentioned in passing.

Any attempt to look ahead must begin with an assessment of the situation today, and what has happened to bring it about. As the Japanese proverb says, you must visit the past to understand the present.

## THE GROWTH IN OIL SUPPLY AND DEMAND

Looking back, the remarkable post-war economic growth was fuelled by oil, and more specifically cheap oil. It may be difficult to recall now that oil was $2 a barrel or less until 1972. Over much of that period oil use grew at a faster rate than economic growth and there were two main reasons for this. Firstly, use of oil displaced large volumes of coal, both for general industrial purposes and for primary electricity generation; and secondly, much of its increased use was in energy-intensive industries such as steel and petrochemicals.

Oil supplied, among other things, the cheap feedstock for the spiralling number of petrochemical plants built to meet the demands of the consumer society that was born in the fifties and flourished in the sixties.

With overall growth rates in chemicals around 15% per annum, some individual products growing even faster, and new markets opening up all the time, the output from these new plants was swallowed up almost as fast as they could be built. Feedstocks, as well as the fuel oil to run the plants, were available at economically attractive prices. The sixties were boom times for all of us and there seemed to be no limits.

Forecasts suggested oil demand would continue to grow at about 6% a year, with Saudi Arabia producing perhaps as much as 1000 million tonnes a year in the 1980's, and every other oil-producing country going flat out to help fuel the enormous oil needs of the consuming countries. Looking back now, with the benefit of hindsight, one can see how unrealistic those forecasts were.

## THE FIRST OIL CRISIS

What really made an energy crisis inevitable was the emergence in the late sixties of the United States as a major importer of oil. After a hundred years of production, the domestic oil industry in the United States finally could no longer keep pace with the huge demand.

This had a fundamental effect on world oil trade generally. The large increase in United States imports added to the already fast-growing import needs of the rest of the developed world. Demand for internationally available oil became, for the first time, greater than the rate at which the oil exporting countries were adding to their production capabilities. Furthermore, governments of consuming countries were taking about ten times as much revenue from oil products as the producers themselves were receiving. OPEC soon realised its underlying strength and flexed its new-found ability to restore this imbalance. It now had all the aces and with the onset of the Yom Kippur war in October 1973, it played them.

From then onwards it became clear that the level of oil availability was not just a matter of price, or reserves or even production potential. The political will to export increasingly became an essential ingredient in the overall picture – as was demonstrated by OPEC's selective supply reductions designed to reduce support for Israel. This led to subsequent misunderstandings about whether 'real' shortages of oil existed then, or were likely to exist in the future.

Most people in the consuming countries saw the 1973–74 shortages as manifestations of physical scarcity and of resources running out, despite the fact that they were short-lived. This perception contributed significantly to the realisation by governments in the consuming countries that some semblance of stability in energy supply could only be restored if dependence upon imported oil could be reduced.

That first oil crisis also saw the quadrupling of oil prices which led to a downturn in economic activity, acute balance-of-payments problems, and recession. As a result, oil demand fell for the first time in many years. At the same time, and as a consequence of the price increases as well as the perceived shortages, consumers began to think in terms of switching to other forms of energy, mainly coal and gas, and of more efficient energy use.

Among those grappling with the higher oil prices was one of the main early warning indicators of economic change – the petrochemical industry. New manufacturing capacity planned towards the end of the 1960s was coming on-stream but new market opportunities were shrinking as the industry moved into maturity and economic growth declined. Greatly increased feedstock and fuel prices added to the problems as more and more products chased stagnating demand.

With demand for oil falling but production returning to its previous levels, it was not long before oil prices began to suffer from the natural laws of the marketplace. For a period in the mid-seventies oil prices actually fell in real terms.

During this period many people, notably in America, came to believe the energy crisis was a myth, and worse, one that was being manipulated by the oil companies. Unfortunately, this swiftly brought signs of a slow-down in the impetus of energy-saving campaigns. The world economy resumed its growth and world demand for oil began to increase once again.


## THE SECOND OIL CRISIS

If this trend had continued it would have led inevitably to the situation where oil demand would have grown to levels that OPEC could not have coped with. Sooner or later, perhaps in the early/middle eighties, crisis point would again have been reached. The events in Iran in 1979 merely brought this crisis forward: they led to another apparent shortage, another round of price increases, and the second so-called 'energy crunch'. Oil prices tripled again, a much worse recession hit many Western countries and oil demand plummeted. Now with fuel switching and energy efficiency even more firmly entrenched in consumer minds, it is inconceivable that oil demand will ever again attain the high growth rates of the sixties and early seventies.

If this look at the past has seemed somewhat lengthy, I make no apology. It is essential that we understand quite clearly what has happened, and what is happening now, before looking at future possibilities and how these may affect us.

## THE MECHANISMS OF OIL SUPPLY AND DEMAND

I referred earlier to the wrong perceptions as to the causes of the shortages of oil in 1973-74. This is a cardinal point because the risk of sudden shortages of oil will remain in the future, liable at any moment to turn our world upside down again. It is important to appreciate the mechanisms that are involved.

The availability of a commodity such as oil will differ depending upon the timescale one is using. In the short- or medium-term the amount of oil available is determined firstly by the capacity of the oil-producing countries to produce and deliver; secondly by their 'political' will to do so; and thirdly, as now, by their economic need to do so. Reserves are not, or are not likely to be, the real problem. So 'availability' will not be constrained by the reserves of oil in the ground. The problem will be the ability and the will of the producing countries to invest in, to install and to utilise new production capacity.

Over the long-term, on the other hand, as reserves are depleted the cost of extraction will tend to increase and this will lead to higher prices. This in turn will reduce demand, so making the reserves last longer as well as providing fresh impetus for the development of alternative energies.

Today's situation provides a good example of the inherent uncertainties facing the oil industry. There is an international surplus of crude and prices have fallen, and continue to fall, as a consequence. This is the market economy at work following the excessive price increases of 1979-80. There has been a collapse in oil demand and individual producing countries are now trying to maintain their exports by adjusting prices downwards to protect their overall levels of incomes.

A key role in this is played by Saudi Arabia with its refusal to reduce its output artificially in order to maintain over-high OPEC prices. What is happening now was, indeed, forecast by Saudi Arabia when some members of OPEC took advantage of the 1979 crisis to push prices up. The present oil suplus and the associated fall in prices is a function of the steady fall in world oil demand, and the difficulty that OPEC countries have in tailoring their production to match it. How long this supply imbalance and its depressing effect on prices will last is difficult to forecast. Nor is it possible to say that prices will not go down further.

But in the long-term the vital ingredients remain the same: the capacity and willingness of OPEC members to produce, rather than the size of their reserves. So, barring exceptional upheavals — and who can ignore the likelihood of these? — the key to stability for both OPEC and the consuming countries is to keep demand on OPEC oil below the maximum that the OPEC members are willing and able to produce. This is often a delicate and difficult balance, although it might not appear so at the present time.

These arguments lead one to the view that there ought to be a natural partnership between oil-consuming and oil-producing countries. Short-term responses to prevailing market conditions, interspersed with 'political' upheavals, are clearly in nobody's interests. The possibility of another oil crisis, however, will always remain. It is the degree of likelihood, its extent and its timing, that is so difficult to determine.

## FUTURE PROSPECTS

To see what the prospects are for future stability, let us now look at some of the other pieces in the puzzle. First, a brief look at international energy demand.

For the OECD countries it is expected that the growth in overall energy consumption will be very small despite, hopefully, a return to slow but steady industrial and economic growth. Increased conservation, more attention to energy efficiency, and the continued decrease in dependence on energy-intensive industries will all combine to ensure that the growth in energy consumption is no longer a simple factor of the growth in industrial output. We shall all be using less energy, and we shall be using it better.

On the other hand, a sharp increase in primary energy demand is expected in the developing countries in the years ahead. Some 600 million people live in countries that have entered the 'take-off' phase of industrialisation and many are well on the way towards establishing an industrial society, particularly in Latin America, the Far East and the Middle East. If a sufficient number of developing nations can attain the status of newly industrialised countries, this could provide a new engine for a major upturn in world economic growth. But it is expected that some newly industrialising countries will develop indigenous energy resources. Biomass and hydropower will have a part to play for some of those who do not have their own oil or gas. Nevertheless, a substantial quantity of their new and rising energy demand will have to be met by oil, and this will add significantly to the call on world supplies.

Overall, it is expected that coal, gas, nuclear and renewable energies such as hydro-electricity will increasingly take over the base load of energy supply in most countries, leaving oil to serve its most valuable and fundamental purpose of providing fuel for transportation and feedstock for petrochemicals. In the light of these considerations, we expect total energy demand in the world outside Communist areas to increase by about 50% by the end the century.

**Future oil demand**
Turning to oil I believe I should start by saying something about energy efficiency.
It is something of a joker in the pack, because it just is not possible as yet to
separate the effects of recession upon oil demand from the effects of energy
conservation. What is clear, however, is that energy efficiency is becoming
increasingly important as more efficient equipment is developed, for example,
the latest motor car engines. The point here is that these developments take
some time to reach the market. For this reason many experts, in my view, are
still understating the effects such developments will have on future oil demand.

As long as OECD countries maintain the pressure for energy conservation
and oil substitution, their demand for oil could be lower by the year 2000 than
it is today. With higher demand from the developing world, we in Shell believe
that the world demand for oil outside the Communist countries might well be
2½-3 thousand million tonnes a year by the year 2000. This compares with a
1980 consumption figure of some 2½ thousand million tonnes and is substantially below the figures that were being projected in the 1970s, when some estimates
for end-of-the-century oil demand went as high as 9 thousand million tonnes a
year!

So how can the oil-producing countries respond to these projected levels
of demand? Today's proven reserves of technically recoverable oil can meet
these needs to well beyond the year 2000. A theoretical decline might be deemed
to set in somewhere about the middle of the first quarter of the next century,
but it takes no account of production from heavy oils, which are the nearest to
conventional crude, or from tar sands or oil shales. So it is possible that there
may be quite a long plateau before the decline in world oil production actually
sets in.

But this is not the end of the question. We have to consider the will to
produce. Even with a demand of some 3 thousand million tonnes a year, demand
on the OPEC producers could, before the end of the century, again be in the
region of vulnerability where a minor upheaval — political or technical — could
cause a recurrence of the type of crisis we saw in 1979. So here is the oil tight-
rope we shall be walking along for the rest of the century at least. If oil demand
is low we might get the sort of price weaknesses we see today and this will test
OPEC's cohesion to the limits. But if oil demand increases, possibly by not more
than 15-20% and fuelled perhaps more by the needs of the developing than the
developed world, we could get instabilities and perceived shortages. It is with
this in mind that I now turn to the contribution of North Sea oil.

**North Sea oil**
North Sea crude from all national sectors currently represents some 4% of total
world oil production, and North Sea gas some 5% of the world's total gas output.
In the world context this is a small contribution.

However, for the United Kingdom, Norway and, to a lesser extent, Denmark, the North Sea offers the important advantages of security of supply and availability until well into the next century − provided economic conditions set by the various governments are favourable to the substantial exploration and development effort needed to achieve this. In the United Kingdom we are seeking adequate understanding by government of the need for such conditions so that work can continue on locating and developing the new oil and gas reserves necessary to maintain self-sufficiency into the next century. These new fields are likely to be smaller than existing fields but they will require more effort and more money than has been expended in the past, not less. If they are not developed, United Kingdom self-sufficiency could be relatively short-lived.

While the United Kingdom's potential for exports surplus to domestic requirements is limited and may be of short duration, the situation in Norway is comparatively favourable. Here there are already substantial proven reserves; production is currently well in excess of national requirements that are in any case comparatively low; and with very large areas still unexplored there may well be further significant discoveries. But there are political constraints and artificial limits set on total production of Norwegian oil and gas. There is a present annual ceiling set by the previous government and this is expected to be maintained. So the pressure on Norway to explore and develop is very limited. Unless present policies are revised, total annual exports of oil are unlikely to exceed 40 million tonnes between now and the end of the century, and the amount of gas is expected to be broadly similar.

The extent to which the United Kingdom and Norway will be prepared to produce over and above their domestic requirements or production ceilings and make further surpluses available for export, depends on many factors. Revenue requirements and the effect on domestic industries are obvious considerations, but much will depend on general perceptions in regard to the whole question of future global resources. A view of a world running out of irreplaceable resources obviously increases the temptation for a country to try to keep its resources to itself. If this tends to inspire or reinforce similar views and actions in other countries, the general development effort is likely to be drastically reduced and the chances of future supply crises are multiplied. So these perceptions of future scarcity become something of a self-fulfilling prophesy, and fortunately we see signs that governments are beginning to recognise this.

Whatever national policies may emerge in the development of North Sea oil and gas, these resources inevitably influence the prospects of future availability in general, even if only by reducing European dependence on OPEC. As such, this helps to stabilise the international supply and pricing outlook.

Total production in the United Kingdom, Norway and Denmark is unlikely to exceed 175 million tonnes a year. And if this is then set against total European demand, which is expected to lie within a range of 5-6 hundred million tonnes a year, it indicates the possible scale of impact − not overwhelming, but significant.

*Pricing policy and the petrochemical industry*

Furthermore, the influence of North Sea oil is unlikely to result in lower oil prices over the long-term. The North Sea price is governed by the world oil market and I believe the long-term trend of prices remains upwards. But the North Sea does offer security of supply for its European customers and it does have an effect on OPEC pricing policy. The United Kingdom operates a unique pricing system since there is a single price for term contracts, effectively set by the government, through the British National Oil Corporation. This can have a considerable influence on the behaviour of other countries — witness the latest round of oil price moves — but it will probably only affect developments in this decade. It is unlikely to be significant in the longer term.

North Sea oil producing countries could opt for cheaper energy and cheaper feedstock for domestic industries, but there is little sense in their doing so. Apart from the loss of government revenues this would entail, there would be other long-term effects. It would encourage consumption, discourage saving and bring forward the day that imports of oil would once again have to be resumed. There would also be a difficult 're-entry' problem as the economy adjusted to the higher prices of these imports. The United States learnt the lessons of a cheap energy policy the hard way, providing an example for others not to follow. The United States chemical industry is still making life difficult for us here in Europe, because of its access to cheap gas. But this situation will end when gas prices are decontrolled, which should happen in the mid-eighties.

So it seems unlikely that the European petrochemical industry will be singled out as an exception to the rule, particularly as cheaper feedstock would have little direct effect on the major problem facing Western countries today: unemployment.

However, the ready availability of ethane and natural gas liquids for the first time from North Sea gas will become increasingly important, as they displace naphtha as a basic feedstock and release it for alternative uses. But while this is important from the point of view of supply security, there is little case for assuming price advantages over other feedstocks, except possibly close to the point of landing, for example, ethane at Mossmorran in Scotland.

Of course, security of supply is a most important factor; with a very high cost of plant there is a lot at stake. But different companies may attach differing degrees of importance to it. Security is clearly worth some insurance premium so far as price is concerned, but much of the time it cannot be very high and sometimes it will disappear altogether.

## SUMMARY AND CONCLUSIONS

So where does this all leave us? It is justifiable to assume that the question of future availability of oil will not be any real problem. In the short- or medium-term there may be extended periods of surplus, punctuated by shorter periods of disruption leading to sharp price increases. But it is hard to see these increases being of the magnitude that occurred in 1973–74 or 1979–80. Any disruptions are likely to be short lived as some kind of balance is restored. After all, supply and demand will always balance in the end.

Although there should be no difficulties in the availability of crude oil, problems could arise in the supply of naphtha if demand for other petroleum fractions, such as fuel oil, were to be very low. But this could be resolved easily by either the oil industry installing additional conversion capacity, or by the chemical industry converting some of their plant to take heavier feedstocks.

By the end of this century, the indications are therefore that there should be plenty of feedstock for the petrochemicals industry, even if demand for chemical products is two or three times that of present levels — which on current estimates seems most unlikely. Furthermore, this does not take into account present or future reserves of gas, and the likelihood of more and more natural gas liquids being used as potential chemical feedstock.

It is therefore the question of feedstock price, rather than availability, which is the real area of concern. Indeed, for the European petrochemicals industry it is more than that: it is the question of their costs compared to those prevailing in oil-rich industrialising countries whose feedstock costs are virtually nil. There are some countries, most notably Saudi Arabia, where cheap domestic energy and feedstock, much of it from gas or gas liquids, are being priced at low levels in order to develop indigenous chemical industries.

It is from these areas that the real challenges for the European petrochemicals industry are likely to come. These new industries may well be given major cost advantages over the European industry which could further constrain its growth and its profitability. But let us not forget the advantages of logistics to the European industry and the technological expertise that has accumulated from fifty years of experience. The North Sea, as I have said, is important, but it is localised and cannot underpin the European chemical industry as a whole, though it may offer advantages to some countries and affect the geographical balance of petrochemical industry economics.

Some critical questions remain. Will the development of alternative feed-stocks, described in later chapters, improve the competitive position of the industry here in Europe? What effect will the short- to medium-term chances of oil price depression have upon these developments? Will they offer improved security of supply? The answers to these questions will depend upon your views of the future.

# 3

# Coal: World availability and future of the European industry

**Dr Gerhard Ott**, General Manager, National Association of the West German Hard Coal Mining Industry

## INTRODUCTION

Until the second half of this century, coal was by far the most important source of energy for Western Europe. Industry in Europe was built on coal throughout the last century and it was coal that fuelled the growth of industry after the Second World War. During the 1960s, however, coal lost a large part of the market as a result of the influx of cheap oil from the Middle East and oil's relative ease in handling. This change was so rapid that by 1973 the EEC's consumption of oil was two and a half times that of coal.

Then came the first oil crisis, shattering the belief that there would be a continuing abundance of cheap oil. The inherent risks of an excessive dependence on oil thus became obvious to all. In this situation, old 'King Coal' won back its attractiveness and there were even voices praising coal as the one solution to the worldwide energy problem. On the other hand, even a short-term fall in oil prices or a temporary oil surplus in Rotterdam seemed to be good reason for others to discard again very quickly the thought of returning to coal. Two realities are clear, however:

● The problems associated with supplying sufficient energy for the future are not temporary — they will continue, and we must learn to deal with them.
● The energy problem is inherently global — no nation is untouched, nor can any act in isolation.

It is in this context that this chapter presents the future of coal — as a growing industry as well as an important supply source for Western Europe.

Since we are dealing with 'the global situation' in this part, the discussion is confined to the overall aspects without going into too many details. The more specific aspects of the use of coal in the European Chemical Industry are discussed and highlighted by experts in subsequent chapters. I shall try, however, to give a realistic picture of what the European Industry may expect from coal in the future and for this purpose I have structured the paper into four parts.

(1) The role of coal in the global context of the world's energy problem.
(2) Western Europe's energy situation and, specifically, its market potential for coal.
(3) Present and future coal supply sources for Western Europe, be it coal from domestic production or coal from the world market.
(4) Conclusions and long-term perspectives.

## THE ROLE OF COAL IN THE GLOBAL CONTEXT OF THE WORLD'S ENERGY PROBLEM

As already mentioned the energy problem is inherently global and no nation's economy is untouched by eventual imbalances on the global scene. Indeed, we are confronted with strong imbalances on the world energy scene which eventu-

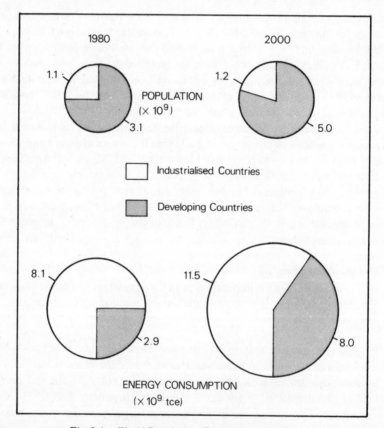

Fig. 3.1 – World Population–Energy Consumption.

ally might even lead to serious and harmful disruptions. It is worth mentioning two of the most serious factors.

(1) As of today one-quarter of the world's population, namely the population of the industrialised countries in the West and in the East, is consuming 75% of the world's total energy production, whereas the other 75% of the world population, namely the population of the developing countries, is left with the remaining quarter of the world's energy production [3.1]. This imbalance certainly shows the need for a better equilibrium. (See Fig. 3.1.)

(2) Another serious factor is the present imbalance between the size of resources of the various fossil energies and the rate at which such sources are consumed:

- Oil and natural gas account for about 20% of currently known geological fossil fuel resources throughout the world, whereas coal accounts for nearly 80% [3.2].
- By contrast, about two-thirds of world consumption of fossil fuels are covered by gas and oil, with coal only accounting for the remaining third. (See Fig. 3.2.)

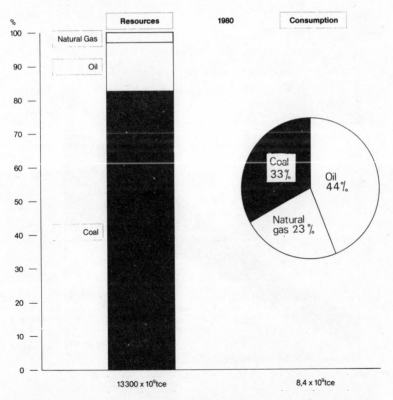

Fig. 3.2 – Total World Resources and Consumption of Coal, Oil and Natural Gas.

In the EEC, coal's contribution is smaller still — only one-quarter. Such an imbalance, it seems, cannot last indefinitely.

As we are faced with a fast growing world population which calls for more and not for less energy, the present imbalanced structure has to be corrected as a matter of urgency if a cutback of the economic and social development also in our part of the world is to be avoided [3.3]. (See Fig. 3.3.)

Naturally, a fundamental change in the structure of energy consumption can neither be achieved overnight nor can it be effected by coal alone. Since, however, all recent studies show that even vigorous conservation, the development and rapid implementation of programmes for nuclear power, unconventional sources of oil and gas, solar energy, other renewable sources, and new technologies will not be sufficient to meet the growing energy needs of the world, a massive effort to expand facilities for the production, transport, and use of coal is urgently required. Otherwise, future energy supply will fall short of requirements. In essence, what is at stake is the future economic growth of the developing, and also of the industrialised world.

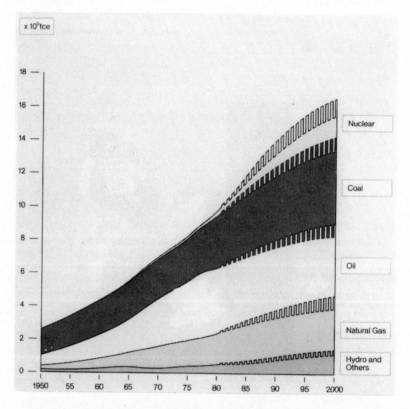

Fig. 3.3 — World Energy Demand–Supply by Sources 1950–2000.

The role of coal is thus destined to become more and more important as being, aside from nuclear energy, the most significant substitute fuel for oil. Numerous reasons indicate that coal will be able to provide an increasing part of the world's energy need:

- First, reserves are abundant. It is estimated that the world has recoverable reserves of 690 billion tonnes, enough for 250 years of consumption at present levels.
- Second, the reserves of coal are widely dispersed, half of them being within the Western sphere of influence.
- Third, coal is in wide use already today. Its technology is known throughout the coal chain, from the mine to the consumer. Coal is a known commodity to the consumer. Over the longer term, new combustion technologies as well as conversion into synthetic gas or liquid fuels will promote greater penetration of coal into other markets.
- Fourth, coal has become more attractive than oil in price and is becoming so in relation to natural gas as well.

On balance, it seems not unrealistic that today's world coal production of some 3 thousand million tonnes could be increased up to 5-6 thousand million tonnes by the year 2000 (Fig. 3.3), if − and this 'if' has to be underlined − action is taken now in order to identify still existing constraints to increased coal production and to develop strategies to overcome them. Some of these constraints are discussed later.

## WESTERN EUROPE'S ENERGY SITUATION AND ITS MARKET POTENTIAL FOR COAL

Clearly, all the factors that characterise the world energy market have direct repercussions on Western Europe, this being particularly true for an industry like the European chemical industry with its numerous worldwide activities and inter-relationships.

The course of action, therefore, seems to be evident also for the EEC. If it wishes to assure its energy supplies in the years and decades to come, a further development of nuclear energy as well as a positive return to coal are essential. However, have we taken this course of action?

- Nine years have now elapsed since the first oil crisis in 1973, yet the EEC's energy supplies still largely depend on hydrocarbons, that is oil and natural gas. They continue to cover two-thirds of total energy requirements. Moreover, the greater part of the oil is supplied by members of OPEC.
- The two major options for reducing such an excessive and dangerous dependence on oil imports are utilised only in a very restrained way. In many

countries of the Community the plans for developing nuclear energy are still being reduced year by year; and the replacement of oil by coal is still moving ahead rather slowly.

Below are given a few key figures on the energy situation of the EEC followed by a closer look into the present and future market potential for coal.

The structure of the primary energy consumption of the Community is characterised by a still very high dependence on imported energies, mainly oil and natural gas:

- Of a total primary energy consumption of 1.3 thousand million tce in 1981, about 48% was covered by oil and about 18% by natural gas.
- About 80% of this oil and more than 25% of the natural gas had to be imported.
- In contrast, only 24% of the total primary energy consumption was covered by coal and only 5% by nuclear energy.
- The overall dependence on imported energies came close to 50% in 1981, compared with 64% in 1973. Certainly some improvement, but not a striking progress.

Interestingly enough, even until the year 2000 the overall dependence on imported energies can hardly be brought below 50%. The main reason for this lies in the fact that the higher contribution expected from nuclear energy and indigenous oil will be balanced by a decrease in domestic natural gas production and increased imports of natural gas and coal.

Since also imported coal, however, can be considered a relatively safe alternative compared with the hazards and risks connected with oil or natural gas imports from OPEC or Eastern Block countries, a higher contribution from coal — domestic and imported — is essential for achieving the aims of at least bringing down the share of oil and covering the additional energy demand by a relatively safe source of energy.

Of course, supply and demand have to be considered together: increased production or imports of coal without markets would make little sense. However, the EEC does offer a large market potential for coal. According to the European Commissions's latest estimates the consumption of coal might almost double between now and the year 2000 (see Table 3.1).

Some of the interesting aspects in more detail are described below.

### The steel industry

The steel industry, although the second largest coal consumer, is a predictable but unexciting market for coal since future growth, if any, will be modest. It should be noted, however, that almost 70% of the coal needed by the European steel industry is provided by domestic production which certainly means a safe supply basis.

**Table 3.1**

Market potential for coal in the European Community

|                      | *1980* million tce | % | *1990* million tce | % | *2000* million tce | % |
|----------------------|------|------|------|------|------|------|
| Total                | 314  | 100  | 360  | 100  | 500  | 100 |
| Power stations       | 184  | 58.6 | 220  | 61.1 | 320  | 64  |
| Steel industry       | 88   | 28.0 | 90   | 25.0 | 95   | 19  |
| Other industries     | 14   | 4.5  | 25   | 6.95 | 55   | 11  |
| Residential, others  | 25   | 8.9  | 25   | 6.95 | 30   | 6   |

*Source:* [3.4]

**Power stations**

Power stations, on the other hand, are clearly a fast growing market for coal which is all the more important since this basic industry is highly vulnerable to any disruptions in the supply of energy. Hopefully, therefore, any imported oil or natural gas will be substituted by coal or nuclear energy in the near future. Although the situation is different from country to country, the prospects look promising (see Table 3.2.)

**Table 3.2**

Electricity generation in the European Community, 1981

|               | *coal* % | *nuclear* % |
|---------------|----------|-------------|
| Denmark       | 87       | –           |
| Great Britain | 74       | 14          |
| Germany       | 59       | 15          |
| France        | 19       | 38          |
| Belgium       | 28       | 25          |
| Ireland       | 16       | –           |
| Netherlands   | 15       | 6           |
| Italy         | 10       | 2           |
| Average       | 43       | 17          |

*Source:* [3.5]

### General heat market

A market with a considerable, but still not sufficiently used potential for coal is the so-called general heat market, namely industries other than steel plus the residential sector. Today, this general heat market, which in many countries accounts for close to 50% of the total primary energy consumption, is still dominated by oil and natural gas [3.6] and Fig. 3.4. This market, therefore, offers a large potential for coal with a variety of applications, such as combined production of heat and power, steam raising, process heat and chemical feedstock.

The fundamental requirement for coal to regain a foothold in this sector is to have available modern and fully developed coal-fired boilers. A considerable proportion of the oil- or gas-fired industrial boilers which are now in service in Europe were installed during the late 1950s and early 1960s. They can no doubt be expected to last another decade or so. The high capital cost of coal-fired boilers is also a disincentive to changing over and, in times of recession and high interest rates, business is not inclined to accept that cost. Thus the desirable rate of replacement will not be achieved without appropriate financial inducements to convert to coal.

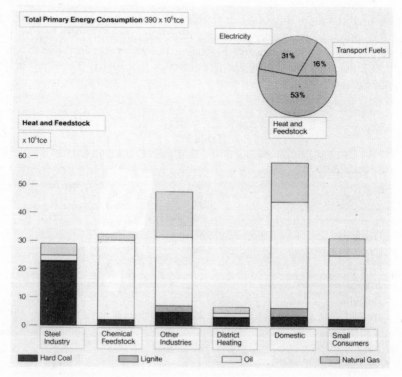

Fig. 3.4 – Energy for Heat and Feedstock Production in the Federal Republic of Germany 1980.

Still, all recent studies forecast that a significant new market may develop during the 1980s and 1990s, and that the use of coal in the general heat market might more than double, from 40 million tce at present to about 80-100 million tce in the year 2000.

**Conclusion**

In conclusion, although the use of coal in the EEC is still limited today, the market potential for a distinctly higher contribution by coal in the future exists, waiting to be developed by coal producers and coal consumers as well. Any potential coal consumer will, of course, want to know whether coal can be considered as a realistic and reliable alternative to oil and gas in terms of availability and price.

**PRESENT AND FUTURE COAL SUPPLY SOURCES FOR WESTERN EUROPE**

The development since the first oil crisis 1973-80, Table 3.3, shows an only modest growth in coal consumption, this being the balance of decreasing domestic production and increasing imports.

Table 3.3
Coal supply sources, European Community

|  | *1973* million tce | % | *1980* million tce | % |
|---|---|---|---|---|
| Community production | 270 | 90 | 247 | 77 |
| Coal imports | 30 | 10 | 74 | 23 |
| Industrialised countries | (14) | 4.7 | (57) | 17.7 |
| Centrally planned economies | (16) | 5.3 | (17) | 5.3 |
| Available | 300 | 100 | 321 | 100 |
| Stock changes | +10 | – | –7 | – |
| Community consumption | 310 | – | 314 | – |

*Source:* [3.7]

**Community production of coal**

The fact that so far Community production of coal has been declining leads to the question of whether and to what extent Western Europe's coal industry will be able to provide a reliable source of supply in the future. The reasons for the decline of Europe's domestic coal production are well known so we will turn to the factors which will determine the future.

*Reserves*
Western Europe has nearly 13% of the world's total reserves, ensuring today's production level for more than 300 years. These reserves are concentrated, as is well known, in West Germany and the United Kingdom, although the smaller reserves of Belgium and France should not be forgotten.

*Coal policy*
On the whole, however, the coal policy of West Germany and Great Britain will be decisive for the future of coal production within the Community. In fact, as a result of the oil crisis, these countries — governments as well as industries — have changed their attitude towards domestic coal.

- The strategy of the United Kingdom coal industry, the largest producer of hard coal in the Community, is based on the National Coal Board's 'Plan for Coal', originally formulated in 1974. The National Coal Board's investment plans aim at an increased coal mining capacity in the years beyond 1985.
- In West Germany, the second largest producer of hard coal in the Community, the coal enterprises have a production, investment and manpower policy designed to enable them to maintain production at its present level and to prepare for an increase in output if required.
- In France, it now seems likely that some increase in the national production target, above the level previously envisaged, will be approved.
- In Belgium, the objective is a modest increase in production above the 1980 figure.

Summing up, there are good prospects that the Community's coal production can be stabilised at the present level on a long-term basis including the possibility of an increase if required by the market.

*Costs*
Talking about the future of coal production in the Community one must not withhold the fact that for reasons of geology, coal in Western Europe can be mined only at higher costs than in many other parts of the world. This, however, should not be regarded as an argument *per se* against Community coal, because the costs of coal mining are not identical with the market prices to be paid by the consumer. Furthermore, Community coal undoubtedly does have a stabilising effect on world coal prices. One could hardly imagine that the replacement of 250 million tonnes of Community coal — or even of a considerable part of it — would not lead to a marked up-surge of world coal prices.

**Coal from the world market**
Even with an indigenous coal production stabilised at the present level or slightly increased, it is evident that more coal from the world market is needed in order to meet the growing demand. Judging from this expected demand, the present

level of coal imports of some 70 million tonnes might well triple by the year 2000.

But how about the actual supply possibilities? From Table 3.4 it seems that with an expected growth of world coal production from some 3 thousand million tonnes today to 5 or even 6 thousand million tonnes by the year 2000, it should be possible to cover an additional Western European demand in the range of some 150 or 200 million tonnes without major problems. There are, however, a number of facts and constraints — economic or political ones — indicating that Western Europe's coal consumers may not expect to get unlimited access to the world coal market free from any uncertainties or risks.

**Table 3.4**
Major coal producing countries, 1980

|  | Production | Exports million tonnes† | Reserves* |
|---|---|---|---|
| United States | 743 | 85 | 190 890 |
| Soviet Union | 606 | 27 | 165 470 |
| People's Republic of China | 606 | 3 | 99 000 |
| Poland | 204 | 33 | 30 600 |
| Federal Republic of Germany | 133 | 22 | 34 536 |
| United Kingdom | 128 | 6 | 45 000 |
| Republic of South Africa | 111 | 29 | 25 290** |
| India | 109 | 1 | 13 134 |
| Australia | 91 | 43 | 36 302 |
| Canada | 33 | 15 | 4 367 |
| Other Countries | 384 | 19 | 42 911 |
| World total | 3 148 | 283 | 687 500 |

*WEC 1980
†Lignite in terms of hce
**Revised estimate: 51 000

*Source:* [3.8]

An analysis of the group of the ten major coal producing countries shows that three of the largest producers, namely the People's Republic of China, the Soviet Union and Poland cannot be expected to offer a large potential for coal exports in the future. For other reasons, the same is true for India. Two more of the major producers, the United Kingdom and West Germany, are Community members.

All analyses indicate, therefore, that the bulk of the growth in exports during this century will be provided by only four countries: The United States, Australia, the Republic of South Africa, and Canada. All of them may be considered as relatively stable, although even for these countries economic or political constraints with respect to higher exports of coal cannot be excluded completely.

## The United States

The United States possesses the largest known deposits of recoverable coal resources in the world, the largest coal-producing industry, relatively the least restrictive environment for entry into all aspects of the coal supply business, and in general one of the safest investment environments in the world today. However, the experiences of 1980 and 1981 emphasised also some weaknesses in the United States coal industry, particularly transportation. Clearly indicated was the need for modernisation and expansion of United States port facilities. But this is a need common to all major competitors in the international market. Other weaknesses tied to United States coal exports, as viewed by the consumer, are the uncertainties of labour stability and, in general, a distinct increase in mining costs.

## Australia

Australia's major strength in the international coal markets rests on abundant coal resources, government and coal industry dedication to the international market, and the country's physical proximity to the Asian markets. Counter-balancing this strength are greater government involvement in the project-approval process and the 'tyranny of distance' the country faces in competing for the European market segment. Also, like the United States, Australia is perceived to possess poor labour–management relations and the need to expand port facilities.

## South Africa

The joint development by government and industry of the Richards Bay coal port and the associated rail facilities have made South Africa a major force in international coal trade. So far, South Africa has proven itself to be a highly reliable, extremely competitive source of coal. Obviously, however, the political uncertainties associated with apartheid rule create an aura of risk that influences the consumer's coal procurement strategies.

Perhaps of even greater influence on the level of South Africa's participation in international coal markets is the government control over exports. South African energy policy places great emphasis on assuring adequate availability of energy to meet future domestic needs. Therefore, coal exports are controlled through export licenses.

Also, South African mining and infrastructure costs will increase significantly. Therefore, the price-related advantages South Africa possessed in the late 1970s will be diminishing.

*Canada*
Of the major traditional international coal suppliers, Canada ranks fourth. As in Australia, coal production in Western Canada (British Columbia and Alberta) is well located to take advantage of the anticipated growth in coal demands in the Pacific Basin, but suffers from a transportation-cost disadvantage relative to the European market. In addition, Canadian mining costs are relatively high, and massive initial infrastructure investment, particularly for rail transportation, is required if coal is to be moved to export markets.

In total, the four major international suppliers are expected to capture about 85% of the incremental growth of total coal trade.

As outlined above, relying on any one of the traditional major international suppliers of coal would involve certain risks and uncertainties, be they the result of labour, politics, or economics. This leads to the conclusion that importers in the 1980s will continue to engage in a risk management strategy of coal procurement that will require diversified sources of coal supplies.

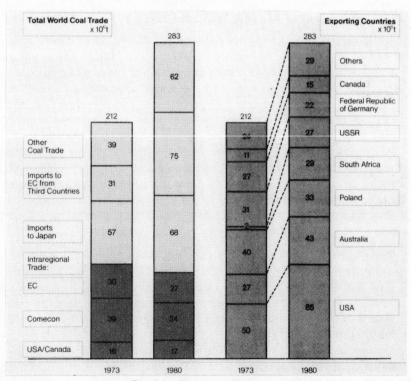

Fig. 3.5 – World Coal Trade.

It should be remembered that world coal trade is still rather limited, [3.9] and Fig. 3.5. At present, only about 200 million tonnes or 7% of the total world coal production are traded intercontinentally. Although an increase in world coal trade over the next two decades up to some 500–600 million tonnes is expected, it must not be overlooked that not only Western Europe, but also other coal importing regions will compete to get their share of this supply. Japan, for instance, draws already today nearly 80 million tonnes from the world coal market. Only recently, it was announced that Japan is planning to quadruple its power station coal consumption.

Finally, the growing demand for coal will also have its effects on coal prices. The development over the last years illustrates this trend. After all, between 1978 to 1981, prices for imported steam coal cif Western Europe increased from about 38 $/t to about 68 $/t, and prices for imported coking coal from about 62 $/t to about 86 $/t.

This leads the discussion back to the importance of Western Europe's indigenous coal industry: there can be little doubt that, in spite of higher mining costs, its importance as a price-stabilising factor will become even more essential, apart from providing a basic supply hardly replaceable by other sources.

## CONCLUSIONS AND LONG-TERM PERSPECTIVES

The world energy outlook has worsened considerably over the past years and we have to face the fact that the world oil market, at best, will be in a very fragile balance over the coming decades. Basic changes in the use of energy are needed if we are to reduce dependence on oil. Stronger conservation measures and determined efforts to accelerate the development of all energy sources are essential to provide the foundation for a sustained and balanced economic growth.

At present, there is a great danger that the effect of the current recession and the temporary surplus of oil will remove the sense of urgency in developing alternative sources of energy. It is critical that the impetus towards reducing the reliance on oil be maintained.

Coal has a particularly important role to play in this process for the following reasons:

● Reserves are abundant and widely dispersed, mostly within the industrialised countries.
● Coal is already widely used for electricity generation and large industrial installations. In the long-term, new combustion technologies and the conversion of coal to synthetic gas and liquid fuel will lead to an even greater market penetration of coal.
● In many areas of the world, coal is now economically competitive with oil in major energy consumption sectors. The competitiveness of coal is likely to improve in the future.

- Many individual decisions must be taken along the chain from coal producer to consumer to ensure that the required amounts are available when needed. Delays at any point affect the entire chain. This emphasises the need for prompt action by consumers, producers, governments, and other public authorities.

Coal demand in the Community, which might double up to the year 2000, can only be met by a combination of indigenous and imported coal. Neither source would be adequate by itself; the two have to be regarded as complementary:

- Western Europe's industry, therefore, will be well advised to stick to indigenous coal as its base of coal supply, at the same time keeping its eyes open for all possibilities of securing additional supplies from the growing world coal market.
- At the same time, Community energy policy must provide for the optimum development and utilisation of both sources of supply, with priority being given to the Community coal mining industry, since no one can fail to realise that the world coal market is subject to risks, as recent experience has shown. We cannot afford to make the same mistake with imported coal that was made with oil, on which Europe became over-dependent.
- The Community coal mining industry is able and willing to make its contribution to greater security of energy supply in the EEC. Its own production, manpower and investment policies are designed to this end, as are its efforts to speed up technical development in coal winning and coal processing and its involvement in commercial coal processing projects and in the world coal market. It will, in this way, be able to play a part in assuring some external coal supplies as a complementary source to indigenous production.

## REFERENCES

[3.1]  United Nations, Department of Economic and Social Affairs *Population Studies* No. 60, World Population Prospects as Assessed in 1977, New York 1977, and Conservation Commission, *World Energy Balance 2000–2020,* London 1981.

[3.2]  World Energy Conference, *Survey of Energy Resources 1980,* and Gesamtverband des deutschen Steinkohlenbergbaus, *Steinkohle 1980/81,* Essen 1981.

[3.3]  Gesamtverband des deutschen Steinkohlenbergbaus, *Steinkohle 1980/81,* Essen 1981.

[3.4]  Commission of the European Communities, Doc. COM (82) 31 final, 10 Feb. 1982.

[3.5]  Statistical office of the European Communities, *Statistical Telegram,* 1982.

[3.6]   Gesamtverband des deutschen Steinkohlenbergbaus, *Steinkohle 1980/81,* Essen 1981.

[3.7]   Commission of the European Communities, Doc. COM (82) 31 final, 10 Feb. 1982.

[3.8]   World Energy Conference, *Survey of Energy Resources 1980,* London 1980, and Statistik der Kohlenwirtschaft eV, *Der Kohlenbergbau in der Energiewirtschaft der Bundesrepublik Deutschland im Jahre 1980,* Essen und Koln 1981.

[3.9]   Gesamtverband des deutschen Steinkohlenbergbaus, *Steinkohle 1980/81,* Essen 1981.

# 4

# Gas: Europe's supplies and the prospects for future growth

**G. Kardaun**, General Manager, Nederlandse Gasunie

## IMPORTANCE OF EUROPEAN GAS INDUSTRY

Natural gas covers an important part of the West European energy requirements (Fig. 4.1). In 1980 natural gas production in Western Europe satisfied 14% of total energy requirements, of which 5% came from the Netherlands and 7% from the other West European countries; 2% of the total West European energy demand was covered in the form of natural gas imports.

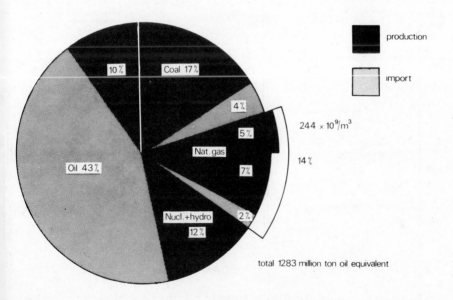

Fig. 4.1 – Gas in West European energy supply, 1980

The future development of the European gas market depends on the following factors:

(1) available reserves and technical production constraints;
(2) the effect of efficiency improvements by existing gas users and the competitiveness of natural gas for potential customers;
(3) export potentials and preparedness to export of countries outside Western Europe;
(4) net-back price to producers in relation to the general energy price level; here the efficiency of the gas use has an important function;
(5) government behaviour in the consuming and producing countries of Western Europe.

A brief analysis of some of these factors follows.

## RESERVE POSITION OF WESTERN EUROPE

### Proven reserves

Figure 4.2 gives our expectations of the total remaining reserves in Western Europe. Until 1974 there was an upward trend in the remaining proven reserves. Since that year, however, the figure has dropped, for one thing because few new fields were discovered, for another because a number of estimates had to be adjusted downward.

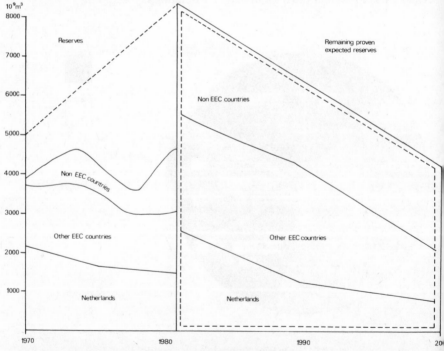

Fig. 4.2 – Development proven and total expected natural gas reserves of Western Europe.

Partly due to the strong energy price increases in 1979 and 1980, the proven reserves could increase again because some already existing reserves became now commercially recoverable. Apart from that there were some impressive findings in the Norwegian part of the Continental Shelf.

In mid-1981, the total remaining proven and recoverable reserves were estimated at approximately $4600 \times 10^9$ $m^3$ of which more than half was located in the North Sea and adjoining waters.

### Additional reserves

There are other reserves which will eventually turn out to be economically recoverable. Our estimate of these reserves, which is based on various sources including the British Brown Paper†, the report of the Norwegian Petroleum Directorate‡, the Netherlands National Geological Service § and others, adds up to nearly $3300 \times 10^9$ $m^3$. About two-thirds of this amount is located in the North Sea (including the Irish Sea). This means that the total quantity of proven recoverable reserves plus the not-proven reserves of Western Europe would

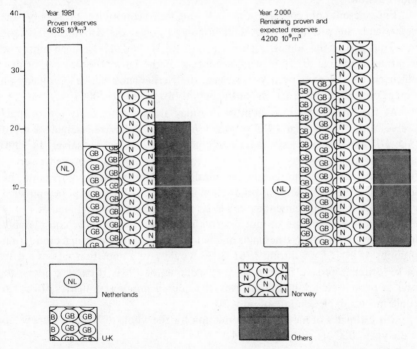

Fig. 4.3 – Shares of main natural gas countries in gas reserves of Wester Europe.

† 'Development of the Oil & Gas Resources of the United Kingdom 1982', report to Parliament by the Secretary of State for Energy.
‡ Annual Report Oljedirektoraten Stavangar.
§ Data collection by Rijks Geologische Dienst as published by the Ministry of Economic Affairs, Natural Gas & Oil in the Netherlands & in the North Sea.

nearly be $8000 \times 10^9$ m$^3$. As Fig. 4.3 shows, total remaining reserves could still amount to $4200 \times 10^9$ m$^3$ in the year 2000, assuming production rates to be as discussed below.

## POTENTIAL PRODUCTION OF NATURAL GAS

Taking into account the total of proven plus additional reserves, as well as a reasonable finding rate for additional reserves, a gas-production for Western Europe of $200 \times 10^9$ m$^3$ in 1990 and $145$-$165 \times 10^9$ m$^3$ in 2000 seems possible.

A somewhat higher production especially towards the year 2000 could be possible. However, technical, financial and policy aspects will limit the rate of depletion of Europe's reserves.

We believe it is not in the interest of the main gas-producing countries in Western Europe to deplete their reserves too fast. Both economic reasons and conservation will play an important role in depletion policy. One of the economic reasons is to spread the financial benefits of the national gas production over a longer period.

For reasons of conservation and long-term continuity of supply, the Netherlands are prudent in not depleting in particular the Groningen field too fast. Thus part of the domestic gas reserves of the Netherlands has been conserved as strategic reserve. That is why, according to the latest Gas Marketing Plan which covers the next twenty-five years, the Netherlands will be producing on average only about $60 \times 10^9$ m$^3$ in the period from 1981 to 2000.

As for Norway, in that country one may also expect a fairly low output/ reserves ratio. In the year 1985 production will not even have reached a level of $25 \times 10^9$ m$^3$, which is less than 2% of that country's proven reserves. In 1990 production might rise to $35 \times 10^9$ m$^3$, which is hardly 3% of proven reserves. The main cause for this lies in technical/economical reasons. As regards the period after 1990, the level of production will depend on the gas policy pursued by the Norwegian government as well as on technical and financial factors.

In Great Britain, the output/reserves ratio is estimated to be considerably higher than that of the Netherlands and Norway. An estimated annual production of about $45 \times 10^9$ m$^3$ during 1981-2000 would give a depletion rate of nearly 6% of Britain's proven reserves at the beginning of 1981. If one considers the total of proven plus potential reserves, the quoted production figures of Britain would mean a depletion rate of only 2%.

Our estimates of natural gas production for the whole of Western Europe up to the year 2000 are given in Table 4.1.

### Table 4.1
Estimates for natural gas production in Western Europe

| 1985 | approx. $200 \times 10^9$ m$^3$ | about | $6.7 \times 10^{18}$ joules |
|------|------|------|------|
| 1990 | approx. $200 \times 10^9$ m$^3$ | | $6.7 \times 10^{18}$ joules |
| 2000 | $145$-$165 \times 10^9$ m$^3$ | | $5.2$-$5.9 \times 10^{18}$ joules |

Figure 4.4 gives the contribution of the different countries to these production rates. Figure 4.5 shows a comparison of forecasts for the West European gas
production by oil companies which are operators in the North Sea.

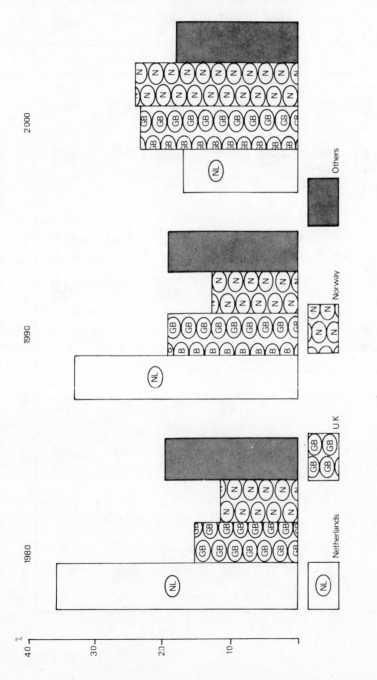

Fig. 4.4 – Shares in natural gas production Western Europe.

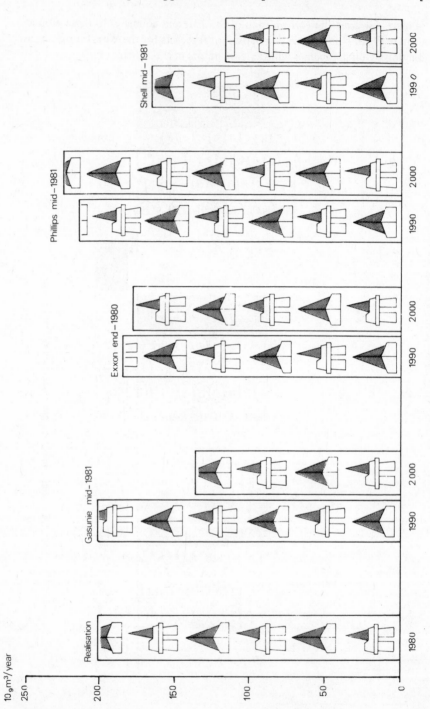

Fig. 4.5 – Natural gas production Western Europe, 1990 and 2000.

**Gas production from coal**
From about 1990 one may reckon with an increasing production of gas from non-conventional sources. Unfortunately, in trying to make a sound estimate of gas production from these sources one finds that many things are much more uncertain than in the case with expected production from proven and additional reserves of natural gas. A number of factors have an influence on the production of gas from coal. These include:

(1)  the need for gas;
(2)  the extent of the current natural gas infrastructure;
(3)  coal availability and cost in relation to other fuels;
(4)  costs of the conversion technologies.

Against this background we have drawn up a very tentative estimate for Western Europe of possible production of coal gas. Both in the United Kingdom and in the Federal Republic of Germany a great deal of effort is going into studies about producing gas from coal. Total production of coal gas may be:

*1990*    3-5 × 10⁹ m³ derived from 6-10 million tons of coal
*2000*    15-20 × 10⁹ m³ derived from 30-40 million tons of coal.

**TOTAL SUPPLY OF GAS IN WESTERN EUROPE**
By analysing the trends and figures we have come up with an approximate fore-cast for the overall gas production of Western Europe (see Table 4.2).

**Table 4.2**
Total gas production in Western Europe

|  | natural gas $10^9$ m³ | gas from coal $10^9$ m³ | total gas $10^9$ m³ |
|---|---|---|---|
| 1985 | 200 | 0.5-1 | 200 |
| 1990 | 200 | 3-5 | 205 |
| 2000 | 145-165 | 15-20 | 160-185 |

This forecast shows that the expected decline in the indigenous natural gas production will, in the 1990s, be compensated 25-40% by the production of coal gas. The share of coal gas in the total gas production could then increase to about 10%.

**Imports of natural gas from outside Western Europe**
The figures in this section should merely be considered to represent an estimate of the potential. In fact, imports will strongly depend on the pipeline price of

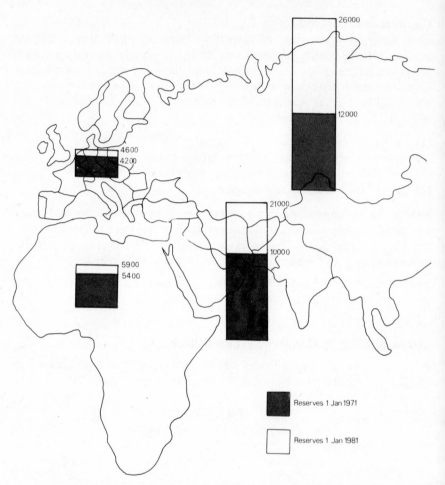

Fig. 4.6 – Proven natural gas reserves in and outside Europe, 1971-1981.

natural gas and on economic and strategic considerations of producing and consuming countries.

The regions from which natural gas is imported or might be imported in the future are: North Africa, West Africa, Soviet Union, The American continent, and The Middle East.

*North Africa*

Taken together the proven reserves of Algeria, Libya, Tunisia and Egypt amount to over $4200 \times 10^9$ m$^3$. Among these countries Algeria has by far the greatest export potential. We estimate the reserves of that country to be over $3300 \times 10^9$ m$^3$, of which about 80% is non-associated gas. It is assumed that about two-thirds of these proven reserves are available for export. This could lead to a potential production of about $75 \times 10^9$ m$^3$/year and a potential export of $50 \times 10^9$ m$^3$/year. Present export commitments for Western Europe amount to approximately $31 \times 10^9$ m$^3$/year and approximately $6 \times 10^9$ m$^3$/year for the United States, apart from the $10 \times 10^9$ m$^3$/year possibly still destined for the customers of El Paso. How far these existing commitments will be extended in the coming years will mainly depend on the natural gas export and price policy of the Algerians and the preparedness of the importing countries to meet the recent Algerian conditions. However, it remains true that from a geographical point of view the West European market is the most logical export market for Algerian gas.

As regards the rest of North Africa, current views are such that no significant expansion of gas exports to Western Europe can be expected. At the moment the contracted export quantity from Libya is over $3.6 \times 10^9$ m$^3$/year. However, the Libyan deliveries to Italy are still interrupted.

### West Africa

The exploration perspectives for this region may be considered as promising. There are vast exploration activities which are now going on in West Africa. Besides Nigeria we should also mention Cameroon and Ivory Coast. A factor which may stimulate exports of natural gas from the most important gas country in the area, Nigeria, is the anti-flaring legislation, which will take effect from 1984. Up to now large quantities of associated natural gas are flared.

On the basis of proven and recoverable reserves now known, one may reckon with an export potential of about $25 \times 10^9$ m$^3$/year from the second half of the nineties. Nevertheless, we fear unavoidable delays for LNG export now the Bonny LNG project in Nigeria is dead. We expect gas from Nigeria will not flow to Europe before 1990. The perspectives for LNG from Cameroon do not look much different. A project of up to $5 \times 10^9$ m$^3$/year LNG exports might be starting in the 1990s.

### Soviet Union

In 1980, $22 \times 10^9$ m$^3$ natural gas was supplied to Western Europe by the Soviet Union. At the moment new contracts have been initiated or signed with Germany, Italy and France for additional supplies. With some other countries negotiations are still going on. In total the Soviet Union currently plans to export to Western Europe an additional volume of $40 \times 10^9$ m$^3$/year. The aim is to start up deliveries by the mid-eighties.

Together with possible expansions within the framework of existing contracts, one may conclude that in the 1990s the Soviet Union could supply about $70 \times 10^9$ m$^3$/year natural gas to Western Europe. The potential however may be higher, perhaps $80\text{-}90 \times 10^9$ m$^3$/year. However, this will also depend on the growth of the giant Russian domestic market of $375 \times 10^9$ m$^3$ in 1980 and the extent to which the desired inland oil consumption of the Soviet Union can be continued to be covered by inland oil production. Assuming an annual gas production increase of 3% per year, and at the same time an annual gas consumption increase of 2.75% (including the other East European countries), an export potential of more than $90 \times 10^9$ m$^3$ for outside Eastern Europe could result at the turn of the century.

Whether Western Europe will be prepared to depend on Eastern Europe for more than 25% of its total natural gas supply and some individual countries for a still greater part, is a different matter altogether. Whether new options will be acquired and taken up will probably strongly depend on prospects offered by other regions elsewhere in the world.

### The American continent

In view of the results of recent exploration work, notably in the north of Canada, a substantial export potential would appear to become available in the 1990s.

If one considers the gas policy pursued by Canada and the United States, exports of Canadian gas in the form of LNG to other markets than the American market may be a possibilty. The distance by sea from the Arctic area of the North American continent to Western Europe is about the same as to the east coast of America. Imports of gas by Western Europe from Mexico however seems out of the question because of the vicinity of the American market to which gas from Mexico can be transported by pipeline. As to possible West European imports from other parts of the American continent, a LNG deal from Trinidad can be considered as a remote possibility only.

### The Middle East

Due to the relation between natural gas reserves in this area and current and expected inland consumption of gas by those countries, the Gulf countries may be considered important future suppliers of gas to Western Europe and Japan. At present the only countries in the area which are actually exporting natural gas, or in which plans for exports exist, are Abu Dhabi and Qatar.

## IMPACT OF GENERAL PRICE LEVEL AND GAS PRICES

The above forecasts of production and import of natural gas depend of course also on the assumed price relationships and the general energy price level. The main assumption for these forecasts was that the natural gas price in the market place will enable a certain further penetration of natural gas in several market sectors. This means that natural gas should remain oriented to the value of the alternative fuels.

For Western Europe this philosophy was generally followed since the beginning of the gas market development. From the very beginning of natural gas sales in the Netherlands, natural gas was considered a very valuable source of energy for several kinds of utilisations. Also for ammonia and methanol production, natural gas is seen as a premium feedstock.

Why do we, as well as other countries, follow this policy of market value and why do we not follow a cost-oriented price policy? The main reason is that if the price were low, there would not be enough gas to fulfill our aim to cover with locally produced indigenous gas all the gas requirements for a given period. There would not exist a strong drive to be efficient with this type of energy. This aim requires a selective use of gas which generally implies that the use of gas has to be restricted to those uses where gas has the highest value for the final user. This means that natural gas should command a certain premium compared with the next best fuel. This premium can be raised by introducing still more efficient gas-using techniques, for instance the high efficient condensating gas boiler. Since we believe the prospects of further increasing supplies at competitive prices are less bright than over the last two decades in most West European countries, this policy will lead via a selective marketing policy to a shift in the natural gas sales towards the residential and commercial market, and at the same to a loss of considerable volume in the power-generation sector and the big industrial steam boiler market. However, this results in a worsening of the market load factor. Of course this sales policy may slightly differ from country to country since the supply conditions as to capacity and flexibility are rather different in the different countries.

If by interference of governments, or because of other constraints, natural gas is not allowed to reach its true market value in the premium sectors the gas market will eventually decrease because it would make new gas projects for exporting countries or producing companies less profitable. Lower market values for gas make some projects in remote areas hardly or not viable, especially now that real oil prices are declining. For this reason and because of the heavy investment, and the financial risks for the financiers, we believe that some of the new gas projects under consideration are becoming uncertain. If c.i.f. prices for gas are requested which, including all costs such as distribution and transportation, exceed the market value of gas, the gas market will further decrease by lack of demand.

So we believe the gas market in Western Europe can only develop further when the gas prices remain competitive, to a certain degree in relation to the next best oil products, taking into account the premiums for natural gas that already exist or can be achieved by new appliances in relation to those products.

## HIGH EFFICIENCY RESIDENTIAL BOILER

The clean fuel natural gas has specific advantages above other fossil fuels like oil and coal. As the total sulphur content of natural gas is much lower than that of oil and coal, the flue gases of gas-fired High Efficiency Residential Boilers can be cooled down below dew-point without causing severe corrosion.

Recently a project-group†, with the aim of examining the effects of flue gas condensate disposal from gas-fired High Efficiency Residential Boilers on the drain system and the environment, has published their report.

The conclusion of these studies were:

- The constructing materials (plastics) currently in use for the sewerage inside and outside a dwelling are corrosion-resistant against flue gas condensate.
- If the flue gas condensate is discharged with domestic sewage water, the impact of flue gas condensate on the drain system is negligible.
- Disposal of flue gas condensate does not harm sewage treatment plants or the surface water.
- From the environmental point of view the gas-fired High Efficiency Residential Boiler is more friendly than conventional ones.

## SECURITY AND FLEXIBILITY OF SUPPLY

A final point is related to the matter of imports: that is the security and flexibility of supply of natural gas.

We have seen above that the countries in Western Europe will gradually become more dependent on gas imports from remote sources and under more stringent conditions. This applies especially to Belgium, West Germany, France and Italy, that is countries which are now mainly importing Dutch gas and using their own indigeneous gas.

More gas from afar may trigger demand for seasonal storage and also for strategic storage. It can be anticipated therefore that the value one can put on security of supply will increase considerably. The problem of security of supply can be met to an only limited extent by providing underground storage because of the high cost.

Apart from this, storage capacity is usually designed to cope with load factor adaptations within a year only, and does not cater for a buffer in case supplies should be interrupted over a longer period. As we have seen before, the future use of natural gas in Western Europe in the public distribution sector is likely to grow more rapidly than in the other sectors, so that the share of the residential and commercial market in the total gas demand in Western Europe will increase. At the same time we have to cope with an increasing share of imports from outside Western Europe. In 1980 such imports accounted for about 14% of the total gas supply. By the year 2000 this percentage might increase to about 50% (see Fig. 4.7). These supplies from remote areas will be imported with a load factor which exceeds by far, the load factor of the residential and commercial market. In general these markets have no dual fuel facilities. The

† **Task** force consists of representatives of: Vereininging von Nederlandse Gemeenten; **Rijksinstituut** voor Zuivering von Afvalawater; Stichting voor onderzoek, beoordeling en kearing van materialen en constructie KOMO; NV Nederlaanse Gasunie.

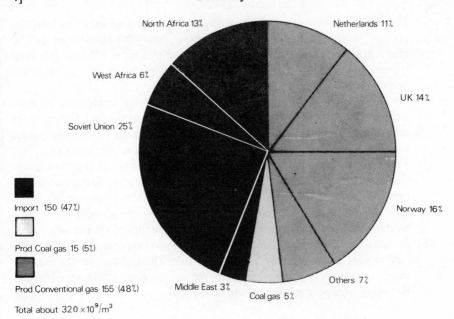

North Africa 13%

Netherlands 11%

West Africa 6%

UK 14%

Soviet Union 25%

Import 150 (47%)

Norway 16%

Prod Coal gas 15 (5%)

Prod Conventional gas 155 (48%)

Middle East 3%

Others 7%

Coal gas 5%

Total about $320 \times 10^9/m^3$

Fig. 4.7 – Natural gas supply Western Europe in the year 2000, $10^9$ m³.

solution of selling interruptible gas to industrial customers and power plants becomes too expensive. This means that security and flexibility of supply will become more and more necessary.

The Groningen field may very well continue to play an important role in this, helping to solve the problem of security and flexibility of supply in the future. As pointed out earlier, this field is playing a central part in the Dutch policy of conservation. Groningen is now – and due to this conservation policy it will remain in the future – the only large field in Western Europe which could perform a load-factor balancing and levelling function. The flexibility in supply which this field offers goes far beyond that achieved by peak shaving. Incidentally, this is being demonstrated already, because our foreign customers tend to make the fullest possible use of the flexibility provided for in our current export contracts.

## SUMMARY

(1) West European natural gas production will remain more or less stable till 1990 at about $200 \times 10^9$ m³/year and then gradually decline to 145-165 $\times 10^9$ m³ in the year 2000.

(2) Gas from coal might compensate by the end of the century for about 25-40% of the decrease in natural gas production within Western Europe.

(3) The share of gas in the residential and commercial market has to increase, because this gas can generate the highest premium value (which can be raised again in these markets by introducing still more efficient gas-using techniques). Volume loss will occur in the power-plant and steam-boiler markets.

(4) On the other hand this trend to the residential and commercial markets creates a serious supply problem: the need for storage and flexibility in general. This problem is worsened by the developments on the supply-side, that is the trend to high load-factor gas imports from remote countries.

(5) Taking all this into account, including the saving potential in gas consumption, we believe that the total West European market for gas could lie between 300 and 350 $\times$ 10$^9$ m$^3$ by the year 2000. This would imply a slight increase in the share of gas in the total primary energy requirements of Western Europe from 14% in 1980 to 15% by the turn of the century.

(6) In spite of the gradual decline of Dutch gas exports after 1990, the Netherlands could still play a significant role. This implies that the risks for other countries importing gas from remote regions, where security and flexibility of supply is lower, could be reduced.

5

# Nuclear energy: The situation and prospects in Europe and the world

**M. R. Carle**, Power Plants Design and Construction Director,
Electricité de France

## WORLD ENERGY REQUIREMENTS

For a quarter of a century, between 1950 and 1975, the world economy experienced a period of economic growth which was exceptional, not only as regards its rate of increase, but also its regularity.

During this period when the exponential function was popular, the population of the planet almost doubled, passing from 2500 to 4000 million inhabitants, and the consumption of energy more than trebled from 2000 million (2 billion) tonnes of oil equivalent (toe) to 6500 million tonnes.

Providing all this energy was apparently a simple matter, as the capacity of the oil industry to meet our needs seemed almost limitless. Its performance is, indeed, remarkable: from an annual production of 500 million tonnes of 'black gold', we have arrived, almost naturally, at more then 2800 million tonnes. Supplies have been multiplied by almost six. Coal, after reigning as sole master of the energy sector during the last century, was suddenly dethroned and condemned almost everywhere to recession.

Since 1974 — the first oil crisis — the situation has, unfortunately, become less simple. The petroleum exporting countries, by altering the prices of oil, which in five years has increased ten-fold in current dollars, remind us that oil reserves, even if they are great, are still limited. These resources must be used sparingly if we are to provide for the needs of the population of the world, which will not easily halt its growth. Within 20 years, by the year 2000, there will be more than 6000 million inhabitants on earth; and 20 years later there will be 8000 million.

The World Energy Conference (WEC) has particularly studied this problem, which is essential for humanity: what will our future energy requirements be, and what resources will it be necessary to use to meet them? The periodic examination of this question is the task of the Energy Conservation Commission.

The last study to date was at the 11th Conference held in Munich in September 1980.

There is a fact (shown in Tables 5.1 and 5.2) of which we are relatively unaware: the disproportion in energy consumption per inhabitant between the

### Table 5.1
Energy geopolitics,
evolution 1950–1975

| | Population (thousand million) | | Energy consumption (toe × 10⁹) | | Average consumption per capita (toe/inhabitant) | |
|---|---|---|---|---|---|---|
| | 1950 | 1975 | 1950 | 1975 | 1950 | 1975 |
| Industrialised countries | 0.8 | 1.1 | 1.8 | 5.0 | 2.3 | 4.5 |
| Developing countries | 1.7 | 2.9 | 0.2 | 1.5 | 0.09 | 0.5 |
| World | 2.5 | 4.0 | 2.0 | 6.5 | 0.75 | 1.6 |

*Source:* World Energy Conference, 1981

### Table 5.2
Consumption levels per inhabitant

| | 1950 (toe) | 1975 (toe) |
|---|---|---|
| *Industrialised countries* | | |
| USA | 5 | 8 |
| UK | 2.9 | 3.7 |
| Germany (GFR + GDR) | 1.7 | 3.7 |
| France | 1.3 | 3.1 |
| EEC (six) | 1.3 | 3.5 |
| Japan | 0.4 | 2.7 |
| USSR | 1.1 | 3.6 |
| Average | 2.3 | 4.5 |
| *Developing countries* | | |
| South America | 0.25 | 1 |
| India | 0.07 | 0.14 |
| Africa | 0.05 | 0.4 |
| China | 0.05 | 0.6 |
| Average | 0.09 | 0.6 |

developed countries (4.5 toe/year on the average) and the Third World (0.5 toe/year for all types of energy; but in fact only 0.3 toe/year if non-commercial forms of energy, such as wood, peat, animal and vegetable waste, are not included). Between 1950 and 1975, the growth trends in commercial energy consumption have indeed differed between industrialised countries and the third World: 4.5% per year for the former, 9% per year for the latter. There is a movement at present in favour of an elementary justice and this will necessarily be continued. If, therefore, for the countries of the Third World, the growth trends we have experienced over the past 25 years continue, the result – coupling this parameter with demographic data – gives an energy requirement for the year 2000 of over 5000 million toe (see Table 5.3). This is near present total world energy consumption. But per inhabitant, the average level of consumption would still only be 1.1 toe/year, much less than that of the average Englishman . . . in 1860!

**Table 5.3**
Energy geopolitics,
prospects 1975–2020

|  | Population (thousand million) | | | Energy consumption (thousand million of toe) | | | Average consumption per capita (toe/inhabitant) | | |
| --- | --- | --- | --- | --- | --- | --- | --- | --- | --- |
|  | 1975 | 2000 | 2020 | 1975 | 2000 | 2020 | 1975 | 2000 | 2020 |
| Industrialised countries | 1.1 | 1.3 | 1.5 | 5.0 | 7.6 | 9.7 | 4.5 | 5.7 | 6.6 |
| Developing countries | 2.9 | 4.8 | 6.8 | 1.5 | 5.3 | 10.4 | 0.5 | 1.1 | 1.5 |
| World | 4.0 | 6.1 | 8.3 | 6.5 | 12.9 | 20.1 | 1.6 | 2.1 | 2.4 |

*Source:* World Energy Conference, 1981

The WEC experts conclude that the industrialised countries must make a serious effort to save energy and slow down the growth in their requirements, to limit their share of world resources so that the Third World may have its fair share. Despite the fact that their estimation was made with this in mind, it nevertheless leads to world energy requirements which will double between now and the year 2000, and treble between now and 2020.

As regards resources, the main fact to be taken into consideration is that it would be unreasonable to count on a substantial growth in oil production (see Fig. 5.1). Unconventional resources (oil shale and sand) can only be utilised progressively; they will help to maintain production more or less at its present level and, at the beginning of the next century, fill the gap left by the inevitable drop in production from conventional oil fields.

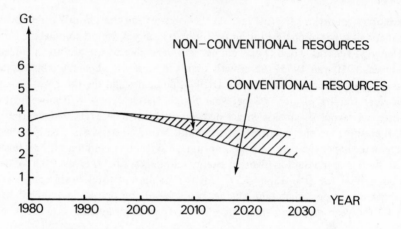

Fig. 5.1 – Maximum oil production capacities according to WEC.

Even if we utilise hydraulic resources still available, and operate renewable energy intensively – giving priority to the sun and the biomass, principally in the Third World – to put them in 2020 in a similar position to that occupied today by coal, if the consumption of natural gas is doubled, and if massive use is made of coal so that its production is raised to three times that existing at present, WEC experts estimate that world production will not be sufficient to meet our requirements. The gap between the two will, in 2020, exceed our present world consumption of oil and gas together! (See Table 5.4 and Fig. 5.2.)

**Table 5.4**
World energy supply, WEC targets in thousand million toe

|                      | 1976 | 2000 | 2020 |
|----------------------|------|------|------|
| *Hydrocarbons*       |      |      |      |
| oil                  | 3.0  | 3.3  | 2.4  |
| gas                  | 1.3  | 2.8  | 2.8  |
| non-conventional     | –    | 0.2  | 1.3  |
| *Solid fuels*        | 1.9  | 3.7  | 5.6  |
| *Renewable energies* |      |      |      |
| Hydro                | 0.4  | 0.8  | 1.3  |
| others†              | 0.7  | 1.0  | 2.1  |
| *Nuclear*            | 0.1  | 1.2  | 4.5  |
| Total                | 7.4  | 13.0 | 20.0 |

† Including non-commercial energies

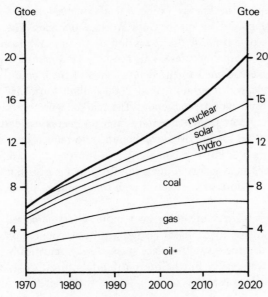

* Including non-conventional resources

Each target assigned to the different energy sources is not in itself binding, but deviation from any source target must be compensated by production from another source in order to maintain the total supply.

### Shares of different sources (%)

|                   | 1976 | 2020 |
|-------------------|------|------|
| Nuclear           | 1    | 22   |
| Renewable energies| 15   | 17   |
| Coal              | 26   | 28   |
| Gas               | 18   | 14   |
| Oil               | 40   | 19   |
| Total             | 100  | 100  |

*Source:* World Energy Conference

Fig. 5.2 – World energy supply, target by energy source.

## THE NEED FOR NUCLEAR ENERGY

Nuclear energy is the only form of energy capable of making sufficient contribution. It appears as the indispensable solution, which, if we are to avoid a catastrophe, it is impossible to ignore; and the initiative lies of course with industrialised countries who alone have the necessary technological capacity.

The studies of the Energy Conservation Commission have also underlined the special part played by oil in Third World economies. The dependence of these countries on oil (with the exception of China) at present exceeds 55% (only 48% for Western countries). This can only be diminished gradually, as oil is particularly well adapted to the requirements of these countries. In the year 2000, they could consume more than 2000 million tons, that is to say about two-thirds of total world production. The market will necessarily experience certain tensions. Apart from elementary humanitarian considerations, this is an additional incentive for industrialised countries to reduce their dependence on this source of energy. But this complicates still further the task, for we are not only to introduce new energy sources to satisfy new needs, but just as much to make a vigorous effort to replace oil with other sources of energy for present uses.

The possible scarcity of oil has been more or less perceived for some time. Indeed, it is also true that nuclear energy was being counted on, consciously or unconsciously, to progressively replace fossil fuels. Scientifically and technically this transition was ready. But before putting it into effect, it was commonly thought that there was sufficient time before us to wait for market conditions to take over.

The two successive oil crises, as well as disorganising our economies and pushing them still further towards a crisis situation, have served to clarify the situation and to accelerate a clear apprehension. The break in oil price increases now makes it essential to carry out, as rapidly as possible, a difficult modification in the way of energy consumption to which we have become all too easily accustomed. This change, for which we were not prepared, must be made in the worst possible conditions, in the context of a general economic and moral crisis. But the financial weight of our oil supplies has become so heavy that in replacing oil, we set up the conditions for restabilisation of our economies.

The provision of 1200 or 4500 million toe in the years 2000 and 2020 using the atom, means that on these dates the installed capacity in nuclear power stations will need to be approximately 850 and 3300 GWe respectively (see Fig. 5.3). At present, this capacity is only 150 GWe.

### Uranium resources

It is normal for us to wonder whether the resources in fissile material which are available are suffient to meet the demands of such rapid growth. A study was made in this respect between 1978 and 1979 under the auspices of the International Nuclear Fuel Cycle Evaluation Conference (INFCE) created by President Carter in 1977. The hypotheses of high- and low-level development of nuclear power in the world chosen at the time agree well with the objectives set by the WEC (see Fig. 5.4). The work of the INFCE can therefore serve as a basis for a precise reply to the question.

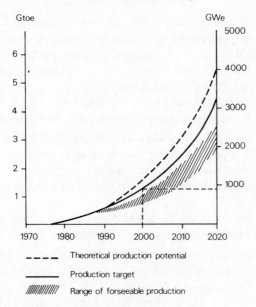

Fig. 5.3 – Nuclear energy, WEC targets.

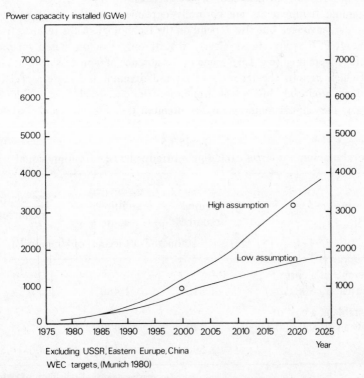

Excluding USSR, Eastern Eurupe, China
WEC targets, (Munich 1980)

Fig. 5.4 – Revised INFCE estimates relative to the installed nuclear power capacities in the world†

The uranium resources which exist in various countries of the world are regularly checked. The results of this investigation are given by the Nuclear Energy Agency(NEA) which has published results regularly since 1965. The last revision to date was in December 1979, and on this occasion it was made in very close collaboration with INFCE group No. 1.

Estimations regarding uranium resources (see Tables 5.5 and 5.6) are classified into three distinct categories corresponding to different degrees of certainty:

(1) Reasonably assured resources are those located in fields which are known at present, and which could be recovered using extraction and treatment techniques which have already been tested.

(2) Estimated additional resources are those whose location is based on direct geological data.

(3) Speculative resources are those whose existence is assumed on the basis of indirect indications and geological extrapolations.

These three categories of resources are of course additive. The first two categories are also subdivided into sections based on estimated costs of extraction. Using only these first two categories, to avoid any overestimation, it may reasonably be estimated that the world will have some 5 million tonnes of metal uranium at its disposal during the next four or five decades.

Uranium requirements are naturally dependent on the range of nuclear programmes envisaged but also depend on the technology of the reactors used.

The INFCE study shows clearly that if only reactors based on present technology were used (e.g. LWR), and in the absence of reprocessing of irradiated fuels (which permits the recovery of unused uranium and more especially of plutonium produced during the irradiation period, which is a usable fissile material), the annual requirements of uranium (see Figs. 5.5 and 5.6) would

### Table 5.5
World uranium resources excluding centrally planned economy countries

|  | Reasonably assured resources | Estimated additional resources | Speculative resources |
|---|---|---|---|
|  | [thousands of tonnes of Uranium] | | |
| Recoverable at a cost of less than 80 $/kg U | 1 850 | 1 480 | |
| Recoverable at a cost of between 80 and 130 $/kg U | 740 | 970 | |
| Total | 2 590 | 2 450 | 6 600–14 800 |

Decreasing degree of reliability of estimates

*Source:* NEA, December 1979

**Table 5.6**

Estimated uranium resources by continent

| Continent | Reasonably assured resources | | Estimated additional resources | | Speculative resources |
|---|---|---|---|---|---|
| | Up to $80/kg U | $80 to $130/kg U | Up to $80/kg U | $80 to $130/kg U | |
| North America | 752 | 224 | 1 145 | 759 | 2.1–3.6 |
| Africa | 609 | 167 | 139 | 124 | 1.3–4.0 |
| Australia | 290 | 9 | 47 | 6 | 2.0–3.0 |
| Europe | 66 | 325 | 49 | 49 | 0.3–1.3 |
| Asia | 40 | 6 | 1 | 23 | 0.2–1.0 |
| South America | 97 | 5 | 99 | 6 | 0.7–1.9 |
| WOCA total (rounded) | 1 850 | 740 | 1 480 | 970 | 6.6–14.8 |

*Source:* NEA/IAEA 'Uranium resources, production and demand', 1979

exceed the maximum production capacity from known resources, and this before the year 2000, whatever the hypothesis of growth used, high or low, for the development of nuclear energy. In cumulative values (Fig. 5.7), the exhaustion of existing resources would be such that the total of reserves of the first two categories would be used up between 2010 and 2020.

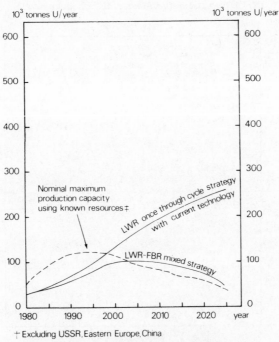

† Excluding USSR, Eastern Europe, China

‡ Reasonably assured and estimated additional resources

Fig. 5.5 – Comparison between annual world supply and demand† of uranium up to 2025, INFCE low assumption (*Source:* INFCE group No. 1).

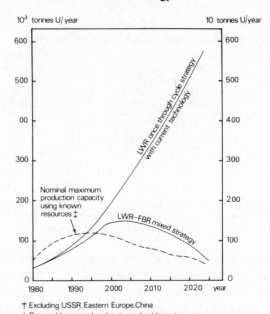

Fig. 5.6 – Comparison between annual world supply and demand† of uranium up to 2025, INFCE high assumption (*Source:* INFCE group No. 1).

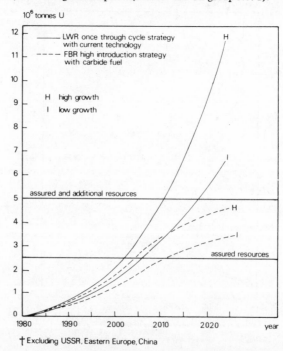

Fig. 5.7 – Cumulative world† uranium demand, 1980–2025, INFCE assumptions.

**The Fast Breeder Reactor**

In order for nuclear energy to be more than a 'flash in the pan' in the history of the world, the treatment of irradiated fuels appears necessary. This is the essential conclusion of the INFCE. Plutonium recycling in light water reactors also is not a sufficient solution to the problem. The use of plutonium in fast reactors, including breeder reactors, is the only way of fundamentally changing the picture.

Great efforts to develop the fast breeder reactor system have been made everywhere for many years. Prototypes and even demonstration power stations are operated in France, the United Kingdom, the Soviet Union, the United States and Japan. Europe — and France in particular — leads the field in this respect. The Phoenix reactor in Marcoule (233 MWe) has now been in operation for more than eight years; it has already produced more than 10 thousand million kWh. A new step will be taken in 1984 when the Super Phoenix, at present under joint construction by EDF (France), ENEL (Italy) and RWE (Federal Republic of Germany), will be commissioned, proving the industrial feasibility of this system. Studies are being actively undertaken with a view to taking at the right moment the next step, which is its commercialisation.

The original feature of fast breeder reactors is that they consume neither natural uranium nor any other rare natural resource, but only depleted uranium, which is discarded by enrichment plants and which has no other use. This is the most important aspect of this system, and perhaps the only one which justifies

**Table 5.7**

Comparison of PWR and FBR strategies from the
point of view of uranium demand,

Balance of material inputs and outputs for 1 GWe in t/year

|  | PWR | Fast breeder reactors Creys-Malville | 1500 MW project |
|---|---|---|---|
| *Inputs* | | | |
| natural uranium | 157 | — | — |
| depleted uranium | — | 26 | 10 |
| *Outputs* | | | |
| depleted uranium | 132 | — | — |
| reprocessed uranium | 22 | 24 | 9 |
| fission products (= energy) | 0.7 | 0.6 | 0.6 |
| plutonium | 0.28 | 0.12 | 0.05 |

The difference between inputs and outputs represents the losses at the various stages of the fuel cycle (fluorisation, fabrication, reprocessing).

the development of new nuclear technology when we already have improved systems at our disposal.

This characteristic can easily be illustrated (see Table 5.7) by comparing the input–output balance of the fast breeder reactor system with that of the light water reactor systems, on the understanding that these balances take into account all the actual operations which the nuclear fuel undergoes between leaving the ore concentration plant and leaving the irradiated fuel reprocessing plant.

The PWR system consumes 157 tonnes of natural uranium per GW each year; it only destroys 700 kg (0.5%) by fission, producing at the same time 280 kg of plutonium. It discards 132 tonnes of depleted unusable uranium at the enrichment plant.

It is this depleted uranium which the fast breeder reactor system consumes, at a rate of 10 tonnes/GW each year in the reactors at present under consideration, once the initial load of plutonium has been introduced. Of these 10 tonnes, only 600 kg are destroyed by fission. Therefore a generating system made of PWRs is able to feed a fast breeder reactor system 13 times larger, without the need to extract or purchase a single extra ton of natural uranium. If this ratio were exceeded, a factor of 10 could be gained in addition by recycling the depleted uranium recovered from the reprocessing of irradiated fuels.

In short, it may be said that the fast breeder reactor is fed by the waste products of the light water system; by this means, uranium ceases to be a scarce resource.

Therefore the limited reserves of uranium can no longer be considered as a simple stand-by for the provision of energy for mankind; they take on a completely new dimension (see Fig. 5.8). Whereas they are equivalent to only 45 thousand million tonnes of coal when used in light water reactors, they have a

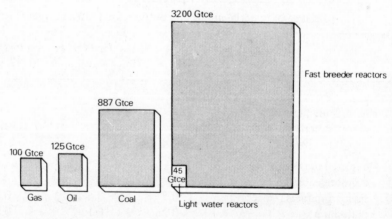

Fig. 5.8 – World energy resources, excluding centrally planned economy countries. (*Source:* WEC, 1980, 'Inquiry into energy resources' and NEA, Activity Report 1980).

potential, when used in fast breeder reactors, of more than $3 \times 10^{12}$ tce. This represents much more than 4 times the total world reserves of coal and 3.5 times the total world reserves of fossil fuels (oil, gas and coal).

Mankind is therefore assured of adequate provision of energy for centuries. We have all the time needed to make breakthroughs in technology and master new energy sources: renewable energies, dry rock geothermics and nuclear fusion.

## FORECASTS OF NUCLEAR POWER STATION INSTALLATION

Since the arrival of electro-nuclear power stations, the atom has already enabled a saving of the equivalent of 1000 million tonnes of oil to be made. At the end of 1981 there were 276 units in operation in the world (23 countries), which had produced during that year more than 800 thousand million kWh, that is more than three times the electricity consumption of France and about 10% of electricity consumed in the world (see Table 5.8). Nearly 200 Mtoe have thus been economised. More than 300 other units, with a total power about double the present installed power, were under construction or ordered (Table 5.9). It may be estimated therefore that in 1990 the nuclear power station capacity in operation will be approximately 450 GWe. This means that the substitution of oil has really been started.

However, the atmosphere is hardly optimistic when we speak today of the development of nuclear energy. Programmes everywhere have been seriously retarded with respect to initial forecasts; and most countries are experiencing great difficulty in removing the obstacles which are blocking the continuation of their projects. So, in fact, the contribution we today expect from the atom is far below what we hoped for only a few years ago, in 1972 just before the Yom Kippur War.

It is useful to examine how estimations have diminished over the years. This is a general phenomenon in all Western countries. But even if lack of data prevents us from studying evolution in Eastern countries, it would appear that they have not been spared either.

In the United States (Table 5.10), the stage has been reached where the estimation made today for 1990 gives an installed power level which was anticipated in 1972 for 1980. Thus in eight years the foreseen programmes have fallen ten years behind! And for the year 2000 the most optimistic estimations hardly correspond to the minimum announced in 1972 for 1985.

A similar pattern is valid for Japan (see Table 5.11). In the short term, the delay seems to be less, only five or six years, as the estimation — an overestimation, it is true — made today for 1990 gives approximately the same figure for installed power as the figure given in 1973 for 1985. But for the year 2000, the present estimations are similar to those made in 1973 for 1990, which means a delay of ten years in the programmes, as in the United States.

**Table 5.8**
World nuclear power plants in operation, situation at 31 Dec. 1981

| Country | Gross installed capacity (GWe) | Number of units | Production (TWh) | |
|---|---|---|---|---|
| | | | *1980* | *1981* |
| USA | 59.9 | 75 | 265 | 289 |
| France | 22.5 | 30 | 61 | 105 |
| Japan | 17.3 | 25 | 83 | 84 |
| USSR | 16.0 | 37 | 65 | 75 |
| West Germany | 9.1 | 14 | 44 | 53 |
| UK | 9.0 | 32 | 38 | 39 |
| Sweden | 6.7 | 9 | 27 | 38 |
| Canada | 5.8 | 11 | 41 | 43 |
| Finland | 2.3 | 4 | 7 | 14 |
| Taiwan | 2.3 | 3 | 8 | 11 |
| Switzerland | 2.0 | 4 | 14 | 15 |
| Spain | 2.0 | 4 | 5 | 9 |
| Belgium | 1.8 | 4 | 13 | 13 |
| East Germany | 1.8 | 5 | 12 | 12 |
| Italy | 1.5 | 4 | 2 | 3 |
| Bulgaria | 1.3 | 3 | 6 | 6 |
| Indian Union | 0.9 | 4 | 3 | 3 |
| Czechoslovakia | 0.8 | 2 | 5 | 6 |
| Yugoslavia | 0.7 | 1 | – | – |
| South Korea | 0.6 | 1 | 4 | 3 |
| Netherlands | 0.5 | 2 | 4 | 4 |
| Argentina | 0.4 | 1 | 2 | 3 |
| Pakistan | 0.1 | 1 | – | – |
| Total (World) | 165.3 | 276 | 710 ~175 Mtoe | 828 ~200 Mtoe |

*Source:* EDF Construction Department

Table 5.9
Nuclear power plants in the world, situation at 1 Aug. 1981

| Country | Units in operation | | Units under construction or ordered | | Total | |
|---|---|---|---|---|---|---|
| | Number | Power (GWe) | Number | Power (GWe) | Number | Power (GWe) |
| South Africa | — | — | 2 | 1.9 | 2 | 1.9 |
| East Germany | 5 | 1.8 | 12 | 5.3 | 17 | 7.1 |
| West Germany | 14 | 9 | 16 | 18.7 | 30 | 27.7 |
| Argentina | 1 | 0.4 | 2 | 1.4 | 3 | 1.8 |
| Belgium | 4 | 1.7 | 4 | 4 | 8 | 5.7 |
| Brazil | — | — | 3 | 3.3 | 3 | 3.3 |
| Bulgaria | 3 | 1.3 | 2 | 1.4 | 5 | 2.7 |
| Canada | 11 | 5.8 | 15 | 11.1 | 26 | 16.9 |
| South Korea | 1 | 0.6 | 8 | 7 | 9 | 7.6 |
| Cuba | — | — | 2 | 0.9 | 2 | 0.9 |
| Spain | 4 | 2.1 | 16 | 15.7 | 20 | 17.8 |
| USA | 73 | 57.5 | 98 | 113.4 | 171 | 170.9 |
| Finland | 4 | 2.3 | 1 | 1 | 5 | 3.3 |
| France | 29 | 21.8 | 29 | 33.1 | 58 | 54.9 |
| Hungary | — | — | 5 | 2.8 | 5 | 2.8 |
| India | 4 | 0.9 | 4 | 0.9 | 8 | 1.8 |
| Italy | 4 | 1.5 | 5 | 4 | 9 | 5.5 |
| Japan | 23 | 15.7 | 14 | 12.5 | 37 | 28.2 |
| Libya | — | — | 1 | 0.3 | 1 | 0.3 |
| Mexico | — | — | 2 | 1.4 | 2 | 1.4 |
| Pakistan | 1 | 0.1 | — | — | 1 | 0.1 |
| Netherlands | 2 | 0.5 | — | — | 2 | 0.5 |
| Philippines | — | — | 1 | 0.6 | 1 | 0.6 |
| Poland | — | — | 2 | 0.9 | 2 | 0.9 |
| Rumania | — | — | 2 | 1.1 | 2 | 1.1 |
| UK | 32 | 9 | 10 | 6.6 | 42 | 15.6 |
| Sweden | 9 | 6.7 | 3 | 3.2 | 12 | 9.9 |
| Switzerland | 4 | 2 | 3 | 3.1 | 7 | 5.1 |
| Taiwan | 3 | 2.3 | 3 | 2.9 | 6 | 5.2 |
| Czechoslovakia | 2 | 0.8 | 11 | 5.4 | 13 | 6.2 |
| Turkey | — | — | 1 | 0.4 | 1 | 0.4 |
| USSR | 35 | 14.6 | 49 | 49.7 | 84 | 64.3 |
| Yugoslavia | — | — | 1 | 0.7 | 1 | 0.7 |
| World | 268 | 158.4 | 327 | 314.7 | 595 | 473.1 |

*Source:* EDF Construction Department

**Table 5.10**

Prospects for nuclear power development (in GWe), evolution of forecasts from 1972 to 1981, USA

| Total installed capacity planned for | Forecast year | | | | | | | | |
|---|---|---|---|---|---|---|---|---|---|
| | 1972 | 1974 | 1975 | 1976 | 1977 | 1978 | 1979 | 1980 | 1981 |
| 1980 | 127–144 | 85–112 | 85–112 | 60–71 | 60–71 | 62–66 | 56.7 | 55 | 53.3 |
| 1985 | 256–332 | 231–275 | 231–275 | 127–166 | 127 | 100–123 | 95–120 | 90–120 | 98.3 |
| 1990 | 412–602 | | 410–575 | 195–290 | 195–210 | 158–193 | 129–155 | 125–135 | 115–144 |
| 2000 | | | | 380–620 | 300–400 | 256–396 | 200–300 | 185–250 | 120–230 |

*Source:* AEC(F), Programs Department

**Table 5.11**

Prospects for nuclear power development (in GWe), evolution of forecasts from 1972 to 1981, Japan

| Total installed capacity planned for | Forecast year | | | | | | | | |
|---|---|---|---|---|---|---|---|---|---|
| | 1973 | 1974 | 1975 | 1976 | 1977 | 1978 | 1979 | 1980 | 1981 |
| 1980 | 32 | | | | | | | | 14.5 |
| 1985 | 60 | | 49 | 35 | 33 | | 25–30 | | 22.8 |
| 1990 | 100 | | | | 60 | | 53 | | 36–51 |
| 2000 | | | | | | | | | 65–100 |

*Source:* AEC(F), Programs Department

**Table 5.12**

Prospects for nuclear power development (in GWe), evolution of forecasts from 1972 to 1981, Western Europe

| Total installed capacity planned for | Forecast year | | | | | | | | |
|---|---|---|---|---|---|---|---|---|---|
| | 1972 | 1974 | 1975 | 1976 | 1977 | 1978 | 1979 | 1980 | 1981 |
| 1980 | 94 | 92 | 84 | 74 | 70 | 58 | 48 | 48 | 41.9 |
| 1985 | 204 | 220–226 | 198–206 | 167–181 | 123–130 | 100–109 | 98 | 98 | 95.3 |
| 1990 | 410 | | | | 201–239 | 160–193 | 151–171 | 140–160 | 131–156 |
| 2000 | | | | 312–344 | | 266–398 | 257–335 | 240–300 | 229–302 |

*Source:* AEC(F), Programs Department

**Table 5.13**

Evolution of installed capacity forecasts, market economy countries

| Total installed nuclear capacity (GWe) | Forecast date (year end) | | | | | | | | |
|---|---|---|---|---|---|---|---|---|---|
| | 1972 | 1974 | 1975 | 1976 | 1977 | 1978 | 1979 | 1980 | 1981 |
| For 1980 | 267–290 | 225–290 | 200–230 | 165–176 | 158–169 | 146–151 | 128 | 126 | 115 |
| For 1985 | 580–650 | 570–630 | 530–600 | 370–428 | 315–329 | 259–293 | 245–270 | 230–260 | 229 |
| For 1990 | 1050–1250 | | | 692–819 | 536–608 | 429–532 | 385–441 | 355–405 | 332–364 |
| For 2000 | | | | | | 800–1126 | 680–972 | 590–850 | 480–640 |

*Source:* Atom Energy Commission (France), 1980

In Western Europe (Table 5.12), the programmes have been cut back even
further, as the power today anticipated for the year 2000 is, under the best
possible conditions, three-quarters of that announced in 1972 for 1990.

For all the countries governed by a market economy the installations in
1980 respresent, in fact, less than half what was announced in 1972 for that date
(see Table 5.13). The contribution today estimated for 1990 is about a third;
and for the year 2000 only a half of what was announced for 1990 is anticipated
today.

To date, for the world as a whole, including Eastern countries, the installed
power capacity at the end of the century (see Table 5.14) could be between 750
and 1000 GWe, and excluding Eastern countries between 500 and 750 GWe.
The WEC objective of 850 GWe is therefore far from being achieved.

**Table 5.14**

Nuclear energy, prospects of development for installed capacity in the world
(in GWe)

|                            | 1980  | 1985  | 1990      | 2000     |
|----------------------------|-------|-------|-----------|----------|
| *Industrialised countries* |       |       |           |          |
| EEC (ten)                  | 32.4  | 73.8  | 107–123   | 165–225  |
| Europe outside the EEC     | 9.5   | 21.5  | 29–33     | 41–59    |
| USA                        | 53.3  | 98.3  | 125–135   | 155–195  |
| Canada                     | 5.5   | 10.4  | 15        | 20–27    |
| Japan                      | 14.5  | 22.8  | 36–51     | 65–100   |
| Other market economy       |       |       |           |          |
| countries                  | –     | 1.8   | 3         | 5–15     |
| USSR                       | 10.8  | 31.6  | 62–83     | 165–200  |
| Comecon countries          | 3.3   | 13.2  | 22–35     | 78–94    |
| Subtotal                   | 129.3 | 273.4 | 399–478   | 694–915  |
|                            |       |       |           |          |
| *Developing countries*     |       |       |           |          |
| North Africa               | –     | –     | 0.3       | 6–12     |
| Middle East                | –     | –     | 0.4       | 3–10     |
| Southern Asia              | 0.7   | 1.3   | 2–3       | 7–15     |
| South-east Asia            | 1.8   | 8.5   | 15–19     | 32–79    |
| Latin America              | 0.3   | 2.8   | 4–6       | 13–32    |
| Subtotal                   | 2.8   | 12.6  | 21.7–28.7 | 61–148   |
|                            |       |       |           |          |
| World (rounded)            | 132   | 286   | 420–500   | 750–1050 |

*Source:* AEC (F), Programs Department, situation at 1 July 1981

Within this figure, the ten countries of the EEC will represent about 200 GWe (see Table 5.15). This is exactly what in 1974, after the first oil crisis, the EEC Commission had as the objective to be achieved by 1985 for the nine member countries (excluding Greece).

**Development of the French nuclear industry**
Going further into the analysis, we note that France alone represents about half this estimate.

### Table 5.15
Prospects of development for nuclear installed capacity in Western Europe
(GWe)

|  | 1980 | 1985 | 1990 | 2000 |
|---|---|---|---|---|
| *EEC* | | | | |
| West Germany | 8.6 | 17.7 | 25.4 | 50–58 |
| Belgium | 1.7 | 5.5 | 6.5 | 6.5–10.5 |
| Denmark | – | – | – | 0–1 |
| France | 12.8 | 36.8 | 56 | 86–106 |
| Greece | – | – | 0.6 | 1.5–3.5 |
| Ireland | – | – | – | 0.5–1.5 |
| Italy | 0.6 | 1.5 | 5.5 | 13–20 |
| Luxembourg | – | – | – | – |
| Netherlands | 0.5 | 0.5 | 0.5 | 0.5–3.5 |
| UK | 8.2 | 11.8 | 15.1 | 23–39 |
| Subtotal | 32.0 | 74.0 | 110.0 | 180–240 |
| *Europe outside EEC* | | | | |
| Austria | – | – | – | – |
| Spain | 1.1 | 7.4 | 15.5 | 22–30 |
| Finland | 1.8 | 2.2 | 3.2 | 3.2–5.2 |
| Norway | – | – | – | – |
| Portugal | – | – | – | 1–4 |
| Sweden | 4.7 | 8.4 | 9.5 | 9 |
| Switzerland | 1.9 | 2.9 | 3.8 | 4.8 |
| Yugoslavia | – | 0.6 | 0.6 | 2–6 |
| Subtotal | 9.5 | 21.5 | 33.0 | 40–60 |
| Total Western Europe | 42.0 | 95.0 | 143.0 | 220–300 |

*Source:* Atomic Energy Commission, July 1981 and UNIPEDE's 'Electricity sector programs and prospects'.

The French electro-nuclear policy deserves careful consideration. Carrying out with remarkable constancy and regularity a wide-ranging programme of construction of nuclear power stations, at a rate of 5 to 6 GWe each year, France would appear, indeed, an example unique of its kind.

In 1976, hardly five years ago, the French nuclear power system, composed mainly of NUGG (Natural uranium-gas-graphite) reactor units, had a capacity less than 3 GWe; and electricity of nuclear origin, 14 TWh, only represented a share of about 8% of the national production.

At the end of 1981, the total capacity of nuclear power stations in operation was about 22 GWe and already included twenty-two PWR units of 900 MWe. Of this total, only five units — located at Fessenheim and Bugey — are a result of installation decisions made before the oil crisis. It may therefore be said that France has already begun to reap the harvest of the efforts made in 1974 to break away from oil. During only the period between March 1980 and June 1981, fifteen months in all, fifteen units were added to the network. Electricity generation of nuclear origin reached almost 100 TWh in 1981, which represents a share of 40% in the nation production of electricity. France has thus, over a period of a few years, become the world's second producer of nuclear energy. The country remains, it is true, well behind the United States, which today has a generating nuclear system almost three times as big.

At the same date, the end of 1981, France had twelve units of 900 MWe and thirteen of 1300 MWe, in all more than 27 GWe, under construction. Taking into account installation decisions already made for 1982 and 1983 (7.5 GW in all), it is indisputable that France will have at its disposal in 1990 a nuclear power capacity of between 55 and 60 GWe (see Table 5.16). This is about half of what the United States today estimates for that date. By 1990 France's dependence on foreign countries for their energy supply will have fallen to 54%.

**Table 5.16**

French energy programme, primary energy supplies in 1990

|  | Eighth Plan | | Two Year Plan | |
|---|---|---|---|---|
|  | Mtoe | % | Mtoe | % |
| Coal | 31 | 13 | 35–40 | 15–17 |
| Oil | 73 | 30 | 70–75 | 30–32 |
| Natural gas | 40 | 17 | 31–40 | 13–17 |
| Renewable energies | 11 | 4 | 10–14 | 4–5 |
| Hydraulic | 14 | 6 | 14–15 | 6 |
| Nuclear | 73 | 30 | 60–66 | 26–29 |
| Total | 242 | 100 | 232 | 100 |

The decision of the French government, at the beginning of 1974, to make massive use of nuclear energy is easy to explain. Of a total consumption of 177 Mtoe at the time, oil alone represented 116 Mt and the degree of dependence on other countries for energy supplies reached the unbearable level of almost 80%. Only nuclear energy could in time enable the situation to be re-established.

At the same time, nuclear energy seemed an opportunity literally within hand's reach. Indeed, since 1945, thirty years of effort have provided the ways and means to promote the atom for peace. Step by step, all the elements of a well-framed nuclear industry covering the whole of the chain had been patiently put into place, in the field of construction and operation of reactors and at all stages of the fuel cycle, upstream and downstream of the power station. At the beginning of the 1970s, the French nuclear industry had thus reached maturity. The ambitious nature of the French electro-nuclear programme has often been underlined, and even criticised.

It is more correct to consider it as the final stage of a long period of effort. The nuclear firms concerned with the construction of a series of nuclear reactors in 1974 were, in general, well prepared for this change.

The first realisations and then the order for a series of nuclear boilers with identical characteristics (thirty-four units in the 900 MWe bracket, and to date, eighteen units in the 1300 MWe bracket) have permitted a solid industrial group to be created and good selling prices to be obtained. Contracts for a period of years which were made with the industry explain clearly the fact that the costs of nuclear power stations have not increased as much in France as in other countries. They permitted, indeed, the costs of studies on a large number of units to be paid and, at the same time, allowed a planning of provisions and constructions in workshops to be made over a number of years.

The programme was carried out, on the whole, without any problems. The only real difficulty encountered was, as regards the sites necessary for the installation of the power stations, in obtaining, in due time and despite protests, all the vital administrative authorisations. To avoid any delays which would endanger the continuation of the programme, it was decided to place a maximum of four units† on the sites available, even at the expense of some over-concentration of power in certain regions. On the other hand, there were benefits from a favourable effect concerning the length and cost of construction. Also, even more important, a clear gain has been made in the performance of installations during the first months of operation due to better training of personnel. The authorities have supported EDF entirely to help them solve these difficult questions.

† Gravelines is an exception, with six units by the sea.

## FACTORS AFFECTING THE GROWTH OF NUCLEAR POWER

How can the almost universal slowing down in nuclear programmes be explained? The question needs to be carefully examined, given its importance for the future of the world and of society.

The principal explanation of this phonomenon of deceleration is without doubt the economic crisis which has existed since the end of 1973, the date of the first oil crisis. It is clear that in this depressed situation, the growth of electricity consumption has slowed down considerably everywhere (see Tables 5.17 and 5.18). All estimations concerning middle-term perspectives of demand, which influence the installation programmes for new power stations, have had to

**Table 5.17**

Domestic electrical energy consumption including losses, (TWh)

|  | 1960 | 1965 | 1970 | 1975 | 1980 |
|---|---|---|---|---|---|
| France | 72 | 102 | 140 | 181 | 249 |
| Germany | 113 | 165 | 235 | 290 | 351 |
| Italy | 56 | 81 | 117 | 141 | 180 |
| UK | 130 | 184 | 233 | 253 | 265 |
| USA | 846 | 1 157 | 1 640 | 2 010 | 2 380 |
| Japan | 111 | 184 | 344 | 454 | 570 |
| Canada | 109 | 144 | 201 | 266 | 335 |
| USSR | 292 | 505 | 692 | 1 030 | 1 220 |

*Source:* EDF annual statistics

**Table 5.18**

Average annual growth rates in electrical energy consumption

|  | 1960-1965 | 1965-1970 | 1970-1975 | 1975-1980 |
|---|---|---|---|---|
| France | 7.2 | 6.5 | 5.3 | 6.6 |
| Germany | 7.8 | 7.3 | 4.3 | 3.9 |
| Italy | 7.6 | 7.6 | 3.8 | 5.0 |
| UK | 7.2 | 4.8 | 1.7 | 0.9 |
| USA | 6.4 | 7.2 | 4.1 | 3.4 |
| Japan | 10.6 | 13.3 | 5.7 | 4.6 |
| Canada | 5.7 | 6.9 | 5.8 | 4.7 |
| USSR | 11.6 | 6.5 | 8.3 | 3.4 |

*Source:* EDF annual statistics

be reduced over the years. It is significant to note that it was in 1976, after a particularly difficult year for the world economy, that nuclear installation programmes underwent their most serious reductions through projects being abandoned or even orders being cancelled.

For the nine countries of the EEC, electricity consumption between 1974 and 1981 only developed on average at a rate of 3% per year, as against 7.2% per year just before the crisis (see Table 5.19). Estimations are about six years behind what was announced in 1974. The case of Germany (Table 5.20) corresponds approximately to this pattern. In the United Kingdom (Table 5.21), where electricity also faced hard competition from North Sea gas, the situation was still worse, putting the country almost ten years behind. France (Table 5.22) is lucky enough to be an exception; electricity consumption — helped, it is true, by the commissioning of the EURODIF gas diffusion isotopes separation plant — is only about two years behind. It is now unfortunately probable that the recession which followed the second oil crisis and which we are at present living through, will lead, as in 1975, to a reduced re-evaluation of installation programmes. But it is still too early to predict the extent of this re-evaluation.

The second explanation is the action of antinuclear pressure groups, who have effectively exploited the fear reflex linked to the military connotations of nuclear energy and to the mysterious and dangerous physiological effects of radioactivity.

In some countries — Germany, Austria or Sweden, for example — nuclear energy has thus become a bone of contention between political parties. The result is a blocking of the situation which it is difficult for governments to dominate. In Austria, an unlucky referendum in November 1978, concerning commissioning of the first national nuclear unit, at Zwettendorf, not only led to the project being shelved, but also put a stop to any new installation for years to come. In Sweden, a referendum was necessary, in March 1981, for a series of twelve reactors, which were almost ready, to be commissioned; but the government was obliged, in order to obtain this, to guarantee that no others would be built. Germany, who had launched before France a wide programme of nuclear power stations, was paralysed by a violent argument concerning operations at the end of the fuel cycle (stockage of irradiated fuels, reprocessing and stockage of waste from reprocessing). A real moratorium took place, which since 1977 has blocked almost every decision concerning new operations. And under pressure from coal producers who were cleverly able to turn the situation to their advantage, German electricity producers were forced to sign a long-term contract to absorb — at the expense of the electricity consumer — all the nationally-produced coal which cannot be used otherwise (30 Mt/year). It was therefore necessary to make the decision to construct new coal-fired power stations, which has diminished even further the possibility of a return to nuclear.

Subject to pressure from political opponents and certain sections, sometimes violent, of their opinion, the governments have come to give priority to their

**Table 5.19**

Prospects for expansion of electricity consumption (in TWh), evolution of forecasts from 1974 to 1981 for the nine EEC countries

| Consumption forecast for | Forecast year | | | | | | | |
|---|---|---|---|---|---|---|---|---|
|  | 1974 | 1975 | 1976 | 1977 | 1978 | 1979 | 1980 | 1981 |
| 1974 | 974.4 |  |  |  |  |  |  |  |
| 1975 | 1 108 | 967.4 |  |  |  |  |  |  |
| 1976 | 1 187 | 1 147 | 1 030.8 |  |  |  |  |  |
| 1977 | 1 270 | 1 228 |  | 1 066.1 |  |  |  |  |
| 1978 | 1 360 | 1 315 |  |  | 1 127.3 |  |  |  |
| 1979 | 1 454 | 1 410 |  |  |  | 1 186.7 |  |  |
| 1980 |  | 1 507 | 1 300 | 1 294 | 1 262 | 1 250 | 1 192.3 |  |
| 1981 |  |  | 1 390 |  |  |  | 1 278 | 1 193.9 |
| 1982 |  |  |  | 1 450 | 1 475 |  |  |  |
| 1983 |  |  |  |  |  | 1 510 |  |  |
| 1984 |  |  |  |  |  |  | 1 508 |  |
| 1985 |  |  |  |  |  |  |  | 1 454 |

[ ] Achievements

*Source:* UNIPEDE's 'Electricity sector programs and prospects'

**Table 5.20**

Prospects for expansion of electricity consumption (in TWh), evolution of forecasts from 1974 to 1981 for the Federal Republic of Germany

| Consumption forecast for | Forecast year | | | | | | | |
|---|---|---|---|---|---|---|---|---|
| | 1974 | 1975 | 1976 | 1977 | 1978 | 1979 | 1980 | 1981 |
| 1974 | 296.3 | | | | | | | |
| 1975 | 332 | 289.6 | | | | | | |
| 1976 | 356 | 345 | 312.4 | | | | | |
| 1977 | 382 | 370 | 326 | 319.6 | | | | |
| 1978 | 410 | 397 | 346 | 351 | 332.3 | | | |
| 1979 | 440 | 425 | 368 | 373 | 377 | 349.0 | | |
| 1980 | | 455 | 391 | 390 | | 370 | 351.4 | |
| 1981 | | | 415 | 419 | | 390 | 379 | 352.0 |
| 1982 | | | | 450 | | 410 | 400 | 394 |
| 1983 | | | | | 447 | 430 | 415 | 409 |
| 1984 | | | | | | 450 | 430 | 424 |
| 1985 | | | | | | | 445 | 439 |

☐ Achievements

*Source:* UNIPEDE's 'Electricity sector programs and prospects'

**Table 5.21**

Prospects for expansion of electricity consumption (in TWh), evolution of forecasts from 1974 to 1981 for the United Kingdom

| Consumption forecast for | Forecast year | | | | | | | |
|---|---|---|---|---|---|---|---|---|
| | 1974 | 1975 | 1976 | 1977 | 1978 | 1979 | 1980 | 1981 |
| 1975 | 283 | 237 | | | | | | |
| 1976 | 300 | 287 | 239.3 | | | | | |
| 1977 | 317 | 304 | 258 | 246.3 | | | | |
| 1978 | 334 | 321 | 272 | 266 | 267.3 | | | |
| 1979 | 351 | 338 | 287 | 279 | 272 | 278.7 | | |
| 1980 | | 355 | 302 | 293 | 283 | 280 | 264.9 | |
| 1981 | | | 319 | 306 | 296 | 287 | 276 | 258.7 |
| 1982 | | | | 320 | 306 | 289 | 279 | 268 |
| 1983 | | | | | 318 | 300 | 283 | 272 |
| 1984 | | | | | | 305 | 289 | 278 |
| 1985 | | | | | | | 292 | 283 |

☐ Achievements

*Source:* UNIPEDE's 'Electricity sector programs and prsopects'

**Table 5.22**

Prospects for expansion of electricity consumption (in TWh), evolution of forecasts from 1974 to 1981 for France

| Consumption forecast for | Forecast year | | | | | | | |
|---|---|---|---|---|---|---|---|---|
| | 1974 | 1975 | 1976 | 1977 | 1978 | 1979 | 1980 | 1981 |
| 1974 | 179.8 | | | | | | | |
| 1975 | 197 | 180.7 | | | | | | |
| 1976 | 211 | 207 | 196.4 | | | | | |
| 1977 | 226 | 222 | | 206.8 | | | | |
| 1978 | 243 | 240 | | | 220.8 | | | |
| 1979 | 261 | 263 | | | | 235.6 | | |
| 1980 | | 285 | 270 | 265 | 260 | 255 | 248.7 | |
| 1981 | | | 290 | | | | 267 | 258 |
| 1982 | | | | 303 | 313 | | | |
| 1983 | | | | | | 325 | | |
| 1984 | | | | | | | 340 | |
| 1985 | | | | | | | | 320 |

▢ Achievements

*Source:* UNIPEDE's 'Electricity sector programs and prospects'

short-term tranquility rather than long-term national interests. Electricity distributors, forsaken on all sides, are finding it more and more difficult to find new locations, open new sites or even complete operations already begun some time ago, which costs a great deal in financial terms. Tired of disputes, they have decided a little too easily to renounce the normal expansion of their market. This effect, added to the delays due to the economic crisis, contributes to limiting even further the need for new installations.

In the face of ever-increasing security restrictions demanded by public opinion, the authorities responsible for the definition of safety regulations for nuclear power stations have become hesitant, over-severe or too careful, which in every case results in an extra burden on installation costs. In the United States, the Three Mile Island accident, in March 1980, with the inevitable questions and reconsiderations it brought about, caused a severe halt in programmes for commissioning and constructing installations, from which the country is only just recovering.

Our countries must, unfortunately, undergo the changes in the economic situation. On the other hand, it is contrary to their interests to accept without question the hold-ups which are imposed by minorities in the name of considerations which are usually wrong or debatable.

### The cost of nuclear-generated electricity

The example of France is there to witness that nuclear energy, when properly used, is by far the most economic means of production of electricity. Even if, as shown in Table 5.23, the cost of the nuclear kWh has considerably increased since 1974 — in fact, by about 70% in constant money terms — principally due

**Table 5.23**

Evolution of costs of nuclear kWh

| Economic conditions | Costs in cF of constant 1981 value | | | | |
| | 1974 | 1976 | 1979 | 1980 | 1981 |
|---|---|---|---|---|---|
| Date of commissioning of the plant | 1979 | 1985 | 1985 | 1990 | 1990 |
| Capital | 5.25 | 6.33 | 7.80 | 7.80 | 8.92 |
| Operation | 2.06 | 2.34 | 2.56 | 2.97 | 3.19 |
| Fuel | 2.65 | 3.76 | 4.86 | 4.97 | 4.50 |
| Total (rounded) | 10.0 | 12.4 | 15.2 | 15.7 | 16.6 |

*Source:* Electricité de France

to more and more severe safety restrictions imposed for the construction and exploitation of reactors, it remains today nearly a third cheaper than the thermal kWh generated by imported coal (see Table 5.24). Comparison with production from fuel-oil, which in 1970 was the most economic option, has become useless, and has lost all meaning.

**Table 5.24**

Comparative costs of nuclear and fossil fuel electricity, for a base-load power plant commissioned in 1990, 1981 conditions

|  | Nuclear cF/kWh | Coal cF/kWh | Fuel cF/kWh |
|---|---|---|---|
| Capital | 8.8 | 6.8 | 5.9 |
| Operation | 3.2 | 3.2 | 2.9 |
| Fuel | 4.5 | 15.5 | 44.7 |
| Total | 16.5 | 25.5 | 53.5 |

Ratio of kWh nuclear: kWh fossil fuel:

|  |  |  |
|---|---|---|
| 1972 |  | 0.97 |
| 1975 |  | 0.55 |
| 1980 | 0.61 | 0.31 |
| 1981 | 0.65 | 0.31 |
| Flue gas desulphurisation (supplement) | 3 | 3.7 |

Discount rate 9%

The utilisation of nuclear energy is therefore, for the electricity consumer, the best guarantee of provision at the cheapest price. At the same time, the nuclear option protects him against possible price increases which coal producers would one day impose and which would have widespread implications on the price of the thermal kWh. Uranium is certainly not protected against such increases, but the cost structure of the nuclear kWh is such that the incidence on the final cost would be extremely limited: fuel only represents a quarter of the production costs and uranium only a third of the fuel costs, that is, a twelfth of the total. The nuclear option brings finally the possibility of an easy guarantee against possible ruptures in provisions, which are always to be feared where imported fuel is concerned, for after all it costs relatively little to form safety stocks of natural uranium to cover several years of operation of power stations.

In fact, nuclear energy brings to the electricity sector the domination of costs and assurance of provisions. This is particularly important given the key role — often unfortunately, ignored — which electricity is called upon to play in 'energy redeployment' which our countries have to operate.

As, indeed, it is a question of substituting other energies — in fact practically coal or the atom — for oil all non-specific uses, it must be recognised that, except for a few rare cases where the direct use of coal is still possible and acceptable, it is necessary to pass by the intermediary of heat or electricity distribution. The special conditions of heat distribution necessarily limit the cases of application. The essential role as a 'vector' of oil-substitute energies falls therefore to electricity; and electricity has a good chance of conquering these new markets.

Firstly, on a physical level, electricity is ready to be used in many ways. For example, where heating is concerned, the Joule effect can be utilised (ovens), as can heating methods of the mass by induction (conducting body) or dielectric effect (insulating body), rays, electric arc or plasma. What oil can accomplish by means of heat, electricity can do by other means. If, for example, it is a question of concentrating certain solutions in food and agriculture industries, rather than transforming the surplus water into steam by heating all the mass, it is possible, by mechanical action, to separate the ingredients by passing them selectively through a membrane (osmosis). In any case, electrical procedures also benefit from a considerable advantage thanks to the extraordinary possibilities of regulation which only electricity can permit. In total, one kWh of electricity can substitute (see Fig. 5.9) for a number of oil-product thermal units, which, although variable according to the cases, are on weighted average approximately 3 or 4 and thus above that needed (2.25) to produce the same kWh in a thermal power station†. Electricity therefore appears as the best method of using energy rationally, permitting the energy savings so highly needed to be made. The obstacle to be surmounted is that it is necessary to rethink entirely the conventional procedures of utilisation and therefore to replace existing plants. This will require time and money; and the problems caused by the economic crisis do nothing to facilitate the change.

Secondly, electricity has the advantage of its price with regard to competitive energies, if nuclear energy ever gets its right part in the generating capacities. In the case of France (see Figs. 5.10 and 5.11), where, in 1990 when the generating system will have recovered its optimal composition, nuclear should provide 75% of the total electricity production (nearly 40% in 1981), it is estimated that the price of the kWh should reach — in constant francs — its level before the first oil crisis. Henceforth, and despite the fact that EDF is only just beginning, over the last three or four years, to reap the harvest of its nuclear commitments, the

---

†The thermal unit used in Fig. 5.9, the 'thermie', is the amount of heat required to raise one tonne of water through $1°C$ ($4.18 \times 10^6$ joules).

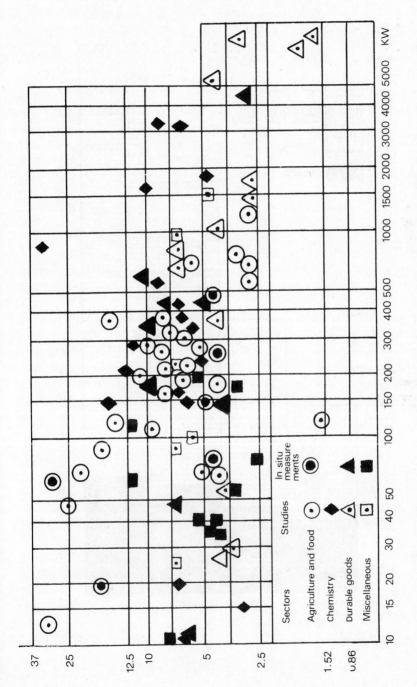

Fig. 5.9 – The coefficient $\gamma$: How many thermies can replace 1 kWh (*Source:* EDF (DER and SEPAC).

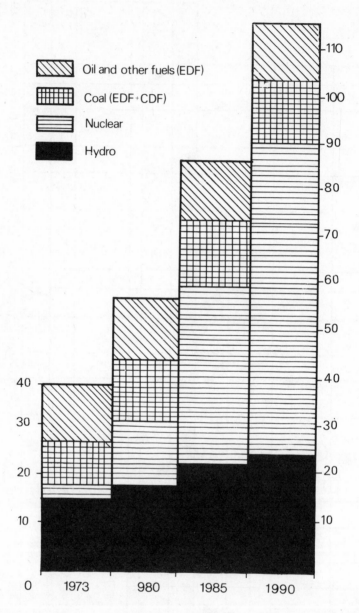

Oil and other fuels (EDF)

Coal (EDF + CDF)

Nuclear

Hydro

Fig. 5.10 — Evolution of total installed capacity in France (in GWe), 450 TWh outlook.

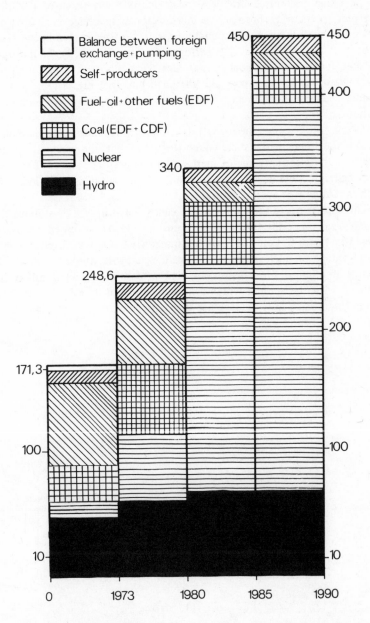

Fig. 5.11 − Evolution of the structure of electricity generation in France, (in TWh) 450 TWh outlook.

average selling price of electricity is today only 9% above that of 1970. (To be more precise, there has been a 28% increase for high voltage and a drop of 20% for low voltage.) Progress in productivity, a factor well known in the electrical industry, is the reason for this performance (see Fig. 5.12). In the opposite direction, fuel prices (oil, coal or gas), after increasing greatly as we know, still tend towards an increase now and in the future (see Fig. 5.13). There is no example where such differences in price have not, in time, led to massive substitutions.

Taking into account the many implications which have just been explained, and also the fundamental stake which they represent, the incidence of nuclear energy on the electricity sector merits a long exposition, indeed. Nevertheless, nuclear energy presents for our economies other advantages no less essential which should now be at least mentioned.

The constraints of the commercial balance constitute for most industrialised countries, and especially for European countries which have few energy resources, an essential factor not only of economic insecurity but also of decreasing growth. The advantage of nuclear energy is here considerable: the amount of foreign currency it requires for construction and provision of fuel is relatively low compared with the amount which can be economised by substituting it for oil or imported coal. Thus, by relaxing the constraint of the external balance, it

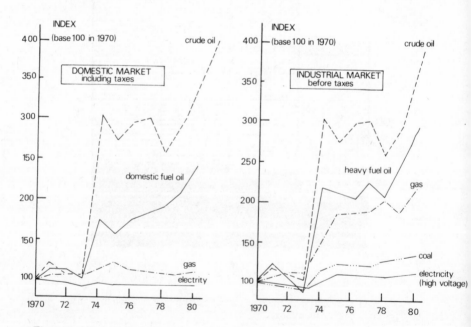

Fig. 5.12 – Evolution of energy prices in France in real value in the consumers sectors (*Source:* Commissariat Général au Plan (F) 1980).

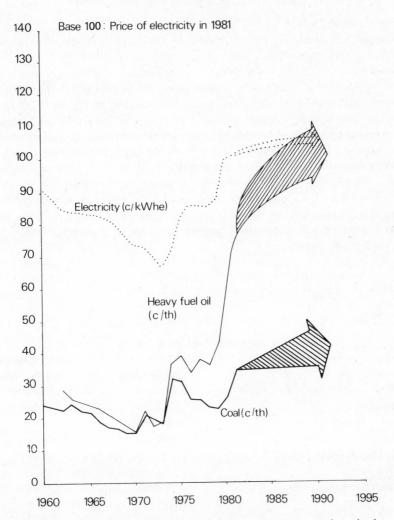

Fig. 5.13 – Energy prices in France for large industrial consumers in real value (customers consuming more than 100 toe per year)

can be estimated in the case of France that nearly 1% growth increase may be achieved. According to calculations, this corresponds, over a period of five years, to the indirect creation of 400 000 jobs.

To this indirect effect on employment should be added the direct effect, also of benefit, linked to specifically nuclear activities. More 'capitalistic' than a thermal power station, a nuclear power station requires more work for construction; and the fuel cycle service leads to a substitution of national employment for importations.

**Environmental Factors**

From an environmental point of view, finally nuclear energy also presents particular advantages with regard to fossil fuel energies (see Table 5.25). This is true as regards polluting waste. Nuclear waste certainly exists; it has been much criticised. In fact, it cannot be stressed enough that the possibility of isolating, in a very small volume which is then totally controllable, all the waste linked to consumption of energy constitutes, on the contrary, for the safety of our planet, an immense advantage which should be considered as an asset of nuclear energy.

After a period of great confusion, it would seem that all these truths are being understood; and we find more and more responsible leaders who support them. For all these reasons, it seems that we may today be reasonably optimistic for the future of nuclear energy. The long-awaited recovery of the economy, even at a lower rate of growth than that which we knew before the crisis, will, provided we can convince public opinion to accept it, give nuclear energy the opportunity to move ahead.

**Table 5.25**

Environmental effects of a 5 GW power plant

|  | Coal fired power plant | Nuclear plant |
|---|---|---|
| *Rejects* |  |  |
| $CO_2$ | 40 000 000 t/year | 0 |
| $SO_2$ | 200 000 t/year | 0 |
| $NO_x$ (expressed in $NO_2$) | 50 000 t/year | 0 |
| Ashes | 30 000 t/year | 0 |
| (without precipitators) | (2 000 000 t/year) |  |
| *Waste* |  |  |
| Flying ashes | 2 000 000 t/year |  |
| High activity wastes | 0 | 70 m$^3$ |
| Medium and low activity wastes | 0 | 2 500 m$^3$ |
| *Radioactivity* |  |  |
| Air for radioactivity dilution† | 50 λ 1500 Gm$^3$/year | 400 Gm$^3$/year |
| Water for radioactivity dilution‡ | 0 | 60 m$^3$/s |
| *Transportation* |  |  |
| Fuel supplies | 120 000 wagons of 100 t each | A few trucks |

† Volume of air required for diluting radioactive rejects.
‡ Maximum water throughput required for diluting radioactive rejects.

## CONCLUSION

In the meantime, our duty is to prepare the ground for this renewal of nuclear programmes.

The rules of the game in our democratic countries is that the decision must always be made by the people; nothing can be done without the agreement of the majority.

This means that we must take great pains to persevere in informing the public of the problems posed by energy and in explaining to them why recourse to nuclear energy is really inevitable, what its nature is, how installations operate, the risks involved, how they compare with other forms of energy. In other words, we must be convincing.

Whether we like it or not, the subject of energy has ceased to be a purely technical field reserved for engineers and economists. The 'energy debate' which began in 1973 is of real interest to the people; it should be pursued with renewed vigour and new methods. Today it constitutes a privileged field for the development of democracy. The stakes are such that we cannot avoid the new efforts implied by the change which must be brought about in the mind of the public.

# 6

# Environmental and political constraints on future energy/chemical industry policy

**Kenneth D. Collins**, Member of the European Parliament

## INTRODUCTION

In 1800, near enough to the beginning of the industrial revolution in Europe, the population of Paris was 547,000 (2% of the French total) and that of London was 850,000 (8.1% of the United Kingdom total) and their experience throughout the nineteenth century and well into the twentieth was typical. By 1950 the figures were 6.7 million (15.7%) and 10.9 million (27.4%). Urban growth took place at unprecedented rates, sustained by industrial expansion and overseas trade with a world colonised by Western Europe.

There were a few protestors about the poverty, living conditions and the squalor of urban industrial slums (Marx, Engels, Rowntree, etc.) but there were very few indeed concerned about growth itself. The slave trade was condemned but there were few protests about the rights of Western Europe to manufacture raw materials extracted from the colonies. Only people like Ruskin or Howard were against the trend. Growth, expansion, exploitation and pollution were the rule and few thoughts were given to the urbanisation of green farmland.

It is inconceivable that a conference such as this could have been held then. Industrial expansion and business confidence inspired few doubts. The environment was a God-given gift that should be used for man's gratification and just as God's love was limitless, so were the resources that he should place at man's disposal. They were boundless and thoughts of conservation, had they occurred, would have been banished as near heresy.

## THE RISE OF ENVIRONMENTALISM

As early public health legislation shows, there was a movement, especially in the hard-pressed cities, to control in a simple way the worst ravages of industrial expansion. Epidemics of cholera, dysentery and typhoid were the common experience of many towns and action was taken to control and purify public water supplies and sewerage systems. It was a similar concern for public health that prompted the laws concerning clean air in cities in this century. This strand of environmental thinking is still important — and it is evident for example in the campaign about lead in petrol and water supply pipes.

There was a second strand too in the realisation that if any degree of efficiency were to be achieved (never mind decent living conditions) then planning was necessary. This theme was taken up with the publication in the late 1960s of the findings of the Club of Rome — a cataclysmic view of the future based on simulating the global use of resources, agriculture, the growth of population, industrial pollution, etc.

In 1970 the United Kingdom set up the Royal commission on Industrial Pollution and United Nations Environment Programme (UNEP) got under way in 1972; these, along with more traditional environmental organisations like the World Wildlife Fund, have had their effects on the wider European and international public and on political opinion†.

The European experience of the rise of environmentalism can be mapped out by the following highlights: the 1972 Paris summit; in 1973 the formation within the European Commission of the Environment and Consumer Protection Service; the First Action Programme on the Environment ('State of the Environment: First Report' Commission of the European Communities, 1977); followed by the Second Action Programme on the Environment ('State of the Environment: Second Report' Commission of the European Communites, 1979, Catalogue number CB-24-78-152-EN-C); now the Third Action Programme is being considered (Draft Action Programme of the European Communities on the Environment 1982-1986 COM(81) 626 final). The trend now is away from early fire-fighting to a more long-term integrated approach which is supposed to link with other EEC strategies on agriculture, energy, etc. But there is a strong economic base as well as more recent concern to recognise that, without long-term resource management policies, the rest of our economic and social efforts will be reduced in their achievements.

## THE POLITICAL CONTEXT

Political authority and political legitimacy for such policies lies with the people themselves and we must look at the political perceptions that may shape the society within which industry must work in the future. We consider the following here:

(1)  whether the environment is a political issue;
(2)  the extent of public control or influence over company policy;
(3)  the scale of the political units that we shall work with;
(4)  the extent of our responsibility to and for the rest of the world.

† See references.

These are all crucial elements in the development of an open, free, and democratic society which is the aim of us all.

### The environment as a political issue

The problem of environment policy (pollution control or wildlife conservation and protection) is that its aims are:

(1) long-term and measurable in time units longer than the political term of office;
(2) based sometimes on a moral philosophical view of man's purpose on earth and his relation to the rest of the living world. Ego- or eco-centric?

These points make it difficult, but not impossible, and there is a growing awareness that today's environmental decisions may have far-reaching consequences for children, and both education and longevity have contributed to this new and more complete view of industrial policy. It is therefore an underlying issue and discussion about it has to be encouraged.

### Company policy

A worrying area for many is the way in which company decisions seem often not to be in the public interest but only in the company interest and here, two points have to be made:

(1) In the Commission's 'Proposal's for directives concerning the assessment of the environmental effects of certain public and private projects' (Document 1-293/80) and on 'Major industrial hazards' (Seveso Directive Document COM/79/353 final) show there is a clear trend towards giving the public a say in decisions affecting industrial location and if this is not given then it will be demanded, for example, Frankfurt airport or nuclear demonstrations. We need to have politicians who are sensitive and aware of public opinion.
(2) In the United Kingdom, we have suffered too long from a class-based society in which worker and management do not work easily together. It is inevitable that in the later part of the twentieth century there will be attempts to have active worker participation, in public and private enterprises.

### Scale of political units

This is a dangerous and difficult world and we need to manage our industries and our economy to maintain and improve the lot of all of the people. To make a British point, it does seem to be the politics and the economics of those suffering from a particularly sad case of tunnel vision to advocate the break-up of the EEC. It would not serve the interests of Britain (except that of the extreme right-wing elements); it would weaken the Community; and it would stifle the possibility of a genuinely wider Europe of the future involving the European Free Trade Association. Nevertheless, such decisions are part of our future and I can only say that as Chairman of the European Parliament Committee on the Environment, it seems to make no sense to risk going back to innumerable conflicting bilateral agreements on environment policy. It makes no commercial sense, but it makes neither political nor environmental sense either and everyone should realise how great a possibility it is and see the problems of the Common Agricultural Policy (CAP) and of the EEC Budget in this long-term sense.

### Responsibility
Finally, it is easy in Western Europe with our common history of dominance in the rest of the world for hundreds of years, to continue with our old patterns of thought. However, as Brandt has pointed out to us, our relations with the Third World in the future must be the relations of interdependence. That must affect our policies of trade and aid but also it must affect the care with which, for example, the environment of the Third World is handled. Should pesticides and herbicides that are banned in Europe be used? What about baby food preparations? Are we always happy with the effects on the poor and illiterate of sophisticated marketing? Who is to control this and what values are to prevail?

## CONCLUSION

Faced with an audience of professionals and senior industrialists, it is always tempting to concentrate on the ideas of the business world. But I am not a businessman — I am a geographer and a politican. I am also a politician of the Left and I happen to believe that the eradication of poverty, the promotion of peace and the conservation of the environment are the three great central issues of our time. How we achieve that depends on how we manage and control our industry, and what goals we set for it and for society.

I have tried to show how environmental awareness has developed as a political issue and I have tried to examine a very few of the problems of the future. These are questions that are not asked only by earnest young environmentalists with cataclysmic views of the future and the fire of bloody revolution in their hearts. They are questions that are being asked all over Europe by ordinary people, and history demands that we should find answers for those who will inhabit the earth long after we have gone (that is, if by our stupidity we haven't destroyed it!). In the words of Mark Antony, 'The evil that men do lives after them, the good is oft interred with their bones'.

## REFERENCES

[6.1] Royal Commission on Environmental Pollution, First Report Cmnd. 4858, February 1971.

[6.2] Royal Commission on Environmental Pollution, Fifth Report 'Air Pollution Control: An Integrated Approach' Cmnd. 6371, January 1976.

[6.3] Royal Commission on Environmental Pollution, Sixth Report 'Nuclear Power and the Environment' Cmnd. 6618, September 1976.

[6.4] Royal Commission on Environmental Pollution, Eighth Report 'Oil Pollution of the Sea' Cmnd. 8358, October 1981.

# Sustainable energy futures

**Professor W. Häfele,** Vorsitzender des Vorstands der Kernforschungsanlage Jülich GmbH, West Germany

## INTRODUCTION

Will there be a sustainable energy future? That is a rather delicate question involving possible long-term solutions, favourable choices, and strategies appropriate to the transition. It requires considerable study in detail and will yield several answers with conditional clauses. Let me anticipate a condensed statement: There can be a sustainable future with indefinite production of energy, without consumption of our resources, and for a world population amounting to 8, or 10, or even 20 billion people. As this might sound perhaps a little fantastic, it will be worth considering a particular solution as a reference for the conception of principal guideposts in the course of the reasoning to follow.

## A REFERENCE SOLUTION

With regard to the energy flows through the energy system (see Fig. 7.1), the usual distinction is made between primary energy which is predominantly supplied to conversion, and the various forms of secondary energy which meet the final energy demand of the end-users. The specific applications of energy in the end-use sectors do not allow any substitution by other forms of secondary energy: a computer cannot be operated with oil; flying an aeroplane needs hydrocarbons or hydrogen. Therefore, a final energy supply either by electricity or by hydrocarbons alone would be impracticable. On a long term basis a second grid is needed in addition to the well-established electricity, or first grid; and hydrogen, as envisaged in our reference solution, would ensure a comfortable basis to complement the functions of electricity; both are extremely clean and can easily be adjusted to the end-use energy requirements.

Electricity and hydrogen would be produced in central conversion facilities utilising the resources of nuclear and solar energy. As for nuclear energy, the resources available are of unlimited extent. This estimate applies to the potential

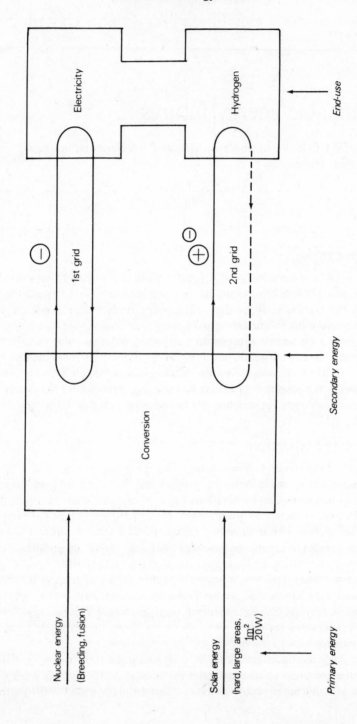

Fig. 7.1 — Unlimited energy without consumption of resources (but investments: Capital (S10 000/per capita), Labour, Resources, e.g. 50 kg/m²)

of breeder-based fission as well as to that of fusion. Breeders will enhance the energy supply potential of the Light Water Reactor by a factor of at least 1000. This accounts for two effects of breeder operation: firstly for the familiar 60-fold higher energy yield per unit of natural uranium; secondly for the enlargement of economically exploitable resources by low grade ores due to the 60-fold lower weight of uranium prices in the fuel cycle cost. The technical realisation of fusion reactors is still far ahead. Based on deuterium-lithium fuel supply, as scheduled for the earliest realisable performance, fusion reactors would offer quantitatively a similar potential as breeders, that is, ensure supply for more than several 10 000 years. The exclusively deuterium-fuelled version would enhance the potential of fusion energy once more by a factor of 1 000.

Thus, it turns out that the profits that can be taken from our immense resources of primary energy, are actually limited by the quality and quantity of the conversion capacities. Their future development and extension will require considerable investments in form of capital, labour, know-how, and natural resources, for example, land and raw material. Hard solar energy conversion for the large-scale production of storable and transportable energy, preferably hydrogen (in this connection local and indirect applications of solar energy may be ignored) is the other long-term energy option rendering possible high power yields from sunshine, but only at the expense of enormous investment burdens. Those large fields somewhere in the arid and sunny zones harvesting solar energy with passable efficiencies of 20 $W/m^2$, even with no restrictions from land use, require substantial amounts of steel, concrete, and glass, with typical steel installation densities of about 50 $kg/m^2$. For the time being, the progressive way to abundant and clean energy will therefore continue to be intense production and processing of raw-material which is necessarily coupled with environmental pollution and also with other undesired side-effects. In contrast, however, to the attitude of an unobjectionable consumption of resources that modern societies are so strikingly reproached with, the indicated way of resource utilisation, albeit troublesome, would essentially be an investive one. Nonetheless, to envisage plain reality, setting up such a system will probably take more than 100 years. This can be visualised from the $10 000 *per capita* which is a fair appropriate figure for the investments immediately involved. Since typically one-third of all investments are for energy, this would correspond to a total capital stock of at least $30 000 *per capita*. In comparison with present $2000 *per capita* world average values, this would be an enhancement by a factor of 15 (about $2^4$); hence it would require four doubling times of gross world product (GWP) for accumulation. The highest GWP growth rates maintainable in the long run, however, will certainly not exceed 3% per year which results in doubling times longer than 23.5 years.

This vision of a principle solution in a possible energy future 100–150 years from now may be completed by some more nearby aspects of a world in transition with problems impending enough to attract present day's sharp attention.

## THE INTERNATIONAL INSTITUTE FOR APPLIED SYSTEMS ANALYSIS ENERGY STUDY

Let me refer to a few very condensed results of a comprehensive 8-year study performed by the Energy Systems Program Group of the International Institute for Applied Systems Analysis (IIASA) in Laxenburg near Vienna. The details are documented in an 800-page book [7.1].

One particular aspect of this world in transition concerns the future development of world population which is expected to increase from 4 billion people today by another 4 billion people during the next five decades (Fig. 7.2). There is some evidence that after 2030 the growth rates could continuously level off thus bringing about a stabilisation at levels of about 10 billion people towards the end of the century, but this will not necessarily occur. In any case, we are confronted with the problem of supplying 8 billion people by 2030, most of whom will live in today's developing countries.

The IIASA approach to this problem has been based on a detailed investigation of demographical, economical and technological structures, and their development potential in a real world showing regional differences and structural imbalances. In particular, the world was subdivided into seven regions and the study considered the resources available, the existing standards of culture and

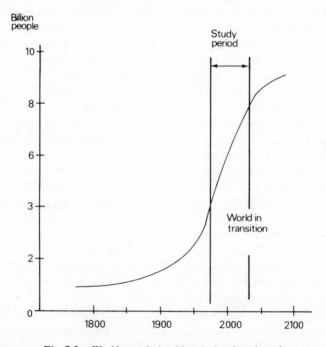

Fig. 7.2 – World population historical and projected.

civilisation, geographical peculiarities and trade relations. On this informal basis two different but self-consistent scenarios were constructed labelled as 'high' or 'low', corresponding to the mean scenario growth rates.

The various reasonings and details of the scenario construction are fully documented in the book, but it is worth emphasising that scenario writing is a technique to effect consistency of basic assumptions and essential restrictions in organising one's thoughts about possible future developments. It proceeds on a quantitative basis checking and readjusting the assumptions iteratively with regard to the resulting implications. It does not aim at the prediction of future events and therefore a scenario is not a forecast.

The global supply of primary energy, specified by source and for the reference years 2000 and 2030, as it turns out from the high and low scenario respectively, is shown in Table 7.1. According to these figures, by 2030 the production of primary energy would have to be enhanced from today's 8.2 TWyr/yr† to 22.4 TWyr/yr in the low, and 35.6 TWyr/yr in the high case. The results in Fig. 7.3 are derived from assumptions that exclude exponential growth, allow to a certain reasonable extent relatively higher growth in the developing parts of the world than in the industrialised countries, and comprise rigorous conservation measures. The global results represent the aggregate of all regions, the primary energy supply of which has been determined on the condition of minimum discounted overall cost by considerate operation of a linear programming routine. In contrast to popular wisdom, the outcome is that the world is not going to move away from oil. The absolute amounts of oil, gas, and especially of

Table 7.1

Global primary energy by source (TW)

|  | 1975 | High scenario | | Low scenario | |
| --- | --- | --- | --- | --- | --- |
|  |  | 2000 | 2030 | 2000 | 2030 |
| Oil | 3.62 | 5.89 | 6.83 | 4.75 | 5.02 |
| Gas | 1.51 | 3.11 | 5.97 | 2.53 | 3.47 |
| Coal | 2.26 | 4.95 | 11.98 | 3.93 | 6.45 |
| LWR | 0.12 | 1.70 | 3.21 | 1.27 | 1.89 |
| FBR | 0.00 | 0.04 | 4.88 | 0.02 | 3.28 |
| Hydro | 0.50 | 0.83 | 1.46 | 0.83 | 1.46 |
| Solar | 0.00 | 0.10 | 0.49 | 0.09 | 0.30 |
| Other | 0.21 | 0.22 | 0.81 | 0.17 | 0.52 |
| Total | 8.22 | 16.84 | 35.63 | 13.59 | 22.39 |

† 1 TWyr = 1 Terawatt year = $10^{12}$ Wyr = 704.5 × $10^6$ toe = 1076 × $10^6$ tce.

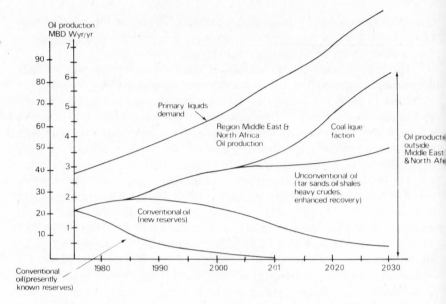

Fig. 7.3 – Oil, 1975 to 2030, World excluding centrally planned economies (high).

coal to be produced will remarkably increase in both scenarios. Decisive reasons for that result derive from the resource situation and from the fact that the recovery rates can be enhanced at relatively attractive cost conditions. The contribution of nuclear energy to supply, by Light Water Reactors and Fast Breeder Reactors, will still be modest in 2030, due to cost-effective restrictions and other constraints in the introduction phase. The thermal energy production of 8 TWyr/yr in the high case – 3 TWyr/yr from LWR, 5 TWyr/yr from FBR – corresponds to an installed capacity of 5000 GWe, or roughly 5000 power stations. Solar energy will still be restricted to local and small-scale energy supplies.

One robust conclusion from these results which was very carefully checked, is that mankind will have to engage in the utmost utilisation of all sources of primary energy; there are no alternative energies available for free choice, there are only additive resources urgently required for supply. The development of the demand for liquid fuels in the Western world in the high scenario case, and the supply situation, indicated in Fig. 7.3, will be an instructive example. The conventional oil from presently known reserves will, as in popular view, gradually draw to an end, roughly around the year 2010. Meanwhile however, new reserves of conventional oil, for instance in Mexico or in Alaska, will be developed. In addition, unconventional oil from heavy crudes, tar sands, oil shales, and from enhanced recovery of course, must serve to meet the increasing demand. If we account furthermore for a continuous and even slightly increasing oil supply from the Arabian countries of the Middle East–North Africa region, there would remain a gap to be filled by coal liquefaction, which would have to start by the year 2000 and increase its production rates up to 2.4 TWyr/yr (34 Mbd) by 2030. From this time scale it turns out that the fundamental technological steps to achieve such production rates must be started now.

Thus it is quite evident that the high increase of demand for coal, as indicated in Fig. 7.4, will be due more to the strategic function of coal in filling the liquid fuel gap than to the expansion of its conventional applications. Figure 7.5 refers to the assumption of autothermal generation of process heat for coal liquefaction in the high scenario case, which indeed reflects present technological tendencies where every pilot plant operates on an autothermal basis. In 2030, the quantities of coal required for the production of liquid hydrocarbons alone would amount to 6.7 TWyr/yr or to 7 thousand million tce/yr. Thus, the total coal consumption of 13.5 TWyr/yr (15 thousand million tce/yr) would still aggravate the problems resulting from restricted potentials of coal production and coal trade, as well as from the emissions of carbon dioxide which are serious enough anyway. Therefore, we have to label autothermal coal liquefaction and gasification with a distinct question mark.

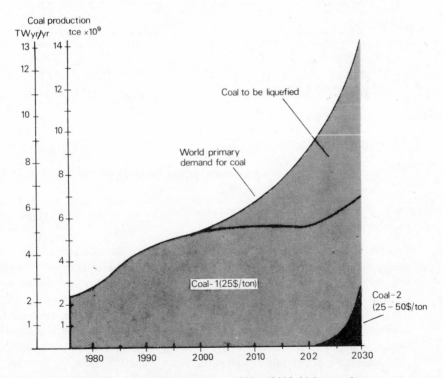

Fig. 7.4 – World coal demand and supply, 1975 to 2030, high scenario.

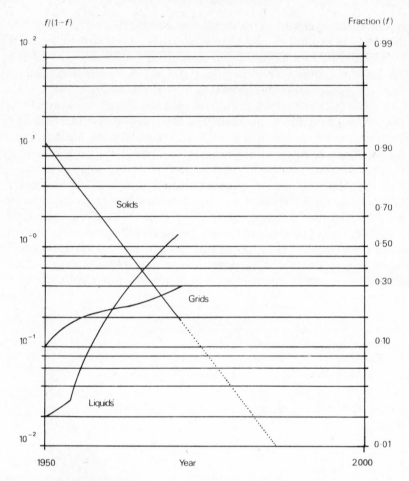

Fig. 7.5 – Secondary energy supply, FRG (1950–1970).

### ENERGY UTILITY

Let me just introduce some basic ideas concerning the qualitative aspects of energy before we enter into a consideration of appropriate technological strategies. It is not energy as such, or its quantity in physical units that is of fundamental interest to the end-user, but its specific utility in the particular end-use applications. The fact that the customers of electric supply companies are not merely supplied with electricity is excellently testified by the term 'utility company'. Hence, in evaluating the proper utility of energy, one will do well to perceive the additional services and burdens attached to its utilisation. Coal, for instance, is not adapted to most of the energy applications in the end-use sectors, nor are wood, oil shale, and tar sand. This is due to the costly, laborious, and uncomfortable operation conditions that all solid fuels show in typical small-scale applications. Automatic control of the energy flow and transformation processes require unusual technical means in comparison with an oil or gas fuelled operation. Moreover, the emissions from small fireplaces caused by coal combustion will seriously affect the environment as clearly indicated by the situation of London in the fifties. Thus, coal turns out to have less utility for a given amount of energy than oil or gas.

### Energy supply grids

It is certainly for these reasons that the development of the market shares of the various secondary energies supplying the end-users show typical and very stable trends, according to which the liquid fuels and the energies conducted by grids, for example, electricity, gas and district heat, take prevalent roles in supply, whereas the importance of solids is continuously fading away. This may be illustrated by the special situation of the Federal Republic of Germany (Fig. 7.5). Many similar figures indicating that this finding generally applies, could be shown irrespective of the particular geographic and economic and technological conditions of the situation.

Hence, coal and other solid fuels as sources of primary energy must necessarily be converted into secondary energy to be of use for supply, and indeed, except for the considerable amounts flowing into the steel production sector (which is a special application affording its own conservation facilities), coal is mostly used for the production of electricity. Many people worrying about the conversion losses that are due to electricity generation, like A. Lovins [7.2] and various members of soft energy parties, call for the abolition of electricity to avoid those losses, because they do not keep in mind the gain of utility due to conversion which really is of factual importance.

### The first grid, electricity

It is true that the demand for electricity has increased with higher growth rates in the past than even primary energy consumption. The growing importance of electricity must be seen in connection with the service functions arranged by the electricity grid, or the first grid, as it is also called (Fig. 7.6). Everybody knows that energy in the form of electricity is transported through the grid from producers to consumers, but there are many additional functions constituting services to both parties. The grid integrates large numbers of different, widely diffused consumers thus rendering possible large-scale operation on the producers' side. The grid ensures increased availability of energy on the consumers' side. Decrease in costs results from the benefits of typically large-scale installations. The high versatility of electric energy in end-use applications, a utility aspect of particular interest for the end-user, is essentially supported by the grid. Furthermore, it effects load levelling on the producers' side through the

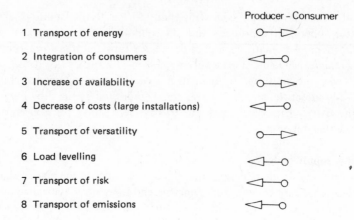

Fig. 7.6 – The first grid, electricity.

aggregation of several fluctuating load characteristics. Finally, in releasing the end-users from anxiety, providing against risk potentials and pollutant emission as would typically occur with stove heating for instance, the grid transfers risk and pollution control to the producer.

The present share of electricity in final supply, on a world average at about 11%, on account of the attractive supply conditions is a surprisingly small but indispensable part. Its potential for significant increase appears to be limited by pertinacious technological or economical constraints, above all by the high utility of hydrocarbons which are well adapted to the typical small-scale applications in the end-use sectors. The power input requirements of typical production units in the energy-intensive industrial sectors are relatively small in comparison with the sectoral power demand. This can be seen from the values given in Fig. 7.7 for the totals of sectoral heat power demand and for average establishments, which refer to the situation in Great Britain in 1975 [7.3]. Qualitatively, one has similar end-use conditions in all industrialised countries. That is the very reason why nuclear energy with its characteristic power units had to engage in the field of electricity generation which operates on the basis of a well established grid.

**The second grid**
Obviously, we need a second grid for the integration of a vast variety of widely diffused end-use applications of high utility secondary energies, in addition to the electric grid. With regard to the operational conditions we already have a well-established second grid. It is distributing the liquid hydrocarbons, not necessarily by means of pipelines, but also by road, railway, and ship. A characteristic feature is that secondary energies are transported and delivered as energy

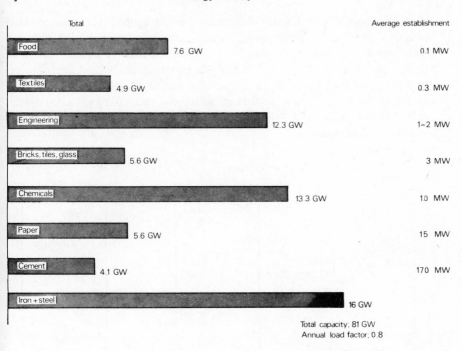

Fig. 7.7 – Total power demand for industrial heat in Great Britain, 1975

packages. It is based on the high utility of liquid hydrocarbons which comprises such advantageous qualities as:

- high energy densities (by volume and by weight) permitting easy storage and transportation at low cost;
- simple chemistry with high degrees of cleanliness and versatility.

Therefore, if we are to advance the utilisation of solid fossil fuels which have a considerable supply potential for the future, we must engage in coal liquefaction and coal gasification.

Referring to the perspective of a largely-extended exploitation of fossil fuel resources that even uses those of low grade, we have to face the environmental concomitants, surely not in the traditional way which looks for additional abatement measures and higher chimneys, but as a principal element of the technical system in question. In Eastern Europe, for instance, they are already burning brown coal with heating values of 1400 kcal/kg, and they are preparing for 900 kcal/kg, with chimneys of 300 m height which merely ensure transferring the dirt from Bulgaria to Turkey, but that is no solution. How much worse will things become with large amounts of shale? Let me raise, therefore, the principle question: How do we increase the utilisation of fossil fuels in connection with the second grid? Figure 7.8 may serve to fix the time available for an appropriate technical solution. We are not going to face the distant future, we will not refer to the investive way of utilising the immense resources of nuclear and solar energy. Nor will we insist on continuing present trends and technologies of more or less well established supply with clean oil and gas, coal utilisation expiring. We shall consider an intermediate situation within the next 20-30 years, of largely extended low grade solid fuel utilisation, and the possibility of an appropriate strategy leading through.

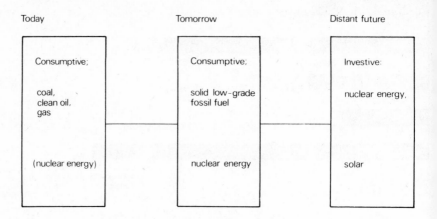

Fig. 7.8 – Primary energy.

To proceed along this line we will consider the following:

*Thesis:*
What the electric generator is to the first grid, the production of synthesis gas ($CO$, $H_2$) is to the second grid.

Synthesis gas is a very important feedstock for chemical synthesis. Almost all chemicals derive from synthesis gas which itself has a potential base in coal. Energy applications, however, will probably constitute a much higher importance for energy supply, because methanol can be directly synthesised from synthesis gas, and hydrocarbons with gasoline specifications either from methanol by the Mobil Oil process, or from synthesis gas by the Fischer–Tropsch process. Besides, energy utility of methanol compares fairly well with that of current liquid fuels. Regarding synthesis gas as a key to methanol, and to higher hydrocarbons, and to the supply potential of coal we may consider the following:

'Zero emission systems'

This means that emissions of pollutants, such as $SO_2$, $NO_x$, $CO_2$ and dust, into the biosphere should really be zero. From an engineering point of view it necessitates qualitatively new solutions, in particular, closed cycles for the mass flows.

Looking for an approach to new technical assignments one often realises that many existing technologies could acquire a breath-taking importance for tomorrow, if they were appropriately modified. One of these technologies is air separation allowing economic production of pure oxygen. Coal combustion with pure oxygen brings about a sharp reduction of the $NO_x$ formation, and additionally, an 80% reduction of the gas flows to be admitted and carried off from the reaction volume. Both effects result in less emission of pollutants even from open flow systems. However, we are interested in mastering the combustion products without a smokestack, and we establish a technical solution as shown by Fig. 7.9. There is a furnace for high pressurised combustion of coal, for instance of low-grade brown coal mined from the ground, supplied with pure oxygen from an air separation facility. The carbon dioxide flue gas is expanded from a pressure of 120 bar to a pressure of 50 bar to drive a gas turbine, thereafter it is conducted through a heat exchanger, where heat is decoupled to be fed to a steam turbine. At 50 bar, 15°C, carbon dioxide is a liquid and a good solvent; therefore it may be used for the transport of the ashes from the furnace

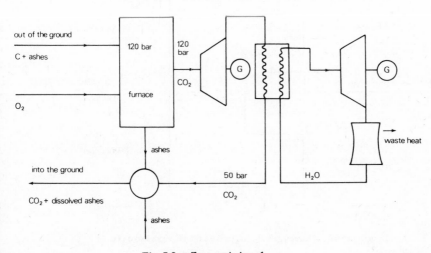

Air separator yields pure $O_2$
 in open combustion of C
   • circulated gas reduced by 80%
   • sharp reduction of $NO_x$ generation
but:
  we are now interested in closed cycles (without stacks)
C + ashes + $O_2$ → electricity + heat + $CO_2$ as dissolvent for the ashes

Fig. 7.9 – Zero emission, 1.

and to be deposited together with the other reaction products in soil or deep sea. Carbon dioxide, is also judged to be of great importance in tertiary oil recovery and is promising big business: US oil companies are considering proposals to produce more oil from the ground by injection of $CO_2$ than had to be burned for its production. This, to some extent, looks like an oil breeding scheme. The particular advantages of either solution have still to be studied.

### Methanol from methane and coal

Switching over from future power plants and electricity to potential second grid units and methanol I would like to start with a conversion facility fed by methane instead of coal. This appears to be a fair approach because coal is a solid, methane a gas, both not directly running short, and also because the reaction $C + CH_4 \rightarrow$ $(-CH_2-)$. (This does not represent an actual chemical equation but symbolises a production scheme of gasoline, certainly a liquid of special interest.)

Methane and water can be converted into methanol and hydrogen by the application of high temperature heat. The overall balances of the sequence of reactions constituting the process and of the energy flows are given in Fig. 7.10. As indicated, we consider allothermal heat supply from the High Temperature Reactor (HTR). If the energy flows are calculated in relative units of power, it

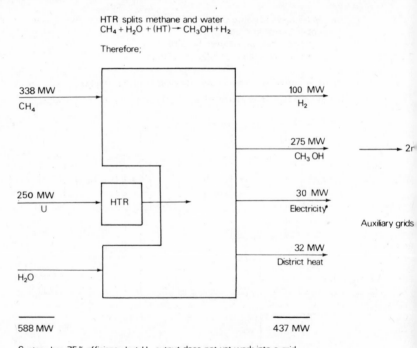

HTR splits methane and water
$CH_4 + H_2O + (HT) \rightarrow CH_3OH + H_2$

Therefore;

| | |
|---|---|
| 338 MW | 100 MW |
| $CH_4$ | $H_2$ |
| | 275 MW |
| | $CH_3OH$ |
| 250 MW | 30 MW |
| U  HTR | Electricity |
| | 32 MW |
| | District heat |
| $H_2O$ | |
| ———— | ———— |
| 588 MW | 437 MW |

System has 75 % efficiency, but $H_2$ output does not yet work into a grid

Fig. 7.10 – Zero emission, 2a.

turns out that the energy input due to methane (338 MW) equals the total of outputs due to methanol (275 MW), electricity (30 MW), and district heat (32 MW). Hence, the power input which is required to split the water and which reappears as chemical energy of the hydrogen output (100 MW), can actually be regarded as a contribution from the HTR whereby, with respect to environment, water shifts to hydrogen without any formation of carbon dioxide. The efficiency, which relates to the total energy input, is 75%. Therefore, the system not only produces methanol as an additional liquid for the existing second grid, thus increasing its flexibility, but also provides the strategically important and, on a long-term basis, essential step to engage in hydrogen technology.

Today, hydrogen is of increasing importance for the conversion of residual fuel oil and, with regard to the utilisation of fossil fuels, it will be so for many decades to come. The infrastructure, however, for its large-scale utilisation as secondary energy is not yet available and will require long development times. Are there other possibilities to integrate hydrogen into a strategy of transition? We can feed the 100 MW hydrogen yield, resulting from the process in Fig. 7.10, to a chemical reactor for the hydrogasification of coal, and take up 237 MW of dirty fossil fuel to obtain 237 MW of methane which is produced in absolute seclusion from the atmosphere (Fig. 7.11). Thus, we are following a consistent strategy of zero emission.

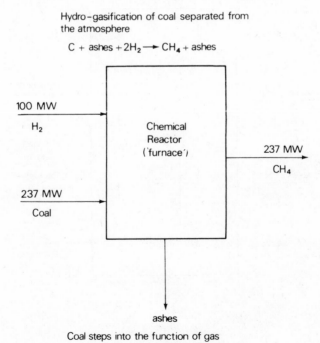

Hydro-gasification of coal separated from the atmosphere

$$C + ashes + 2H_2 \longrightarrow CH_4 + ashes$$

100 MW
$H_2$

Chemical
Reactor
('furnace')

237 MW
$CH_4$

237 MW
Coal

ashes

Coal steps into the function of gas

Fig. 7.11 – Zero emission, 2b.

### Use of the High Temperature Nuclear Reactor

The balance sheet of overall energy flows in a combination of both processes which accounts for the 237 MW methane feedback to the original 338 MW input, is shown in Fig. 7.12. By the application of 250 MW heat power from the HTR, a 101 MW methane input shouldering 237 MW of coal has been transformed into 275 MW methanol of immediate utility for the second grid, and additionally, into electricity and district heat. By the use of coal and the HTR methane can be converted into 2.75 times the amount of liquid fuel (methanol) energy thus fulfilling the function of a utility injector rather than that of an energy injector. One could also say that it facilitates the marriage between gas and coal which is an interesting perspective since there will be available more natural gas than formerly expected.

The pebble-bed reactor at Jülich is capable of performing the functions of an HTR, so far as the nuclear core is concerned. Our research and development efforts are concentrated on problems of material technology. Up to now, we have not yet had the alloys to utilise the heat at the required high temperature

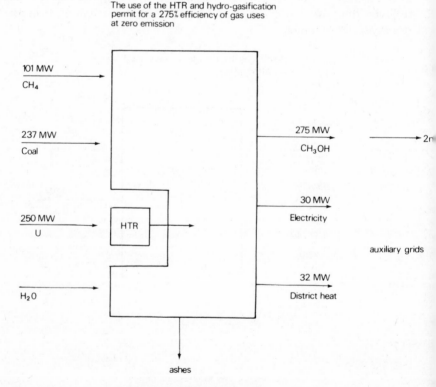

The use of the HTR and hydro-gasification permit for a 275% efficiency of gas uses at zero emission

101 MW
CH₄

237 MW
Coal

250 MW
U

HTR

H₂O

275 MW
CH₃OH

2n

30 MW
Electricity

auxiliary grids

32 MW
District heat

ashes

Fig. 7.12 — Zero emission, 2c.

level, for instance, of heat exchangers, or steam generators or steam reformer channels. Having reached test operation times of 2000–3000 hours with nickel alloys we are now on the right track and we are about to demonstrate this: we intend to construct a 250 MW module in correspondence to the presented scheme. We are working furthermore on the Adam and Eve project where the high temperature reactor heat will be used for the formation of synthesis gas. This is then conducted to the places of final consumption to discharge its energy by catalytical synthesis of methane and thereafter be recycled to the nuclear power plant.

## THE TRANSITION PERIOD

Thus, there are three different modes of using gas in energy supply (Fig. 7.13). It can be used in the consumptive mode to feed immediately the energy services as

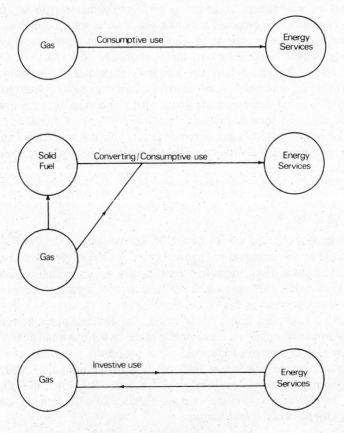

Fig. 7.13 – Different uses of gas.

exercised in today's supply system. In the very long run, it must be used in an essentially investive mode, saving our energy resources and preserving a salubrious environment. In a transition period to be started very soon, gas could already shoulder the fossil solid fuels, thus keeping pace with the demand of the end-users for appropriate secondary energies as well as with the aggravating need for conversion of fossil fuels by clean energy technologies. I think that any reflection upon future energy strategies will be confronted with these transitional aspects.

One is right to object that within the intermediate range, zero emission cannot be achieved because $CO_2$ and $NO_x$ will be released anyway due to the combustion of methanol and hydrocarbons in the end-uses. If we consider however the end-use supply by liquids reduced to the essential applications, only 50% of the future final energy consumption would have to be based on hydrocarbons. Only 50% of that 50% share gives rise to emission of carbon dioxide, the other 50% being contributed by the oxidation of hydrogen producing water. Compared to the situation of today and to present trends, this would constitute an improvement of the energy benefit to pollution ratio by a factor of 3 or 4. Therefore, the proposed scheme gains time, at least a couple of decades which might be decisive. Let us have a look once again at the energy system of the future which I referred to at the beginning. In 100–150 years time, the carbon dioxide emissions into the atmosphere should be limited to those amounts that are compatible with the natural carbon cycle. Our investigations at IIASA indicate that these amounts would correspond to an energy consumption of 5 TWyr/yr or 5 thousand million tonnes of coal/yr or 3.5 thousand million tonnes of oil/yr, to give some reference figures. All the remaining demand for fuels beyond these 5 TW would eventually have to be met by liquid hydrogen in this far future.


## CONCLUSIONS

For the decades to come we must establish and perfect a strategy which makes prudent use of the carbon atom, hydrogen, and nuclear power, being aware of a world in transition from traditional consumptive energy utilisation to a stable energy system that provides an actually inexhaustible supply. A time frame for this transition is given in Fig. 7.14. By the year 2030, electricity should predominantly be supplied from nuclear power. The essential point, however, is to get away from the traditional consumption of the carbon atom and hydrocarbons, and to engage in investive technologies that ensure the possibility of a gradual build-up of a second grid for the final supply with hydrogen after 2030. That is a frame to make our choices. It is my personal conviction that in realising the choices for the future, we might even get a chance to cope with the challenges of energy supply and the environment.

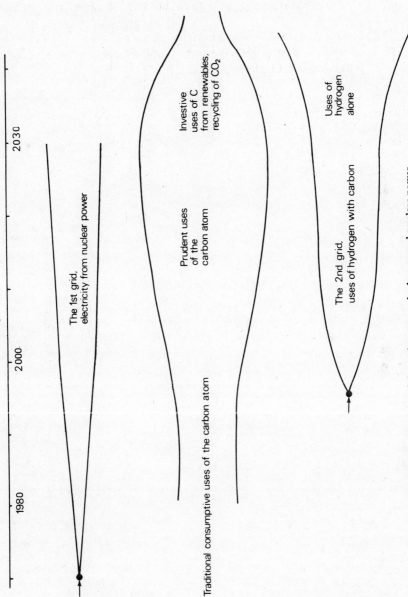

Fig. 7.14 – The prudent uses of the carbon atom, hydrogen and nuclear power.

## REFERENCES

[7.1] IIASA (1981), *Energy in a finite world – A global systems analysis*, Ballinger.
[7.2] Lovins, A. B. (1976), Foreign Affairs 55 No 1 65-96.
[7.3] Llewely, G. T. W. (1975), Proceedings of the European Nuclear Conference (ed) P. Zaleski, Pergamon Press 121.

# Part 2

# FEEDSTOCK FUTURES FOR THE CHEMICAL INDUSTRY

*Chairman:* **P. P. King**, General Secretary,
Society of Chemical Industry

# 8

# Sustainable chemical feedstock futures

**Andrew Stratton**, Consultant

## INTRODUCTION

The basic issue addressed by this conference is whether, in the long term, an economically viable heavy chemical industry can exist in Europe, in competition with that of North America. The chemical industry shares with the economy as a whole its dependence on an adequate supply of energy at an economic price. It is unique, however, in using hydrocarbon fuels with high opportunity cost elsewhere, as feedstock for petrochemicals.

This paper examines some of the technological and resource options open to the West European chemical industry. To do so it is necessary to look well beyond the lifetime of existing plants and processes, into the first 25 years of the next century.

## FUELS AND FEEDSTOCK IN WESTERN EUROPE

Figure 8.1 illustrates the conversion of natural energy resources into fuels for use by the consumer. Nuclear fusion in terrestrial reactors is unlikely to contribute significantly in the time scale; the solar fusion reactor is the source of the so-called renewable energy resources namely direct solar radiation, hydro-electricity, wind and wave power and biomass.

Compared with North America, Western Europe is not well endowed with natural energy resources. North America, in particular, has very large reserves of petroleum, in the form of shale oil and tar sands, and of coal; Many areas, such as California, enjoy an annual level of solar radiation twice that of Western Europe [8.1]. Apart from North Sea oil and gas, West European hydrocarbon resources are basically bituminous coal in the UK and bituminous coal and lignite in West Germany.

Liquid fuels for transport and natural gas for space heating dominate the consumer requirements for hydrocarbon fuels.

Figure 8.2 illustrates chemical usage of fuels for process energy, chemical reactant and feedstock. Typical reactant usages are carbon as a reducing agent in

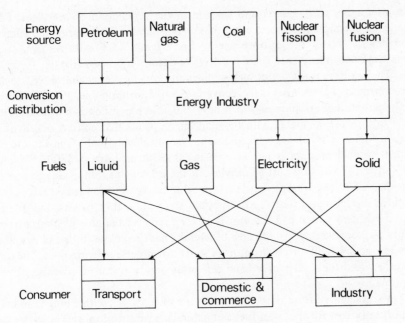

Fig. 8.1 – Conversion of primary energy into fuels.

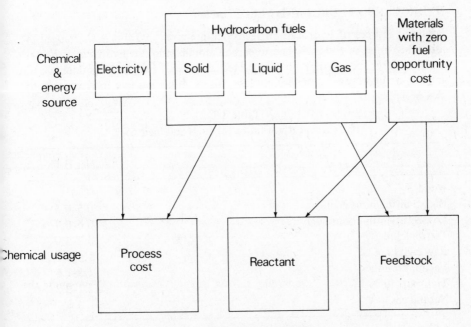

Fig. 8.2 – Chemical usage of fuels.

the production of hydrogen and calcium in the production of acetylene. For inorganic chemical processes, using materials with zero fuel energy, the energy cost linkage is limited to process energy. For sulphur, a fuel with zero (or even negative) opportunity cost, there is a positive output of process energy.

The key issue is the use of high opportunity cost fuels as chemical feedstock. Petrochemical plants essentially convert one hydrocarbon fuel into (an)other hydrocarbon fuel(s), some of which have desirable properties as chemicals. The only difference, compared with fuel conversion plants that convert one hydrocarbon fuel into other desirable and hence more valuable fuels, is in the use to which the fuel product is put as a chemical. With naphtha at $300/t† about 65% of the production cost of ethylene is the price of naphtha; added value is less than 15%. The European petrochemical industry was founded on the use of naphtha at an opportunity cost substantially below that of gasoline [8.2]. Improved plant design and economies of scale reduced the cost of olefins from naphtha. Now the opportunity cost of naphtha has risen close to that of gasoline and low added value offers little or no scope for process improvement. The use of other feedstock such as ethane can offer only a transient advantage over naphtha.

A sustainable future must rest on the use of feedstock with a long-term low opportunity cost for transport fuel and natural gas substitution, and plant design aimed at achieving high thermodynamic efficiency and minimising carbon usage.

## HYDROCARBON FUELS AND FEEDSTOCK

Potential sources of hydrocarbon feedstock are given in Table 8.1, arranged in ascending order of the hydrogen to carbon atomic ratio; the hydrogen is the net available after oxygen has been extracted as water, nitrogen as ammonia and sulphur as hydrogen sulphide. Cellulose, for example, has one oxygen atom to two of hydrogen.

**Table 8.1**
Hydrogen–carbon ratios for fuels and feedstock

| | Net H/C |
|---|---|
| Wood | 0 |
| High S bituminous coal | 0.5–0.6 |
| Low S sub-bituminous coal | 0.6–0.7 |
| Tar/shale oils | 0.7–1.0 |
| Raw coal liquids | 1.4–1.6 |
| Vacuum residue | 1.5–1.6 |
| Transport fuels | 2.1–2.2 |
| Natural gas | 4.0 |

† t = metric ton.

In converting the lower H/C ratio hydrocarbons to transport fuels and substitute natural gas (SNG), carbon is used in three ways:

(1) carbon in the product;
(2) hydrogen production;'
(3) process energy.

The carbon used in hydrogen production and process energy is immediately converted to $CO_2$; the product carbon is eventually converted to $CO_2$. One of the factors that has to be taken into account, in addition to the resource implications of carbon usage, is the uncertainty of the effect of $CO_2$ emissions on climate [8.1].

The consumption of carbon is illustrated in Fig. 8.3 for the conversion of coal to liquid fuels and gas. The carbon usage ratios are for efficiencies that should be obtained in production plants.

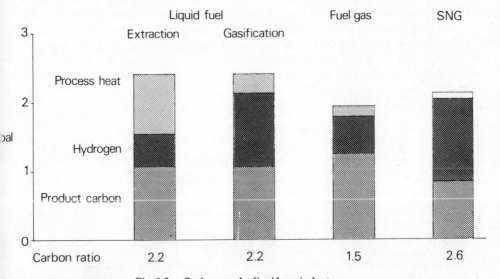

Fig. 8.3 – Coal usage: 1 t liquid equivalent.

The extra hydrogen requirement in liquefaction via gasification, as for any synthesis gas process, arises from the two extra hydrogen atoms required to remove the oxygen from each CO molecule; this reaction generates the balance of the process heat.


## AVAILABLE ENERGY (EXERGY)

The full impact of H/C ratio becomes apparent by examining the power flow in chemical conversion, through the second law of thermodynamics.

The availability of energy, or exergy, is the maximum work that can be obtained during a change of chemical or physical state in a defined environment; for a continuous flow process it is given by equation (8.1).

$$\text{Maximum work} = EX_1 - EX_2 = (H_1 - H_2) - T_0 (S_1 - S_2) \qquad (8.1)$$

where: $EX_1$, $EX_2$ are exergies in states 1 and 2

$H_1$, $H_2$ are enthalpies
$S_1$, $S_2$ are entropies
$T_0$ temperature of environment, $^\circ K$

For fuels, the zero enthalpy and entropy states are referenced to $H_2O$ (l), $CO_2$, $O_2$, as zero, at 298K and 1 atmosphere. It is convenient to express the maximum work as a power flow in MW(e); it is not necessary that electricity should be generated.

Figure 8.4 gives the second law analysis of the production of ($CO + 2H_2$) synthesis gas from methane, using the HHV energy efficiency of 70% from [8.3]. The exothermic reaction (only the energy components are shown) potentially generates 24 MW(e) per t mole per hour of carbon converted. An additional 0.37 t mole/h of carbon is used for process heat, with a potential of 78 MW(e). The power actually generated is taken to be 35% of the total 102 MW(e) available,

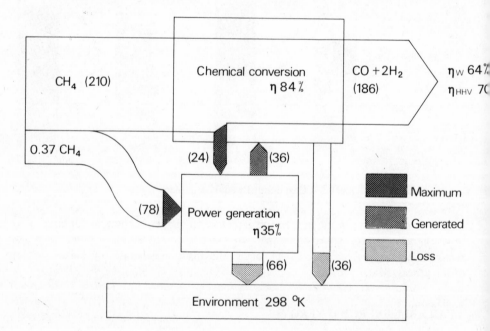

Fig. 8.4 – CO + 2H₂ synthesis gas from methane, power flow MW(e)/t mole/h carbon converted.

which is a normal figure for good power generation practice. The 36 MW(e) of power supplied to the chemical process is lost, irreversibly, giving a chemical conversion efficiency (power out)/(power out + irreversible loss) of 84%.

Figure 8.5 gives a similar analysis for coal feedstock. In this case the endothermic reaction requires a power input of 51 MW(e). At the heat efficiency of 53% [8.3], 2.3 t moles/h of carbon have to be supplied as coal. This input of

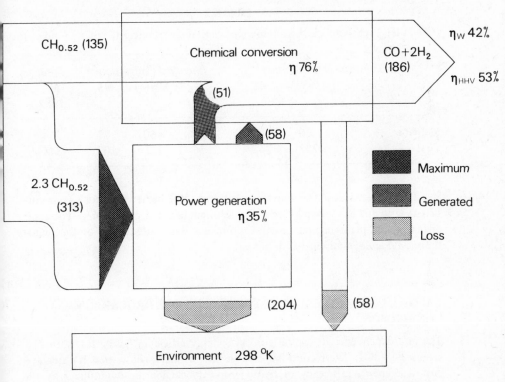

Fig. 8.5 — $CO + 2H_2$ synthesis gas from coal, power flow MW (e)/t mole/h carbon converted.

313 MW(e) converts at 35% to 109 MW(e) of power generated; the 58 MW(e) above reaction requirements is irreversibly lost, giving a chemical conversion efficiency of 76%. There is thus little difference in the chemical efficiency of the two processes. The difference lies in the requirement to generate three times the power for coal conversion to synthesis gas as compared with natural gas. The engineering limitation, today, of 35% efficiency of power generation, lies in the temperature limits of materials used in high pressure steam generation. The availability of suitable gas turbines would raise this temperature limit to that of the gas turbine blades.

The requirement for power generation also reacts on the capital cost of the plant. Table 8.2 takes the capital costs of the plant of [8.3] for producing 150 million CFD of synthesis gas (7.5 t moles/h) and breaks it down into power generation and chemical conversion. The cost assumption used is that a 1 GW(e) power generation plant costs $600/kW with coal and $450/kW with gas; these costs are then scaled in Table 8.2 with a 0.7 power law.

Table 8.2

(CO + 2H₂) synthesis gas; break-down of capital cost, conversion of 7.5 t moles/h

| | Power Generation | | | Chemical Conversion | | | Total |
|---|---|---|---|---|---|---|---|
| | % | MW(e) | M$ | % | MW(e) | M$ | M$ 1982 |
| Methane | 35 | 90 | 83 | 84 | −60 | 71 | 154 |
| Coal | 35 | 273 | 242 | 76 | 128 | 102 | 344 |

The deduced cost for conversion of coal, 50% higher than for gas, is consistent with the extra cost of coal and ash handling and gas cleaning. The 2.2:1 difference in plant capital cost is a fundamental consequence of the higher thermodynamic requirement for power.


## CAPITAL COST AND ENERGY EFFICIENCY OF HYDROCARBON CONVERSION

The interdependence of capital cost and energy efficiency has been reported by Gaensslen [8.4]. The second law analysis in this paper shows that both capital cost and energy efficiency of hydrocarbon conversion are determined by the available power generation technology and, furthermore, that both are a function of the change in available chemical energy (or exergy) in the conversion process. This is illustrated more fully in Fig. 8.6 which gives the chemical exergy, relative to $CO_2$, $H_2O$ (l) and $O_2$ as zero, of some common fuels, feedstock and products. Due to limitations in entropy data and of chemical complexity, crude oil is not shown; reference fuels octane, cetane and benzene are given. The exergy of coal is an approximate estimate.

To convert any hydrocarbon to another with a higher chemical exergy will incur, with present technology, an irreversible loss of about twice the chemical conversion requirement. Conversely, in moving down the scale, two-thirds of the available energy is irreversibly lost. These losses have to be paid for, not only in process fuel, but in capital investment. Processes that go up the scale and then

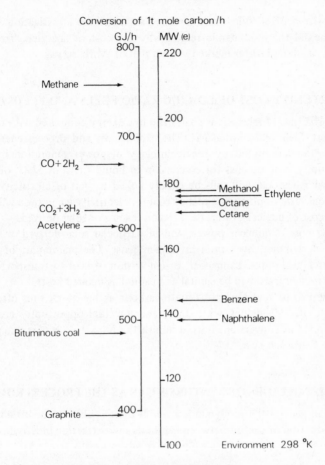

Fig. 8.6 – Available energy (exergy).

come down are thus to be avoided. Some conclusions from Fig. 8.6 on optimum thermodynamic route and capital cost are:

(1) the optimum synthesis gas composition for methanol is a mixture of CO, $CO_2$ and $H_2$;

(2) dehydrating methanol to ethylene is more efficient than dehydrating to an aromatic gasoline;

(3) with the optimum synthesis gas composition there should be little difference between direct synthesis of ethylene and the methanol route;

(4) solvent extraction of coal to give aromatics should have low capital cost and high efficiency; upgrading the product to transport fuels, however, will bring the total process energy loss and capital cost up to that of the optimum synthesis process.

The position of acetylene is interesting and warrants investigation. There is no fundamental thermodynamic reason why the cost of acetylene from coal and natural gas should differ markedly from that of synthesis gas.

## OPPORTUNITY COST OF LOW H/C RATIO FUELS AND FEEDSTOCK

A low H/C ratio fuel, such as coal, has a low exergy compared with the desirable transport fuels and natural gas. The capital cost and process energy for conversion are determined by the technology of power generation by coal. The opportunity cost of coal for conversion to liquid fuels, or SNG, or as a petrochemical feedstock, will thus be closely linked to that of electricity generation from coal. The opportunity cost of coal for electricity generation will be limited by the cost of nuclear power, be it fission reactor or solar derived.

The cost of nuclear power, and in particular solar derived, will be almost wholly determined by capital investment cost. The major part of the cost of electricity generation from coal, or conversion of coal to transport fuels, SNG and petrochemicals, will be capital and capital associated costs.

The use of low H/C ratio hydrocarbons as feedstock thus offers the dual prospect of: a feedstock with stable and low fuel opportunity cost; and high added value in conversion. Both diminish the impact of geographic price difference on feedstock cost.

## THE ADVANTAGES OF SYNTHESIS GAS AS THE PROCESS ROUTE

Whilst on general thermodynamic grounds there is likely to be little to choose in the production of olefins between synthesis and extraction plus hydrogeneration, the synthesis gas route offers a number of advantages:

(1)  independent of the chemical composition of the feedstock;
(2)  potentially highly selective conversion through catalyst development;
(3)  the use of methanol made from low opportunity cost non-transportable fuels (e.g. lignite, high ash coals), as a means of transporting synthesis gas.

For reasons already stated, however, extraction is likely to be the most economic route from coal to aromatics.

## CONSERVING CARBON

Conservation of carbon by restricting consumption to that required in the product makes the best use of indigenous resources, reduces $CO_2$ emission, and may have economic advantages. Thermodynamic analysis has shown, for example, that conversion of one t mole/h of carbon to synthesis gas, requires about 110 MW(e) of power to produce hydrogen and to provide process power.

In principle, 2.3 t moles/h of carbon could be saved by substituting 110 MW(e) of non-fossil fuel generated electricity (nuclear fission or solar derived). With the close link between power generation costs that has been discussed earlier, the economics should be comparable. To achieve the same thermodynamic efficiency, however, requires careful process selection and design, for the production of hydrogen and the use of power.

Whilst thermochemical water-splitting reactions, other than carbon, have received much attention, none so far achieve even a comparable 'first law' efficiency. Hydrogen by electrolysis, however, is an efficient use of available power. An intermediate step that might have process advantages is the possibility of direct electrolysis of coal [8.5] particularly if a $CO + 2H_2$ synthesis gas could be generated directly.

Direct substitution of electricity for process heat is thermodynamically inefficient. The use of heat pumps to produce high temperature heat, starting from the base temperature of, say a nuclear reactor, is one possibility. High temperature electrochemical reactors that generate process power (fuel cells), rather than process heat, would increase the efficiency of power recovery from exothermic processes.

The ultimate carbon conservation, however, obtains by using $CO_2$ as a feedstock with electrolytic hydrogen to produce, say, methanol. This would raise the power requirement (Fig. 8.6) to about 180 MW(e)/t mole/h of carbon, plus irreversible losses. The economics would be highly dependent on the cost of electricity and the opportunity cost of carbon dioxide; restrictions on the emission of $CO_2$ could however make the latter negative.

## SUSTAINABLE WEST EUROPEAN CHEMICAL FUTURES

Returning to the question posed by the Conference, how can the relationships of Fig. 8.2 be changed in favour of Europe? While it lasts, North Sea gas, through synthesis gas, can support the chemical industry. Catalysts exist now, at least in the laboratory, for converting synthesis gas to olefins, either directly or indirectly through methanol. With the price of gas in both North America and Europe rising to the opportunity cost of gasoil for heating, advantage will only obtain through economically unexportable gas from small wells, for example in Alberta. The greater threat to Europe, as for North America, will be Middle East conversion of gas to petrochemicals.

The long-term future options are shown in Fig. 8.7. Electricity has replaced natural gas for heating and hydrocarbon resources are devoted to the production of liquid fuels for transport, in competition with imported oil at a price determined by North American production from shale oil, tar sands and coal.

The use of oil residues as a feedstock for synthesis gas is a transient possibility, but in the long term still leaves the chemical industry coupled to oil economics.

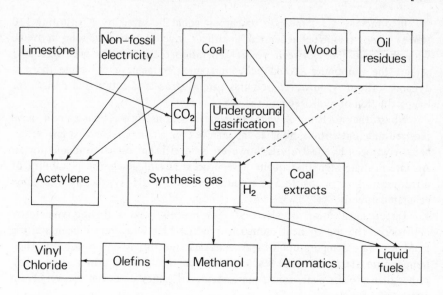

Fig. 8.7 – Sustainable European chemical future.

Studies done on wood by the Alberta Government [8.6] placed it at a disadvantage compared with coal, even in a highly forested region or at a centre of lumber residue. This is basically due to the cost of bringing wood to a conversion centre and the small scale of plant that can be supported. The higher extraction cost of European coal may move the economics more favourably to wood. Concentration on the large-scale production of 'packaged' plants for local use would bring down plant costs. Further investigation is required on this option; in the meantime the dotted connection in Fig. 8.6 signifies a much greater uncertainty than the other options.

This leaves indigenous or imported coal, supported by nuclear fission or solar derived electricity, as the potential feedstock.

Given improved plant thermodynamics, acetylene has a potential role, particularly for vinyl chloride, where the conversion cost is low. Coal extracts are likely to be the cheapest source of aromatics. Production of methanol and olefins would come from synthesis gas, in the production of which underground coal gasification may play a part. Recovery of $CO_2$ from coal combustion provides additional feedstock and limestone represents a virtually inexhaustible reserve. Nuclear energy substitutes for coal in hydrogen production and in process energy. The production of, say, 10 million tons of methanol would require less than 4 million tons of coal carbon and production of 10 million tons of olefins about 11 million tons of coal carbon.

Is such a future sustainable?

## THE CONSTRAINTS

Constraints in technology, resources, and implementation have to be considered.

### Technology constraints
Future research and development objectives that can be identified are:

- Thermodynamic design of process plants
- Materials that will allow heat extraction from combustion at higher temperatures
- Selective catalysts for the conversion of synthesis gas or methanol to olefins
- The development of electrochemical processes and electrochemical cells that will operate at high current density and high temperatures.

Given the will and the resources, all these should be achievable within twenty years.

### Resource constraints
The basic resource constraints are the availability and cost of the hydrocarbon feedstock, 'non-fossil' (that is, nuclear fission or solar derived electricity) and the general economic climate. It is clear from the IIASA studies [8.1] and from the limitations of indigenous coal supplies, that the long-term future of the West European economy is dependent on a massive increase in 'non-fossil' electricity. The investment cost of obtaining this from the solar reactor, whether by radiation, wind, waves, or hydro, will be much higher than for nuclear fission, the supply will be intermittent, installations will cover a large area and present major environmental problems.

If a society decides against nuclear fission, the economic penalty is likely to be high in competition with those who concentrate resources on ensuring the safety of nuclear fission reactors.

Is coal, then, an economically viable feedstock? The situation, today, is summarised in Fig. 8.8, which shows the break-even price for coal versus gas for methanol production [8.7]. A straight pre-tax DCF in constant $ 1979 has been used. The line for 15% DCF coal and 5% DCF gas indicates a typical approach to long-term investment in a new project, wherein a higher return is required for the risk of investing nearly three times the capital to give the same production as an existing plant that has been partly depreciated. Table 8.3, from [8.7], summarises the 10% DCF gas and coal break-even prices; the relevant cost data are given in the Appendix.

Viewed against current gas prices paid by the fertiliser industry in the United States of between $0.5/GJ and $2.0/GJ [8.8] and an average landed price of North Sea gas of $1.6/GJ [8.9], coal is not competitive. However, if natural gas is valued at its opportunity cost as the alternative to gasoil for residential heating, its value is about $6/GJ [8.10]; coal, particularly lignite and sub-bituminous, is then competitive.

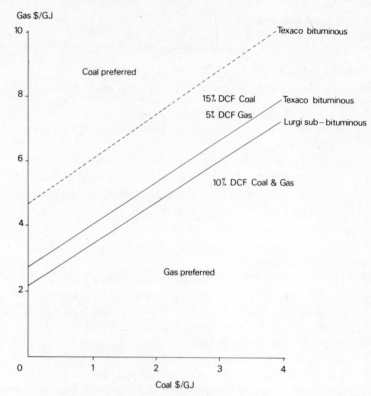

Fig. 8.8 – Methanol break-even price, coal v. gas.

**Table 8.3**
Competitive coal and gas prices for methanol production

|  | *Coal price range* $/GJ (1979) | *Gas price at which coal becomes competitive* $/GJ (1979) |
|---|---|---|
| *Western Europe* | | |
| Imported or indigenous hard coal | 2–3.5 | 4.6–10 |
| UK hard coal | 2.9 | 5.8–9.1 |
| West German Lignite | 0.9 | 2.7–5.3† |
| *North America* | | |
| Bituminous | 1.5–2.5 | 4.0–8.6 |
| Sub-bituminous | 0.5 | 2.6–5.8 |

† This assumes that West German lignite can be gasified at the same cost as the sub-bituminous reference coal of Fig. 8.8, not necessarily by the Lurgi gasifier.

A 50% increase in coal price from \$2/GJ to \$3/GJ adds less than 20% to the cost of methanol. This linkage would be reduced to about 6% if 'non-fossil' electricity provided all the power requirements. When combined with the fundamental thermodynamic reasons for the opportunity cost of coal remaining low relative to oil and gas, the cost of methanol and petrochemicals from coal, whether in Western Euripe or North America, could be determined by capital investment in new technology and not geographic differences in feedstock price. This poses the major issue, that of implementation constraints.

**Implementation constraints**

The scenario outlined requires flexibility of approach to the technological future and investment in new processes. The industry has relied overlong on economies of scale; only recently has it paid attention to efficient thermodynamic design and in doing so has concentrated on short pay-back times. It has largely contracted out of the engineering design of plant. By contrast, catalysts have been developed in the laboratory that show good prospects for the future; investment in pilot plant to develop and test production catalysts has however been lacking.

The present situation of economic recession and a temporary glut of oil does not encourage the long-term approach. Nevertheless, if Western Europe is to be competitive in 2025 there is little time to spare for developing new technology, building and proving pilot plants and implementing a long-term investment strategy that does not penalise new processes, as does the 15% to 5% line of Fig. 8.8.

The chemical industry, however, can only operate within the constraints of government policies on energy and of investment. Maintenance of present lifestyles and future economic growth are dependent on availability and price of fuels. A policy for the chemical industry of concentration on coal (and possibly wood) as carbon feedstock and on nuclear fission or solar derived electricity, is only viable within a long-term committed energy policy to a sustainable energy future in 2025. This will require firm policy decisions on nuclear electricity and large national capital investments.

There is a large difference, currently, between the price and cost of oil. This economic rent is distributed between producer, consumer and government. Government can use its share for current consumption, short-term or long-term investment. Currently only a small proportion of the economic rent is applied by industry or government towards the type of developments that appear necessary to secure a sustainable energy future for Western Europe and for the chemical industry.

**CONCLUSIONS**

A sustainable chemical feedstock future for the West European chemical industry could exist by using low H/C ratio feedstock such as lignite, hard coal and, possibly, wood. On thermodynamic grounds these feedstocks can be expected to maintain a low opportunity cost relative to liquid transport fuels and gas. In the transition, a wider range of feedstocks would be used. Synthesis gas would be the main chemical route, with a possible contribution from acetylene.

Substitution of nuclear generated electricity (fission or solar) for the conventional fossil fuel power generation in existing plants could reduce the carbon requirement, and hence the sensitivity to feedstock cost.

There are no technological constraints that could not be overcome in the time scale. Implementation, however, will require a concerted effort by the industry, within the framework of long-term policies by government on energy and large-scale capital investment.

## REFERENCES

[8.1] Hafele, W., (1981), *Energy in a finite world, paths to a sustainable future,* Ballinger.

[8.2] Stratton, A., (1982), 'Decoupling the industry from oil' in *The Chemical Industry,* Ellis Horwood.

[8.3] See Chapter 12.

[8.4] Gaensslen, H., (1980), 'Thermal efficiency and production economics of chemical plants', *CHEMSA,* June.

[8.5] Coughlin, A. W. and Farooque, M. (1979), 'Hydrogen production from coal, water and electrons', *Nature,* 279, 301–303, 24 May.

[8.6] Alberta Energy & Natural Resources, (1979), 'Energy & chemicals from wood', *ENR Report,* No. 90, March.

[8.7] Stratton, A., Hemming, D. F. and Teper, M., (1982), 'Methanol production from natural gas or coal', Report No. E4/82, Economic Assessment Service, *IEA Coal Research,* December.

[8.8] *Nitrogen,* Jan-Feb, (1981), No. 129, 5.

[8.9] British Gas Corporation, (1980/81), *Annual Report & Accounts,* Her Majesty's Stationery Office, London.

[8.10] Stratton, A., (1980/81), 'Feedstock resources for the world's petrochemical industry', *Process Economics International,* 12, No. 2, Winter, 32–36.

## APPENDIX

**Fuel methanol from coal and natural gas, cost data $ mid-1979 DCF pre-tax**

|  |  | Texaco bituminous | Lurgi sub-bituminous | Gas |
|---|---|---|---|---|
| Fixed capital (CP) | M$ | 3380 | 3140 | 203 |
| Build | yrs | 5 | 5 | 3 |
| Life | yrs | 20 | 20 | 20 |
| Production | SD/yr | 310 | 310 | 333 |
| Feedstock | $10^6$ GJ/yr | 265 | 282 | 35.0 |
| Methanol | Mt/yr | 5.9 | 5.4 | 1.0 |
| By-products | $10^6$ GJ/yr | – | 25 | – |
| Efficiency | HHV % | 51 | 52 | 65 |

|  |  | DCF 10% | DCF 15% | DCF 10% | DCF 5% | DCF 10% |
|---|---|---|---|---|---|---|
| *Operating costs* |  |  |  |  |  |  |
| Annual capital charge | %CP | 15.4 | 23.5 | 15.4 | 9.0 | 14.4 |
| Other non-fuel | %CP | 7.8 | 7.9 | 8.1 | 6.9 | 7.0 |
| Total non-fuel | %CP | 23.2 | 31.4 | 23.5 | 15.9 | 21.4 |

Plants are on a 'greenfield' site USA/Europe.
*Source:* [8.7].

# 9

# Coal liquefaction processes

Professor **Werner Peters** and **Dr Ingo Romey**, Bergbau-Forschung GmbH, West Germany

## INTRODUCTION

Due to the fact that securing the petroleum supply is becoming difficult and costly for countries that are largely dependent on imported oil, investigations into coal technologies has gained substantial interest again all over the world.

For the production of motor fuels in the future, mineral oil can be substituted partly by coal liquefaction processes which are under development or available on a commercial scale.

According to the current state of technology the liquefaction of coal can be accomplished in three different ways:

(1) synthesis;
(2) hydroliquefaction;
(3) solvent extraction.

This paper describes and compares the different processes from technical and economic viewpoints and gives an outlook on further developments in coal liquefaction.

## SYNTHESIS

The synthesis of hydrocarbons has been developed in Germany. F. Fischer and H. Tropsch provided the foundation for the technological Fischer–Tropsch synthesis during the 1920s; the process was subsequently used in commercial operations with coal as a starting material. Prior to 1945, nine installations with a total capacity for producing petrol/gasoline at 750 000 t/year had been constructed. However, the importance of hydrocarbon synthesis was not very great in comparison to that of direct liquefaction of coal.

For the same coal as starting material, the production of motor fuel by direct hydrogenation is more economical than synthesis. Nevertheless, in South

Africa there exist vast reserves of a coal which is cheap but has a very high content of ash and is thus unsuitable for hydrogenation. With the use of this coal as raw material, South Africa is endeavouring to become less dependent on imported crude oil.

### Fischer–Tropsch process

The method for obtaining liquid products from coal differs fundamentally from extraction and hydrogenation; it proceeds indirectly by way of gasification of the coal. In the first stage synthesis gas is generated from the coal. By means of the Fischer–Tropsch synthesis, carburetor motor fuels and diesel oil can be obtained. The synthesis step occurs at the surface of a special iron catalyst, according to the general equation:

$$n(CO + 2H_2) \longrightarrow n\,(-CH_2-) + nH_2O - \Delta H$$

Whereas the process of gasification is strongly endothermic, the synthesis reaction is highly exothermic and much heat must be removed.

Quite logically the production of methanol from synthesis gas can be carried out as a form of indirect liquefaction of coal. In addition to the possible immediate use as carburetor motor fuel, the newly developed Mobil M process has increased the interest in methanol. By means of the latter method petrol/gasoline is produced with a good yield by a zeolite catalyst.

The Fischer–Tropsch process is at present the only technique which is commercially applied for liquefying coal.

Despite the greater economic drawback of synthesis in comparison to hydrogenation, South Africa erected a complex, Sasol I, for synthesis, in 1955; it currently still converts coal to oil at a rate of 10 000 t/day, also producing raw materials for the chemical industry. The decision in favour of synthesis was made in view of the vast reserves of cheap coal with a high content of ash; such coal is hardly suitable for hydrogenation.

A further synthesis plant, Sasol II, commenced operations in 1980. When operating at full capacity it will convert coal at a rate of 40 000 st/day to petrol/gasoline and diesel oil at a rate of 1.5 Mt/year. When the operation of Sasol III begins — presumably in 1983 — South Africa will obtain motor fuel from coal at a rate of about 3.5 Mt/year and thus satisfy about 30% of its motor fuel requirement from its own resources.

### Process description

The overall process of liquefaction (shown in Fig. 9.1) comprises two steps: gasification of the coal; and subsequent Fischer–Tropsch synthesis.

The coal, with an ash content of about 30%, is gasified with the use of steam and oxygen at a pressure of 30 bar; Sasol II thereby employs 36 Lurgi gasification reactors with a diameter of over 4 m. The raw gas for synthesis is subsequently treated in a Rectisol scrubber.

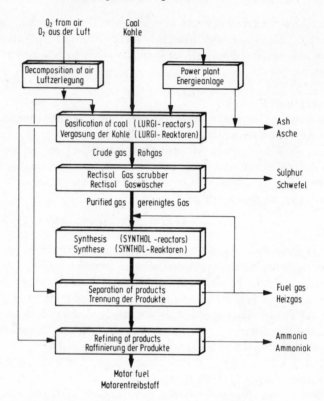

Fig. 9.1 – Fischer–Tropsch synthesis (Sasol II plant)
(Fischer–Tropsch–Synthese (Sasol II Anlage))

The synthesis step is carried out in a series of Synthol reactors (Fischer–Tropsch fluidised bed). The gas for the synthesis is passed through the reactors, together with the circulating gas, at 22 bar and 160°C. Fluidised, powdered catalyst (whose exact composition is kept secret) is continually metered into the gas stream from a stand pipe. The strongly exothermic reaction causes a rapid rise in temperature. By means of vigorous agitation of the gases and intensive cooling by water flowing through interior tubes, the temperature of reaction is kept to 335°C.

After leaving the zone of reaction the catalyst is separated from the product mixture in a precipitation chamber and in cyclones; it is then recycled to the stand pipe.

After indirect cooling the product gas is quenched. After removal of the water the condensates are refined to yield the end products. The remaining gas is partially recycled to the reactor as circulating gas and partially employed as fuel.

# HYDROLIQUEFACTION

## Bergius–Pier process

The development of direct hydrogenation originated with the work performed by F. Bergius from 1911 onward. In 1924 the patents of Bergius were transferred to I. G. Farben. At the plant of present BASF in Ludwigschafen the decisive progress was achieved under the leadership of M. Pier. As a result of this work the hydrogenation of coal evolved into a large-scale industry within the span of twenty years. This eventually comprised twelve plants for producing motor fuel from coal with a total capacity of nearly 4 Mt/year. Independently of I.G., only the British firm ICI succeeded in establishing a plant for hydrogenating coal on a technical scale in Billingham, with a petrol production rate of 100 000 t/year.

Subsequently, the results of the war, the high cost of coal production, as well as the availability of abundant and cheap crude oil, hindered further development and operation of coal hydrogenation plants for almost thirty years.

In the United States too, a large pilot plant was operated on the basis of German technology in Louisiana, Missouri until 1952, but no further work on coal liquefaction was conducted until the beginning of the 1960s.

Four methods of direct liquefaction of coal have been developed to such an extent during recent years that their large-scale technical applicability is now being demonstrated in large pilot plant operations:

(1)  the Kohleöl process (German technology);
(2)  the SRC II process;
(3)  the EDS process;
(4)  the H-Coal process.

Besides these there have been several other quite different developments. However, they have to be thoroughly tested in larger pilot installations before their technical applicability can be appraised.

## Kohleöl process (German technology)

In 1973, after the first major increase in the price of crude oil, the production of oil from coal became a subject of prime importance from the standpoint of energy policy in the Federal Republic of Germany. With both regional and federal government financial support, the worldwide status of coal liquefaction was investigated in detail in the Ruhrgebiet and Saarland. Moreover, the design and construction of pilot plants was begun in both districts.

From the inventory of the various processes of hydrogenation it became obvious that the old I.G. liquid-phase technique presented a feasible alternative to the American developments. Of course, several decisive modifications of the method had to be undertaken.

Since 1976 this modified liquid-phase hydrogenation process is being tested and confirmed at the pilot facilities of Bergbau-Forschung GmbH in Essen

and of the Saarbergwerke AG in Reden. In 1979 the Rheinische Braunkohlen-
werke AG also began operation of a laboratory installation by means of the same
process.

On the basis of results obtained from the new pilot facilities and from the
former large installations, two large-scale pilot plants have been designed during
the interim: Saarberg began operation of a facility for a coal input of 6 t/day
near Volklingen in 1981, Ruhrkohle AG and VEBA Oel AG recently started
operation of a plant for 200 t/day in Bottrop.

Developmental work is currently being condusted on preliminary projects
for large-scale technical facilities at several companies in the Federal Republic of
Germany. It appears possible that one or two commercial coal hydrogenation
plants will be operating by the end of the 1980s, though the construction will be
financed largely by the government.

*Description of the process*
The overall process is illustrated in Fig. 9.2. The coal slurry is prepared from
recycling oil, ground coal and about 2% of catalyst (red waste matter from the
aluminium industry). After the addition of hydrogen and circulating gas, the
slurry is heated and subsequently passes through reaction tubes at a temperature
of 460–480°C and at a pressure up to 300 bar. The products of the reaction, still
under high pressure, are then condensed in precipitators at various temperatures.

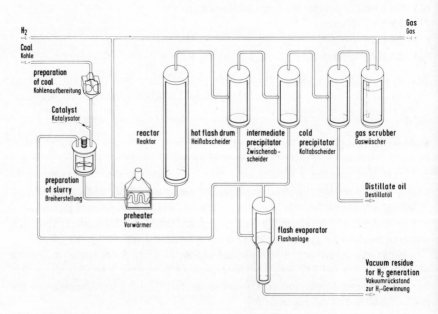

Fig. 9.2 – KOHLEÖL process
(KOHLEÖL-Verfahren)

The solids, together with the high-boiling and non-distillable oil, are removed from the hot flash drum. By means of flash evaporation in vacuum, oil is distilled as long as the residue remains pumpable in the molten state. This pitch-like residue, with about 50% solid matter, is gasified by partial oxidation under a pressure of 80 bar. After conversion the gas satisfies the hydrogen requirement of the hydrogenating facility.

The distillate from the flash evaporation is combined with the oil from the intermediate precipitator and recycled for use as priming oil for preparing the slurry.

In the cold precipitator medium and light oil is collected. This oil is the product oil from the process. It can be further refined to yield fuel oil, motor fuels and raw materials for the chemical industry.

The Kohleöl process differs from the basic I.G. liquid-phase method especially in the following respects:

(1) The formerly necessary elimination of the solid matter by centrifuging and low-temperature distillation of the residue is replaced by a distillation technique.

(2) This measure provides a distillate free of asphalt as a recycling oil for preparing the slurry in the new process. The asphalt substances formerly imposed a considerable additional load on the reactors. Consequently the operating pressure can be decreased from 700 to 300 bar; at the same time the specific throughput of coal can be augmented.

(3) The gasification of the liquid hydrogenation residue under pressure permits operation which is self-sufficient and economically favourable with respect to the generation of hydrogen. Coal was formerly gasified in Winkler generators without the application of pressure.

(4) The only ultimate residue from the new process is the easily disposable, inorganic slag from the gasification. The residue from the older liquid-phase process was a useless distillation waste coke with a high content of both ash and sulphur; its safe disposal was a great environmental problem.

### SRC II process

The SRC II process (see Fig. 9.3) was developed from the SRC I method by Gulf. with participation from Ruhrkohle AG and Steag.

The experimental verification of the data was first conducted in laboratory installations of the Gulf subsidiary Pamco in Merriam, Kansas and at Gulf in Harmarville, Pennsylvania. Furthermore, the 50 st/day SRC I facility of Pamco near Tacoma, Washington was converted for operation by the SRC II technique.

A 6000 t/day demonstration plant in Morgentown, West Virginia under participation of West Germany and Japan, was scheduled to improve the technology in a semitechnical scale. Due to the fact that there was not sufficient funding of the project, further work was cancelled in 1981.

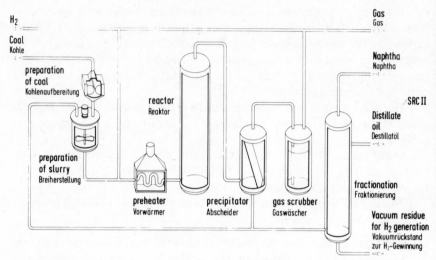

Fig. 9.3 – SRC II process
(SRC II-Verfahren)

## EDS process

The Exxon donor solvent process includes features of both extraction and direct hydrogenation.

The process has been under development by Exxon since about 1970. With the help of two laboratory facilities, fundamental data and specifications were obtained for the construction of a large pilot plant. This installation, which is designed for a coal input of 250 st/day, commenced operation in Baytown, Texas in the summer of 1980.

### Process description

As illustrated by Fig. 9.4, the circulating oil is hydrogenated on a fixed-bed catalyst in a separate reactor. The use of the donor solvent allows the application of comparatively mild conditions in the reactor for coal hydrogenation; consequently, the operation proceeds at about 450°C and 100–140 bar – so far without any additional catalyst.

The products of the reaction are separated by distillation. The residue, which contains the solids, is then transferred in the liquid state to the flexicoker developed by Exxon. In this fluidised-bed coker, fuel gas, fuel oil and low temperature coke are produced; the latter is in turn used for generating hydrogen. Recently, the recycling of distillation residue to the reactor for coal hydrogenation is being carried out; the pressure has thereby been increased to 170 bar. At the expense of the specific throughput of coal, the previously moderate yield of straight-run oil is thereby enhanced. There is a corresponding drop in the amount of low temperature carbonisation products of little value.

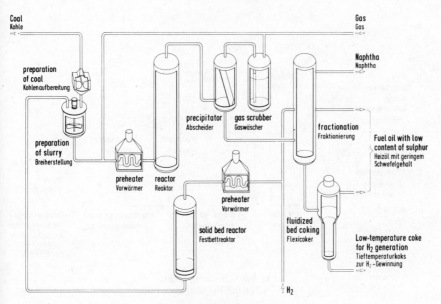

Fig. 9.4 – EDS process
(EDS-Verfahren)

## H-Coal Process

The H-Coal process of Hydrocarbon Research Inc. (HRI), has originated from the H-Oil process, by means of which petroleum residues are desulphurised in commercial installations. The H-Coal process has been tested experimentally by HRI in Trenton, New Jersey, in facilities for 20 kg/day and 3 st/day. In Catlettsburg, Kentucky a large-scale pilot plant with a coal input of 200 st/day (syncrude) or 600 st/day (fuel oil) attained an operational state in mid-1980, after a delay of one year. The department of Energy, the State of Kentucky, as well as Ashland, Conoco, EPRI, Mobil and Standard Oil (Indiana), are participating in the project. As a result of German federal financial support, the Ruhrkohle AG also has access to the experience gained from the operation.

### Process description

The overall process is shown in Fig. 9.5. A special feature of H-Coal process is the ebullated bed reactor, in which a bed of catalyst is agitated and maintained in a state of turbulence by the motion of the circulated reactants. Thus clogging is prevented and a good material and heat exchange is provided. Hence the catalyst remains in the reactor chamber, while the products of the reaction are drawn from the top of the reactor and subsequently refined. Cobalt molybdate on an alumina carrier, in extruded form, serves as the catalyst. For the purpose of maintaining the activity, spent catalyst can be continually removed and replaced by fresh catalyst.

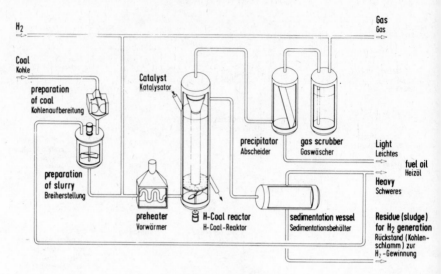

Fig. 9.5 – H-Coal process
(H-Coal-Verfahren)

Typical conditions include pressures around 200 bar and temperatures around 450°C. At a low specific throughout of coal the main product consists of distillates (syncrude process); at a high throughput an oil containing residues is the chief product (fuel oil process). In the fuel oil process the separation of the solids is accomplished not by distillation, but rather by precipitation of the asphalt substances (antisolvent deashing). The residue from distillation or precipitation is gasified for the purpose of generating hydrogen.

## EXTRACTION

### Pott–Broche process
In the Pott–Broche process, the typical method of extraction, the hydrogen is provided exclusively by the donor solvent.

A large-scale pilot plant operated by Ruhröl in Bottrop-Welheim between 1938 and 1944 produced extract at a rate of 80–90 t/day. A mixture of tetralin and cresol as well as middle oil from a pitch hydrogenation plant was used as solvent. The yield of coal free of water and ash was about 75%. The material had a softening range around 220°C and an ash content of 0.15–0.2%. Originally further processing of the extract by means of liquid-phase hydrogenation to gasoline had been intended but was not successful. Therefore the extract was utilised for preparation of high grade electrode coke.

In contrast, the I.G. Uhde process, which evolved at about the same time, utilised hydrogen under high pressure and under hydrogenation conditions in order to produce the extract.

The technology is similar to recent developments: the SRC I processes in the United States and Japan, and the two-stage liquefaction in the United Kingdom.

## SRC I process

The development of the process was started in 1960 by the Spencer Chemical Co., which was later bought by the Gulf Oil Corp. Since 1974 the Pittsburgh and Midway Coal Mining Co., likewise a subsidiary of Gulf during the interim, has operated a pilot plant for a coal input of 50 st/day in Tacoma, Washington. (The installation has been used later for the development of the SRC II process.)

Since 1976, Catalytic Inc. a subsidiary of Air Products, together with other companies in Wilsonville, Alabama, is operating a 6 st/day pilot plant for producing extract by the SRC I method. In 1980 the Department of Energy, Air Products and Wheelabrator-Frye signed a contract for constructing an SRC I plant for a coal input of 6000 st/day. The installation will be built near Newman, Kentucky and should commence operation in 1984/85.

SRC I, the main product of the process, is to be employed as a solid or molten power-plant fuel free of ash and with a low content of sulphur. Corresponding combustion tests on a larger scale have been conducted. As with all extracts from coal, the conversion to special coke for metallurgical purposes is possible too.

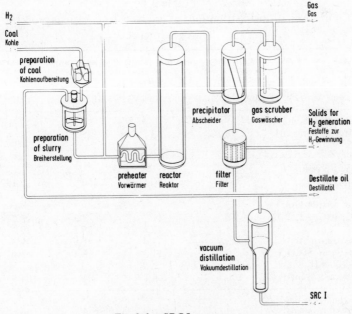

Fig. 9.6 – SRC I process
(SRC I-Verfahren)

Recently a further treatment by means of hydrocracking is under discussion. The intended product is a distillate at temperatures below 450°C, to provide feedstock for making fuel oil and motor fuel.

In the SRC I process (Fig. 9.6), which operates in the usual way, with recycling oil for preparing the coal slurry, the hydrogen is applied at 100–150 bar and about 450°C. The entire liquid product is separated from the solid matter by filtration, or by extractive technique. The circulating oil is subsequently separated by vacuum distillation. The residue of distillation – SRC I – is granulated. Its softening point lies around 200°C. Provided certain types of coal with a high ash content are used as starting material, the operation can be conducted without any additional catalyst.

## COMPARISON OF THE PROCESSES

The yield efficiencies and product distributions of the hydrogenation processes described are summarised in Table 9.1. As a consequence of the more severe conditions, the generation of gas and the consumption of hydrogen are greater for the Kohleöl process than for the other processes. Both the yield and quality (boiling range) of the product oil, however, are substantially better with the Kohleöl process.

The coal which occurs in Germany is more difficult to hydrogenate than that in the United States; this is already evident from the comparative content

### Table 9.1

Comparison of the distributions of products from various processes for hydrogenating coal

| Process | KOHLEOL | SRC II (Gulf) | Exxon Donor Solvent | H-Coal | |
|---|---|---|---|---|---|
| Input coal | Ruhr-Gasflamm | Kentucky | Illinois | Illinois | |
| Volatile constituent in %, free of water and ash | 38 | 43 | 45 | 46.5 | |
| Operating pressure (approx.) in bar | 300 | 150 | 150 | 190 | |
| Product distribution | | | | syncrude | fuel oil |
| gases $C_1$ to $C_3$ | 18.3 | 16.3 | 7.3 | 10.7 | 5.4 |
| oil | 54.2 | 42.8 | 38.8 | 49.7 | 48.7 |
| residue, free of ash | 23.9 | 34.4 | 41.8 | 29.5 | 36.3 |
| $H_2O; NH_3; H_2S; CO; CO_2$ | 9.7 | 11.3 | 16.4 | 15.0 | 12.8 |
| Total (100 + $H_2$ as chemical reagent) | 106.1 | 104.8 | 104.3 | 104.9 | 103.2 |
| Approximate boiling limit of oil from coal in °C | 330 | 450 | 530 | 520 | – |

of volatile matter. However, the German coal with its higher degree of coalification, also generates considerably less water and inorganic gases. The higher pressure prevailing during the Kohleol process also allows the liquefaction of comparatively inert types of coal with a good yield. With the use of a reactive Australian coal or American coal the pressure could be decreased considerably. It is quite probable that the operating pressure of the American processes will be augmented in the future, in order to achieve a greater yield and throughput.

Table 9.2 gives an overview of coal liquefaction plants under construction, or in preparation, for a coal input of more than 100 t/day for the period 1980–1985.

**Table 9.2**

Coal liquefaction plants in operation, under construction or in preparation for construction for coal input more than 100 t/d (1980–1985)

| Process | Location State | Raw coal input t/d | Commencement of operation |
|---------|----------------|--------------------|---------------------------|
| EDS | Baytown, Texas, USA | 220 | 1980 |
| H-coal | Catlettsburg, Kentucky, USA | 550 (fuel oil) 180 (syncrude) | 1979/80 |
| Kohleöl | Bottrop, Ruhrgebiet, W. Germany | 200 | 1981 |
| SRC I | Newman, Kentucky, USA | 5500 | 1984/85 |
| Fischer-Tropsch synthesis | Sasolburg, South Africa | I 10 000 II 36 000 III 36 000 | 1955 1980 1983 |

## CONSTRAINTS ON COAL LIQUEFACTION

A study of the World Energy Conference (Coal Liquefaction Task Force) showed that the timing of the start of commercial-scale coal liquefaction depends on the following factors:

- demand for synthetic fuels
- availability of suitable technology
- availability of coal
- resources for establishing a liquefaction industry
- environmental acceptibility of coal liquefaction technology
- economics of coal liquefaction

In some countries, strategic considerations of security and the use of own resources will be important, but in general coal liquefaction will only become widespread as a response to limitations on the supply of crude oil and the resultant rise in prices.

Countries with large coal production may also manufacture synthetic liquids for export, particularly where coal is of low calorific value so that transport costs of the coal itself are high.

The greatest common interest for synthetic oil is in the transport fuel market. Only Japan foresees also the production of coal liquids for firing boilers. (See Fig. 9.7.)

| Country \ Process | Hydrogenation Solvent extraction | Fischer-Tropsch Synthesis | Methanol Synthesis Mobil-Process | Pyrolysis |
|---|---|---|---|---|
| Australia | ▓ | ▓ | | |
| Canada | ▓ | ▓ | ▓ | ▓ |
| Germany F.R. | ▓ | | ▓ | |
| Japan | ▓ | | | |
| S.Afrika | | ▓ | | |
| U.K. | ▓ | | | |
| USA | ▓ | ▓ | ▓ | |

Fig. 9.7 — Preferred coal liquefaction processes under development

The perceived time of commercial start-up varies from country to country (Fig. 9.8). Only South Africa has liquefaction plants on a commercial scale available today. The earliest time of introduction elsewhere appears to be the late 1980s in Australia, Canada the United States and Germany.

Based on information from national reports the relative contribution that coal liquefaction could conceivably make to the world liquid hydrocarbon supply is shown in Fig. 9.9.

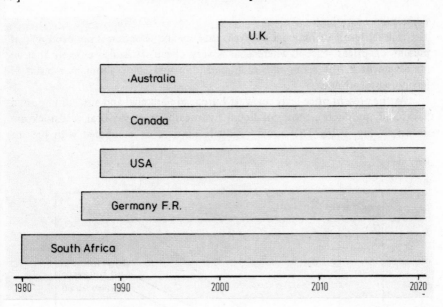

Fig. 9.8 – Estimated times of introduction of commercial-scale coal liquefaction

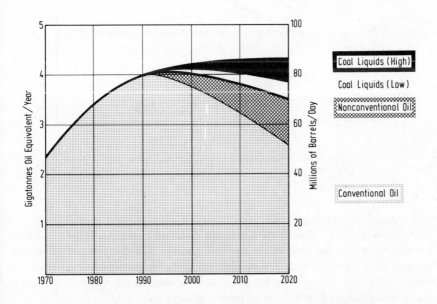

Fig. 9.9 – Prospective production of liquid hydrocarbons from coal compared with conventional and unconventional oil production. (Study World Energy Conference)

**Economic constraints**

The general result of national studies is that coal liquefaction is not economic at present oil prices, except where coal is very cheap. It is also unlikely that an improvement in the economics of liquefaction will come about as a result of technological advance.

At the present price level for West European hard coal and also for imported hard coal, production costs of liquid hydrocarbons versus coal feedstock are shown in Fig. 9.10. The liquid products have to be compared with present

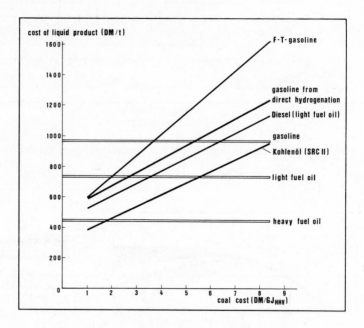

Fig. 9.10 – Production cost of liquid fuels from coal.

market prices. They vary widely among the different countries. Table 9.3 shows a comparison between production costs of gasoline from hydroliquefaction and mineral oil for West Germany. Raw material costs for gasoline from coal are no longer essentially above those for gasoline from mineral oil, but operational or conversion costs are significantly higher for gasoline from coal.

Table 9.4 compares the different types of costs of light fuel oil from mineral oil and coal, based on the German market.

**Table 9.3**
Comparison of the gasoline price from mineral oil and coal

|  | Gasoline from mineral oil (Pf/l) | Gasoline from coal (Pf/l) |
|---|---|---|
| Fuel costs | 39.0 | 55.0 |
| Operating costs | 9.0 | 44.0 |
| Distribution costs | 15.0 | 15.0 |
| Mineral oil tax | 51.0 | 51.0 |
| VAT | 15.0 | 21.0 |
| Gasoline station price | 129.0 | 186.0 |

German conditions, March 1982

**Table 9.4**
Comparison of the light fuel oil from mineral oil and coal

|  | light fuel oil from mineral oil (Pf/l) | light fuel oil from coal (Pf/l) |
|---|---|---|
| fuel costs | 54.1 | 57.7 |
| operating costs | 6.8 | 30.4 |
| distribution costs | 7.8 | 7.8 |
| mineral oil tax | 1.7 | 1.7 |
| VAT | 9.1 | 12.7 |
| total cost incl. delivery to consumer | 79.5 | 110.3 |

German conditions, Jan. 1982

**CONCLUSIONS**

From the technical viewpoint there are no insurmountable barriers which prevent coal liquefaction making a major contribution to future energy supplies. However, a boom in the development of coal conversion technology or a substitution of the resulting products for crude oil cannot be expected under present conditions; there are too many impediments:

(1) The production of coal, which declined severely under economic pressure during the past decade, can be augmented only slowly.

(2) Not only have the prices of oil and gas risen sharply since 1973; coal too has become much more expensive.

(3) The technology of coal liquefaction has to be improved on a commercial scale.

It would appear that the earliest date coal liquefaction is likely to become commercial outside South Africa is the 1990s but that by 2000–2020 coal liquefaction could be a well established technology, making an important contribution to liquid fuel supplies.

# Coal-derived liquids as a source of chemical feedstock

J. Owen, Deputy Director, Coal Research Establishment, National Coal Board.

## INTRODUCTION

The energy crisis of 1973 focussed attention on sources other than crude oil and natural gas. Though shortages are often exaggerated, it is worth noting that in 1978 the United Kingdom chemical industry alone consumed some 15 million tonnes of oil equivalent of which about 55% represented chemical feedstocks. Naphtha accounted for just over 5 million tonnes and natural gas about 2 million tonnes. Assuming a modest growth rate of 1½–2%, as given in recent CEFIC† estimates, this consumption could be doubled by the year 2020. It is hardly surprising then that, with increasing demand for chemicals (and fuels), coal is beginning to replace oil as a source of such materials. Recently, Eastman Kodak have proposed the construction of a plant in the United States to make acetic anhydride from coal [10.1], and other similar projects will no doubt be undertaken in the coming years. As with fuels, it is believed that a gradual rather than a rapid changeover from oil to coal as the major source of chemicals will take place.

## COAL AS A SOURCE OF CHEMICALS IN THE PAST

Before considering the present day methods of converting coal to chemicals (and fuels) it is interesting to recall what has been done in the past.

Coal tar has been used industrially for more than 300 years, one of its first uses being as a preservative for wood and rope. However, the rapid expansion of the coal-tar industry had to await the development of town gas in the early part of the nineteenth century. At first, by-product tar was unwanted, but it was soon recognised that industrially useful materials could be separated from it by distillation. In 1845 Hofmann isolated benzene which he used to prepare nitrobenzene and aniline. About the same time, others demonstrated the presence of

† European Council of Chemical Manufacturers' Federations.

phenol, cresol, naphthalene, anthracene and phenanthrene in coal tar and thus began a chemical industry based on coal. This industry gradually expanded and Bakeland's discovery of phenol-formaldehyde in 1909 added coal-based synthetic resins to it. In the ensuing years, growth in coal tar processing in Britain was gradual and the demands of World War II increased it so that by 1945 the capacity to distil tar (which can represent up to 7–8% of the coal charge) had risen to about 2 million tonnes/year, with benzole, toluole, naphthas, naphthalene, pyridines and anthracenes being the major components isolated (Fig. 10.1). Tar production continued to rise, reaching over 3 million tonnes/annum

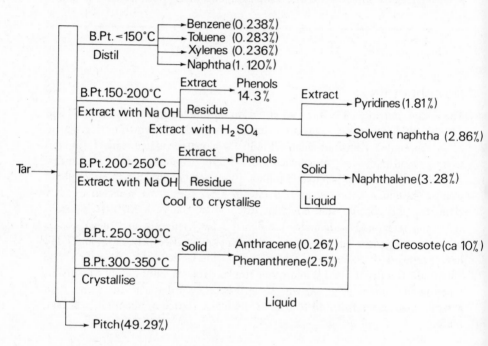

Fig. 10.1 — The major constituents, given in % by weight, of a typical continuous vertical retort coal tar.

by 1957 but, thereafter, declined and is now less than 1 million tonnes [10.2] (Fig. 10.2). The decline was, in part, a result of technical advances in the iron industry which led to reduced coke requirements and hence less tar being made, and also to the fact that benzene, toluene, xylenes, phenol, etc., the prime chemical feedstocks, could be obtained more cheaply from petroleum sources. In any event, the rapidly increasing demand from the 1950s onwards for chemicals could not have been met by the conventional coal carbonisation industry because of the very low yields ($< 5\%$ w/w) of useful hydrocarbon chemicals.

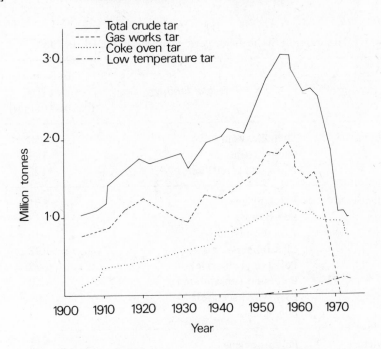

Fig. 10.2 – United Kingdom production of crude coal tar.

In the 1930s, ICI Ltd at Billingham, used coal to make synthetic fuels and chemicals by direct hydrogenation of coal slurries. This was based on the German Bergius I. G. Farben technology. Again, however, when cheap petroleum oil began to flow from the expanding oilfields, after the Second World War, the ICI process too ceased to be competitive and the plant was closed down in the 1950s. Consequently, for the past twenty-five years the principal source of synthetic organic chemicals has been petroleum. The change to the cheaper and more convenient feedstocks caused a rapid growth in what has become known as the petrochemical industry. The principal products are resins, plastics and synthetic rubbers. The United Kingdom productions of these commodities are given in Table 10.1.

New methods of converting coal to liquid (and gaseous) products in high-yield and at acceptable cost are now being developed in several countries but the primary aim is making transport fuels. This is because the principal products from petroleum are gasoline (21%), middle distillate fuels (35%) with the petrochemical industry consuming only about 12% [10.3]. It is almost certain that in future the conversion of coal will be fuel-oriented rather than chemical. However, when that happens, feedstocks for chemicals will also become available and at prices that are likely to be competitive with those then pertaining to petroleum.

**Table 10.1**
Consumption of the major petrochemicals in the UK in 1980

| Feedstock | Principal Product | Consumption in 1980 (× 1000 tonnes) |
|---|---|---|
| Synthesis gas | Phenolic resins | 58 |
| | Amino resins | 123 |
| | Nitrogenous fertilisers‡ | 1122 |
| Ethylene | Polyethylene | 591 |
| | Polystyrene | 153 |
| | ABS terpolymers | 40 |
| | SBR rubbers | 193 |
| | Poly (vinyl chloride) | 372 |
| | Polyesters (unsaturated) | 40 |
| | Poly (vinyl acetate)† | 94 |
| | Polyamides | 16 |
| Propylene | Polypropylene | 225 |
| | Acrylics | 25 |
| | ABS terpolymers | 40 |
| | Nitrile rubbers | 3.8 |
| | Epoxide resins | 13 |
| Butadiene | SBR rubbers | 193 |
| | Polybutadiene | 27 |
| | Polychloroprene | 11 |
| Benzene | Polystyrene | 153 |
| | ABS terpolymers | 40 |
| | SBR rubbers | 193 |
| | Phenolic resins | 58 |
| | Polyamides | 16 |
| | Polyesters (unsaturated) | 40 |
| | Alkyds† | 119 |
| Toluene | Polyurethanes | 84 |
| | Phenolic resins | 58 |

† From *Annual Abstracts of Statistics* 1982
‡ From the *Quarterly Business Monitor* 1981

Table 10.1 — *continued*

| Feedstock | Principal Product | Consumption in 1980 (X 1000 tonnes) |
|-----------|-------------------|-------------------------------------|
| Xylenes | Polyesters (unsaturated) | 40 |
|         | Alkyds‡ | 119 |
| Phenol | Phenolic resins | 58 |
|        | Polyamides | 16 |
|        | Epoxide resins | 13 |

*Source: CSO European Plastics News* Jan. 1982 Edition and *Rubber Statistical Bulletin* Oct-Nov. 1981.

‡ From the *Quarterly Business Monitor* 1981


## FEEDSTOCKS FOR THE PETROCHEMICAL INDUSTRIES

The feedstocks for the petrochemical industries at the present time are natural gas, liquefied petroleum gas, naphtha and other light petroleum cuts [10.4]. Although these are used in large amounts (Table 10.2), the range of primary feedstocks needed for the petrochemical industries is surprisingly small. The most important feedstocks and some of their principal derivatives, already given in Table 10.1, show that most of the plastics, rubbers and resins of today can be made from one or other of the following: benzene, toluene, xylene, butadiene, ethylene and propylene, and many from the olefins, ethylene and propylene

### Table 10.2
Consumption of raw materials by the UK petrochemical industry, 1980

| Product | Consumption (X 1000 tonnes) |
|---------|------------------------------|
| Natural gas | 117 |
| Liquefied petroleum gases | 96 |
| Other petroleum gases | 95 |
| Naphtha | 3 206 |
| Other petroleum products | 344 |

*Source: Digest of Energy Statistics 1981.*

alone. It is therefore the future availability of these key feedstocks, particularly the low olefins ($C_2$-$C_4$) from coal that will be of increasing concern in the coming years. Suitable technology for coal conversion is being developed by adapting existing petroleum processes for (steam) cracking of high boiling fractions [10.5].

Synthesis routes from some of these key intermediates to plastics, rubbers and resins, are given in Figs. 10.3 to 10.5 and typical uses to which these products are put are listed in Tables 10.3 to 10.5. The immense diversity of application is most striking, impinging upon most aspects of our lives and contributing to the high standard of living which we enjoy today.

## CONVERSION OF COAL TO CHEMICALS

Coals vary widely in physical and chemical properties yet they are all essentially carbons (carbon 70-94%) containing hydrogen (about 5%) with smaller amounts of sulphur, nitrogen and oxygen. Various amounts of ash and moisture are also present. The notional components are believed to have three-dimensional aromatic ring structures linked by methylene bridges and cycloaliphatic rings (Fig. 10.6) [10.6]. The conversion of coal to liquid products, therefore, involves the breakdown of these complex (average) molecular structures to lower molecular weight fragments, with simultaneous hydrogenation to increase the proportion of hydrogen to carbon. These requirements can be achieved either by the Direct (degradation) or Indirect (synthesis) routes [10.4] (Fig. 10.7). The main difference between the two lies in the degree of breakdown of the complex coal structure. In the former, breaking at the connecting cycloaliphatic rings and at methylene bridges, leaves most of the original aromatic units intact [10.7]. In the synthesis route, the coal substance is first reduced almost to its component atoms, that is, a mixture of carbon monoxide and hydrogen (synthesis gas) before being catalytically converted to liquid products. Both the degradation and the synthesis processes, the latter now being operated commercially as SASOL in South Africa, originated in Germany where they were extensively used in the production of liquid fuels in World War II.

## DIRECT LIQUEFACTION (Solvent Extraction)

Several degradation processes have been described recently [10.8] but they differ in procedure rather than in principle. In the Liquid Solvent Extraction process developed by the National Coal Board (Fig. 10.8) crushed coal is dissolved at 400°C in hydrogenated recycle oil before being filtered to give a low ash content ($< 0.1\%$) coal extract [10.5]. The extract is catalytically hydrocracked at 400°C to produce oil distillates. These primary liquid products are produced with an overall thermal efficiency of 65-70%.

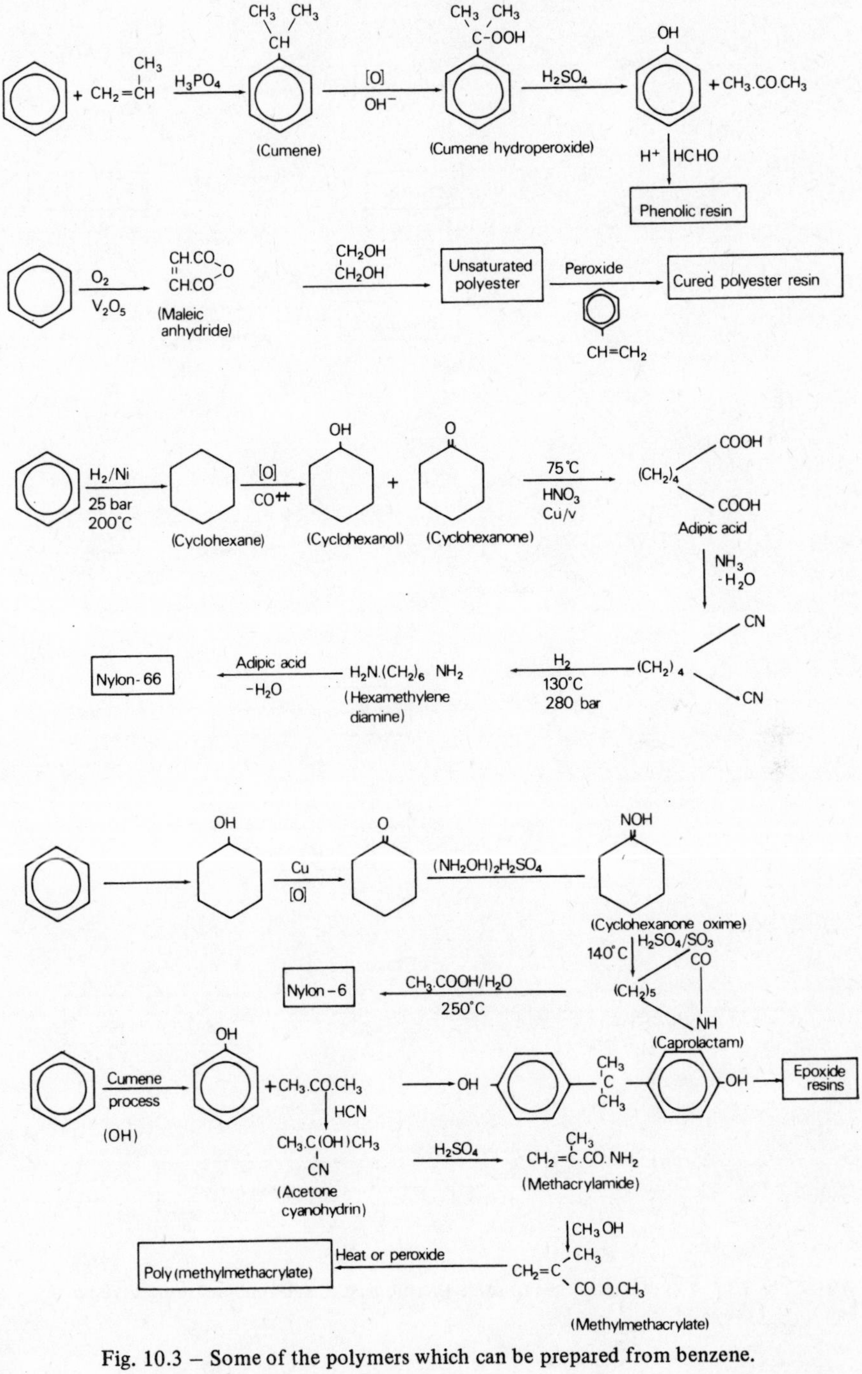

Fig. 10.3 – Some of the polymers which can be prepared from benzene.

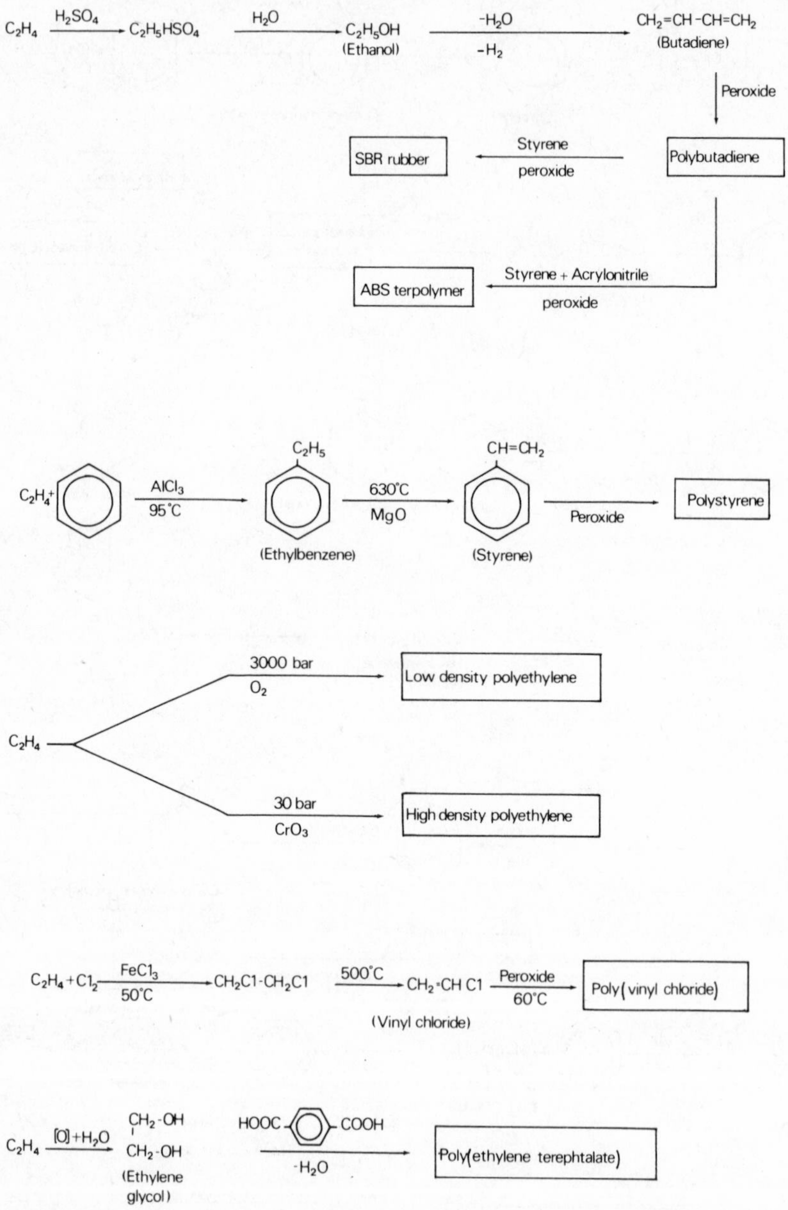

Fig. 10.4 – Synthesis routes to some thermoplastics and rubbers from ethylene.

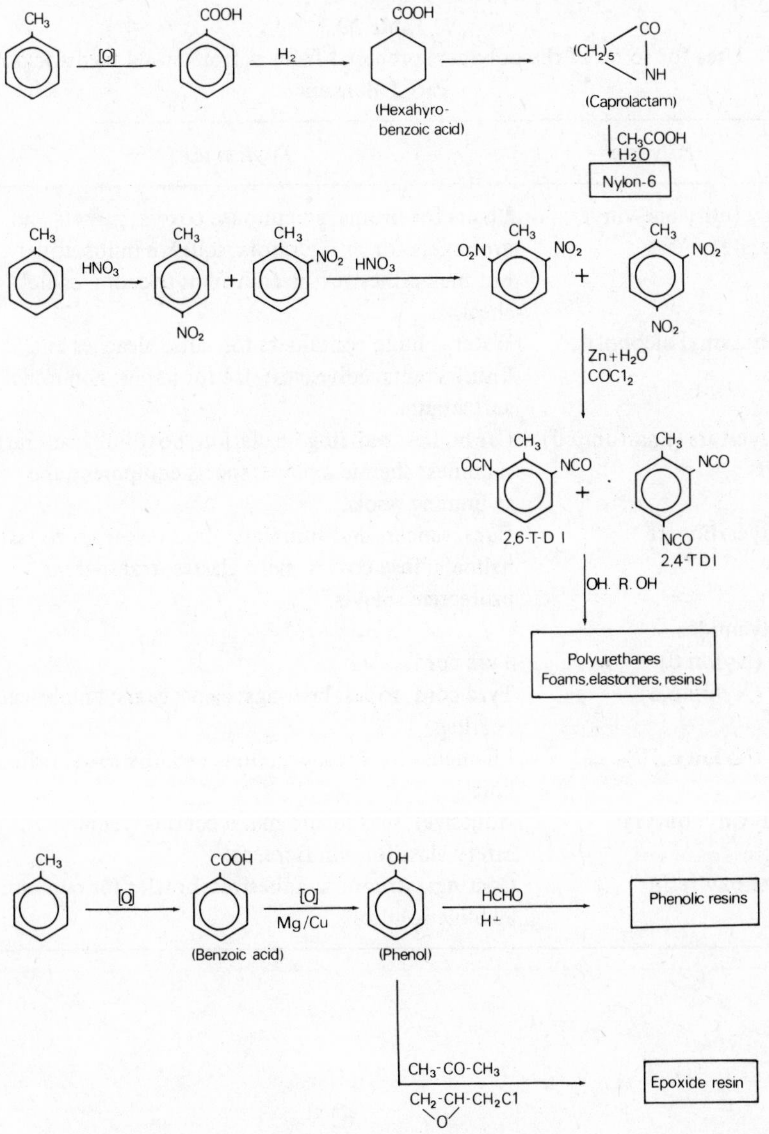

Fig. 10.5 — Routes to some resins, elastomers, and polymers from toluene.

**Table 10.3**
Uses for some of the polymers prepared from petroleum oil feedstocks:
Part 1 Benzene

| Polymer | Typical uses |
| --- | --- |
| Poly (ethylene-vinyl acetate) | Liners for drums; greenhouse covers; gaskets and grommets for refrigerators; squeeze bulbs; toys; hot melt adhesives; stretch film; telecom. cable sheaths. |
| Poly (vinyl alcohol) | Water soluble containers for salts, bleaches and disinfectants; adhesives; size for paper; non-ionic surfactants. |
| Polyesters (unsaturated) | Car bodies; building insulation; boat hulls; aircraft radomes; chemical plant; sports equipment and swimming pools. |
| Polycarbonate | Cups, saucers and tumblers; food mixer bowls; safety helmets; fuse covers; sight glasses; transparent protective shields. |
| Polyamides | |
| (Nylon 6) | Tyre cord. |
| (Nylon 6,6) | Tyre cord; ropes; bearings; cams; gears; unlubricated bearings. |
| (Nylon 6,10) | Filaments for brushes; tennis racket strings; fishing lines. |
| Polyvinyl butyral | Adhesives; textile and metal coatings; laminated safety glass for windscreens. |
| Phenoxy resins | Coatings; structural adhesives; bottles for containing alkaline solutions. |

**Table 10.4**

Uses for some of the polymers prepared from petroleum oil feedstocks:
Part 2 Ethylene

| Polymer | Typical uses |
| --- | --- |
| Polyethylene | Film for packaging; sheets; pipes; tubes; household objects; powders for surface coatings; insulation of electric cables. |
| SBR rubber | Adhesives; tyres; hoses; flexible tubing; covers for rollers. |
| Nitrile rubber | Adhesives; hoses for carrying petrol. |
| Polystyrene | Refrigerator linings; battery cases; radio housings; packaging; thermal insulation; pipes; toys; footwear. |
| ABS terpolymers | Pipes; safety helmets; lawnmower grass boxes and wheels; electrical conduit; housing for electrical equipment; boat hulls; water pump impellers; telephone casings; automobile parts. |
| Poly(vinyl chloride) | Guttering; waste water pipes; window frames; bottles for fruit juices; electric cable insulation; garden hoses; leathercloth; bathroom curtains; washable wallpapers; footwear; beach balls. |
| Poly(vinyl acetate) | Emulsion paints; adhesives, permanent starch; plastic tiles. |
| Polyethylene terephthalate | Summer- and medium-weight clothing; magnetic tapes; drawing office equipment; set squares and rulers. |
| Ethylene/ethyl-acrylate copolymer | Flexible hoses; toys; wire and cable coating. |
| Chlorosulphonated polyethylene | Fabric coatings; protective coatings for other elastomers and white side-walls for tyres. |

**Table 10.5**

Uses for some of the polymers prepared from petroleum oil feedstocks:
Part 3 Toluene

| Polymer | Typical uses |
|---------|--------------|
| Polyurethane | |
| Fibres | Bristles; filter cloths; sieves; swimsuits. |
| Elastomers | Shoe soles and heels; fork lift truck tyres; small industrial wheels. |
| Foams | Fabric backing; cushions; shoulder pads; paint rollers; sponges; draught excluders; thermal insulation. |
| Surface coatings | Gymnasium floors; dance floors; bowling pins; a variety of outdoor and marine use where good weather resistance is needed. |
| Epoxide resins | Castings; pottings and encapsulation in the electrical industry. Flooring; adhesives; solders; foams; printed circuitry; bonding new concrete to old; boat hulls. |
| Phenolic resins | Electrical, automobile, radio and television parts; lamp sockets; missile nose cones; glasscloth and paper laminates; cutlery handles; brake linings; abrasive wheels; sandpaper; plywood. |

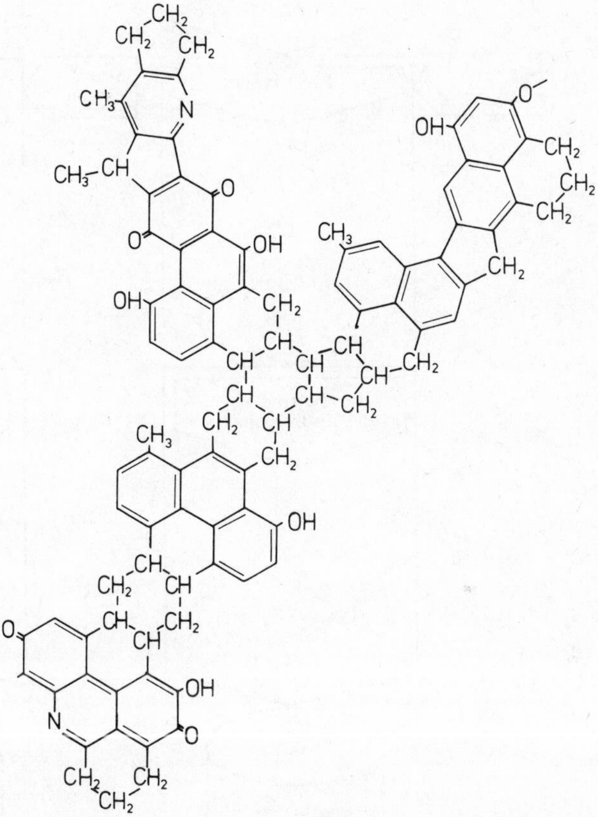

Fig. 10.6 – Proposed molecular model of an 82% carbon vitrinite – the principal component of hard coals.

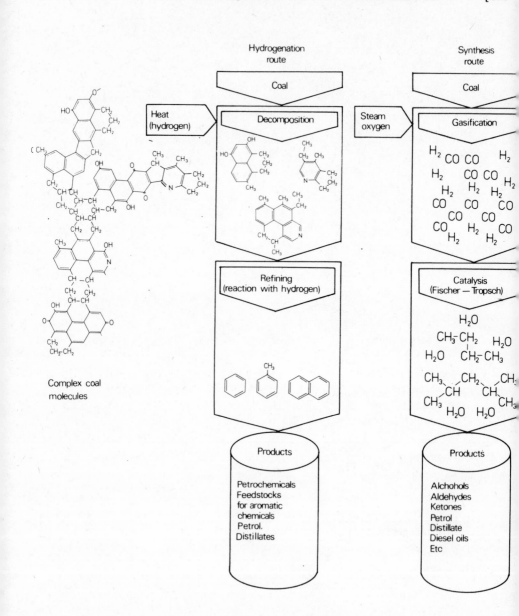

Fig. 10.7 – Hydrogenation and synthesis process routes.

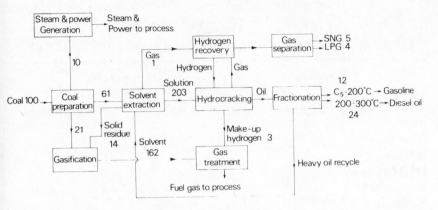

Fig. 10.8 – NCB liquid solvent extraction process.

The naphtha cut (b.p. 80–175°C) containing about 20% of monocyclic aromatics and about 70% cyclo-paraffins (Table 10.6) may be reformed over an appropriate catalyst yielding a product having up to about 80% of aromatics. These are largely benzene, toluene, xylenes and ethyl benzene (Table 10.7).

The mid-distillate fractions (b.p. 175–300°C) of the hydrogenated extract, which are constituted of partly hydrogenated polycyclic aromatic hydrocarbons, provide additional sources of feedstock for petrochemicals (and fuels). Further catalytic hydrogenation, adding about 3% of hydrogen, converts them to their alicyclic counterparts and on steam cracking they yield butadiene, benzene, toluene and xylenes. By this means, ethylene is produced in high yield (about 18%) together with propylene and butylene (12%) (Table 10.8). Thus, most, if not all, of the key intermediates required by the petrochemical industries, can be produced from coal by direct coal liquefaction and subsequent processing.

### Table 10.6
Typical composition (% by weight) of a coal-based naphtha (b.p. 80–175°C)

| C-number | Paraffins | Naphthenes | Aromatics |
|----------|-----------|------------|-----------|
| $C_6$ | – | 25.9 | 1.5 |
| $C_7$ | 2.9 | 24.9 | 7.1 |
| $C_8$ | 2.0 | 18.1 | 11.3 |
| $C_9$ | 0.1 | 5.0 | 1.1 |
| Total | 5.0 | 73.9 | 21.0 |

**Table 10.7**
Typical composition of a reformed coal-based naphtha (b.p. 80–175°C)

| Compound | Amount present (% by weight) |
|---|---|
| Benzene | 22.2 |
| Toluene | 28.1 |
| Xylenes | 11.5 |
| Ethyl benzene | 14.5 |
| $C_9$ aromatics | 4.4 |
| Naphthenes | 4.8 |
| Paraffins | 9.8 |
| Hydrogen | 3.0 |
| $C_1$–$C_4$ gases | 1.7 |
| Total | 100.0 |

Reformer temperature 480°C

**Table 10.8**
Thermal cracking of coal based feedstocks

| Products (% by weight) | Naphtha† | Feedstock Mid-distillates‡ | Hydrogenated mid-distillates‡ |
|---|---|---|---|
| Benzene | 12 | 8 | 11 |
| Toluene | 5 | 6 | 7 |
| $C_8$–$C_9$ aromatics | 3 | 3 | 5 |
| Ethylene | 23 | 9 | 18 |
| Propylene + Butylenes | 13 | 5 | 12 |
| $C_1$–$C_3$ paraffins | 13 | 9 | 14 |
| Butadiene | 8 | n.d. | 5 |
| Total | 77 | 40 | 72 |

† Temperature of cracking 825°C
‡ Temperature of cracking 800°C

## INDIRECT LIQUEFACTION (Gasification/Synthesis Route)

The coal gasification/synthesis route to chemicals has been employed on a commercial scale in the South African SASOL plants for over thirty years. Here, fixed-bed Lurgi gasifiers are used in which sized coal is converted to a mixture of carbon monoxide and hydrogen by oxygen and steam. The crude gas mixture is purified and shifted to the requirements of the particular synthesis reaction. The synthesis gas (together with recycle gas) is passed over a fixed-bed of iron oxide-cobalt catalyst in an Arge reactor or in a fluidised bed in a Synthol reactor to produce, by a Fischer-Tropsch reaction, a broad spectrum $(C_1-C_{30})$ of products. These include straight-chain high boiling hydrocarbons, medium boiling oils, diesel oil and some oxygenates (Figs. 10.9 and 10.10) [10.9].

The poor selectivity in terms of product distribution is due to the nature of the Fischer-Tropsch reactions which involve the progressive addition of carbon atoms to growing oligomer chains to give product distributions like those normally encountered in many polymerisation processes. At the same time, competing reactions occur yielding alcohols and ketones.

Clearly, improving catalyst selectivity is of both technical and economic importance. Alternatively, upgrading the less desirable by-product oils and water-soluble oxygenates to valuable chemical feedstocks would make the process much more attractive. In this connection it has recently been reported by the Dow Chemical Company that the Mobil zeolite catalyst (H–ZSM-5) can be used to convert by-product oils and oxygenates to olefins and aromatics [10.10].

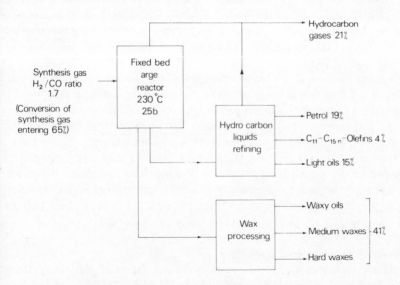

Fig. 10.9 – The principal products, given in % by weight, from SASOL fixed-bed reactors.

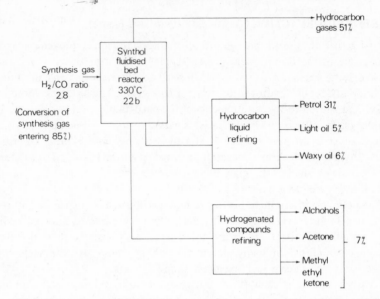

Fig. 10.10 — The principal products, given in % by weight, from SASOL synthol reactors.

By using a different catalyst system (Cu/Zn/Cr) to that commonly used in the Fischer–Tropsch process, synthesis gas can be converted to methanol in high-yield. This alcohol may be catalytically oxidised with air to formaldehyde from which phenol-, urea- and melamine-formaldehyde resins are made. It is also used in the manufacture of poly (oxymethylene) resin — an engineering plastic often known as Delrin (Du Pont), from which gear wheels, cams, etc., are fabricated [10.11].

In Lurgi gasifiers, a small proportion of the coal is converted to tar and oils. The tar, which is similar in composition to continuous vertical retort tars [10.12], could make a chemical feedstock, particularly for the manufacture of phenolic, epoxide, polycarbonate and phenoxy resins. It is likely to assume significance if coal is gasified to supplement and eventually replace natural gas.

The SASOL plants in South Africa have demonstrated that a completely coal-based chemical (and fuel) industry is possible, even though the conversion processes used may not be particularly thermally efficient (40–45%). However, with better product selectivity, such as might be achieved by the downstream use of zeolite catalysts, the position could probably be improved.

### The Mobil Zeolite process

A most notable advance of recent years in the conversion of coal, via synthesis gas, to chemicals (and fuels) has been made by the Mobil Research and Development Corporation. In 1976, the company announced the development of a new and simple catalytic process based on a shape-selective synthetic zeolite (H-ZSM-5), for the conversion of methanol (derived from synthesis gas) to hydrocarbons which were predominantly in the gasoline $C_4$–$C_{10}$ boiling range [10.13]. The hydrocarbons were propane, butanes, pentanes, $C_5^+$ aliphatics, benzene, toluene, xylenes, $C_9$-aromatics and $C_{10}$-aromatics (Table 10.9).

### Table 10.9
Hydrocarbons obtained by passing methanol vapour over a porous aluminosilicate catalyst

|  | NCB | | | Mobil† |
| --- | --- | --- | --- | --- |
| *Reaction conditions* | | | | |
| LHSV ($h^{-1}$) | | 0.3 | | 1.0† |
| Bed temperature ($^\circ$C) | 350 | 400 | 450 | 371 |
| Methanol conversion to water plus hydrocarbons (% by weight) | 85 | 63 | 95 | 100 |
| *Hydrocarbon distribution* (% by weight) | | | | |
| Methane | 2.4 | 15.0 | 11.2 | 1.0 |
| Ethane | 9.1 | 8.8 | 16.9 | 0.6 |
| Ethylene | 7.6 | 5.8 | 10.4 | 0.5 |
| Propane | 6.2 | 0.9 | 7.2 | 16.2 |
| Propene | 3.9 | 1.6 | 4.3 | 1.0 |
| Butanes | 2.8 | 3.7 | 9.4 | 24.3 |
| Butenes | — | — | — | 1.3 |
| Pentanes | 1.1 | 2.1 | 3.0 | 9.1 |
| Pentenes | — | — | — | 0.5 |
| $C_6+$ aliphatics | 4.4 | 4.1 | 7.5 | 4.3 |
| Benzene | — | — | — | 1.7 |
| $C_7$-Aromatics | 5.8 | 17.5 | 7.0 | 10.5 |
| $C_8$-Aromatics | 8.9 | 10.6 | 5.1 | 18.0 |
| $C_9$-Aromatics | 3.2 | 15.0 | 5.9 | 7.5 |
| $C_{10}$-Aromatics | 10.8 | 2.3 | 1.5 | 3.3 |
| $C_{11}$-Aromatics | 9.7 | 3.7 | 0.6 | 0.2 |
| $C_{12}$-Aromatics | 9.9 | 2.1 | 0.6 | — |
| $C_{13}$-Aromatics | 13.9 | 6.4 | 8.9 | — |

† Results obtained by Chang, C. D. and Silvestri, A. J. [10.8] using the Mobil catalyst, ZSM–5.

Mobil workers [10.14] and others [10.15], have assumed that the unique properties of the new zeolite catalyst were due to a strongly proton acidic surface and a crystalline structure with closely controlled pore dimensions (Fig. 10.11).

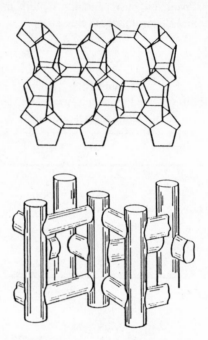

Fig. 10.11 – ZSM-5 structure.

The acidic sites are thought to be responsible for the catalytic reactions which lead to the formation of hydrocarbons while the average pore diameter of about 0.6 nm restricts the size of the largest molecule which could be produced to durene ($C_{10}$). The overall reaction mechanism is not fully understood but the reaction paths from methanol to hydrocarbons can be represented by the scheme below.

$$2\,CH_3OH \rightleftharpoons CH_3\,O\,CH_3 \xrightarrow{-H_2O} C_2\text{-}C_5 \text{ olefins} \longrightarrow \begin{array}{l} \text{Cycloparaffins} \\ \text{Paraffins} \\ \text{Aromatics} \\ C_6\text{-Olefins} \end{array}$$

In an extension of this work it has been demonstrated that by using a composite catalyst consisting of Fe or Zn–Cr and H-ZSM-5, synthesis gas can be used directly to make similar chemicals to those produced using methanol [10.16]. This procedure overcomes the need to produce methanol in a separate stage.

Another synthetic zeolite, recently described by Japanese workers [10.17], is claimed to be suitable for the catalytic conversion of methanol to $C_2$-$C_4$ olefins with high selectivity. It seems feasible to prepare all of the main feedstocks for the petrochemical industry which are listed in Table 10.1 from synthesis gas using zeolite catalysts.

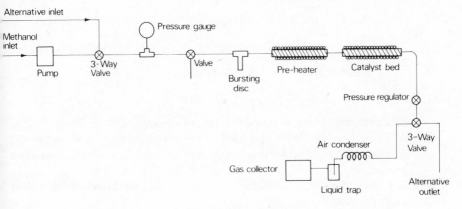

Fig. 10.12 – Schematic diagram of apparatus used for catalyst evaluation.

## Amorphous Aluminosilicates

An interesting extension of the use of aluminosilicate catalysts is being developed at the Coal Research Establishment of the National Coal Board [10.18]. It has been found that when a monolayer of aluminium silicate is deposited over part of the surface of an S-type xerogel (surface area 800 $m^2$/g and pore volume 0.4 $cm^3$) a catalyst is obtained which converts methanol, using apparatus as shown in Fig. 10.12, to a mixture of aliphatic and monocyclic aromatic hydrocarbons in the range $C_1$-$C_{13}$ (Table 10.9). The rate of conversion of methanol to these products can be significantly increased and the molecular size range reduced ($C_2$-$C_{11}$) by using the catalyst in the active H-form. As with the Mobil catalyst, it is assumed that a strongly acidic proton surface is responsible for the catalytic reactions which convert the methanol to hydrocarbons.

## ACETYLENE AS A CHEMICAL INTERMEDIATE

Acetylene has been made traditionally by reacting coke or anthracite with lime and treating the resultant calcium carbide with water, as follows:

$$3C + CaO \longrightarrow CaC_2 + CO$$

$$CaC_2 + 2H_2O \longrightarrow C_2H_2 + Ca(OH)_2$$

Acetylene is made in Germany from natural gas and some lower-boiling petroleum fractions by an electric arc process where the (difficult) technology of production and subsequent processing have been exceedingly well developed. Acetylene is then used in various ways.

(1)  Reaction of acetylene with hydrogen chloride gives vinyl chloride monomer:

$$CH{\equiv}CH + HCl \longrightarrow CH_2{=}CH$$
$$\mid$$
$$Cl$$

(2)  Acrylonitrile is produced by the reaction of acetylene with hydrogen cyanide:

$$CH{\equiv}CH + HCN \longrightarrow CH_2{=}CH$$
$$\mid$$
$$CN$$

Polymers containing acrylonitrile are extensively used as fibres (Orlon, Acrilan). Acrylonitrile is also used in the manufacture of ABS and nitrile rubber, and the homopolymer, polyacrylonitrile, is used for carbon fibre production.

(3)  Reactions of acetylene with carbon monoxide and an aliphatic alcohol gives an acrylic ester:

$$CH{\equiv}CH + CO + C_2H_5OH \longrightarrow CH_2{=}CH$$
$$\mid$$
$$CO.O.C_2H_5$$

There are several commercially available acrylic esters, the polymers and co-polymers of which are used as pressure-sensitive adhesives and thermoplastics.

Acetylene could well become more important in the future as ethylene becomes more expensive to produce. Though the conversion for coal is high in energy demand the hydrogen to carbon ratio is favourable for coal and for the polymers listed above. Much hinges on the cost and availability of hydrogen in all future coal conversion processes.

## MATERIALS OF THE FUTURE

It is likely that primary chemical feedstocks will remain the same in the foreseeable future but the raw materials from which they are derived will probably be quite different. Biomass could possibly become one of the major sources of organic chemicals in some parts of the world, although at the present time there is little incentive to change from the fossil fuels [10.19]. Nevertheless, detailed studies have been carried out from which it has been calculated that about 70 million tonnes of lignocellulose would be needed to produce all of the major resins, rubbers and plastics at 1978 levels in the United States [10.20]. It is recognised, moreover, that while the technology for the changeover from oil to coal is well advanced (and in some cases complete), that for lignocellulose conversion, although feasible in principle, would require a considerable amount of development work.

It is not unreasonable to speculate that some future products of the petrochemical industries could include resins, rubbers and plastics based on silicon and other inorganic elements. We may also be forced to return to using many of the natural products that have been abandoned in favour of (cheap) synthetic materials.

## CONCLUSIONS

This paper takes a wide ranging view of the future supplies for chemical feedstocks.

From laboratory work that has already been done it seems clear that most of the materials for the petrochemicals industry can be made from coal when the need arises, that is, when crude oil supplies become difficult. However, there remain some formidable problems of process development, and these must be solved, before shortages become acute.

## REFERENCES

[10.1] Harris, G. A., Sinnett, C. E. and Swift, H. E. (1981), *Chemtech* December, 764.

[10.2] Gibson, J. and Paul, P. F. M. (1979), *RAPRA Jubilee Conference* July, Shrewsbury, Salop.

[10.3] *BP Statistical Review of the World Oil Industry* (1979).

[10.4] Gibson, J. (1980) *Chemistry in Britain* 16 (1) 26.

[10.5] Davies, G. O. (1979) *The Minerals Engineering Society National Conference* Durham, UK.

[10.6] Gibson, J. (1978) *J. Inst. Fuel* June 67.

[10.7] Larsen, J. W. (1978) 'Organic Chemistry of Coal', *ACS Symposium Series 71*, 126.

[10.8] Rider, D. K. (1981) *Energy: Hydrocarbon Fuels and Chemical Resources* John Wiley & Sons, 148.

[10.9] Hoogendoorn, J. C. (1977) *Proc. Ninth Synth. Pipeline Gas Symposium* Chicago.

[10.10] Storbe, R. A. and Murchison, C. B. (1982) *Hydrocarbon Processing* (1), 147.

[10.11] Brydson, J. A. (1975) *Plastics Materials* Newnes-Butterworths, London, 425.

[10.12] Vaughan, G. A. (1958) *Coal Tar Research Association Report No. 0209.*

[10.13] British Patent No. 1 446 552 (1976), Mobil Oil Corporation.

[10.14] Chang, D. C. and Silvestri, A. J. (1977) *J. Catal.* 47 249.

[10.15] Dejaifve, P., Vedrine, J. C., Bolis, G., and Derouane, E. C. (1980) *J. Catal.* **63** 331.

[10.16] Chang, D. C., Lang, W. H. and Silvestri, A. J. (1979) *J. Catal.* **56** 268.

[10.17] Inui, T., Ishihara, T. and Takegami, Y. (1981) *J.C.S. Chem. Comm.* 936.

[10.18] British Patent Application No. 81/18956 (Coal Industry Patents Ltd.).

[10.19] A. I.Ch.E. Study; reported in (1978) *Chem. Eng. News* **56** (48), 23.

[10.20] Reported in (1979) *Chem. Eng. News* May **57** (19), 25.

[10.21] Gibson, J. and Paul, P.F.M.P. (1979) *Proceedings RAPRA Conference* July, Shrewsbury.

## ACKNOWLEDGEMENTS

It is a pleasure to record thanks to several colleagues, especially to Dr J. G. Robinson, who have contributed thoughts and ideas in discussions. This paper is published by kind permission of the National Coal Board, and the opinions expressed are the author's own.

# 11

# Acetylene: A basic chemical with a future?

**Dr Peter Paessler**, Works Manager, Acetylene Plant, BASF, West Germany

## INTRODUCTION

The question in the title has been raised indirectly over and over again, ever since the changeover from coal chemistry to petrochemistry that took place during the 1940s in the United States and during the 1950s in Europe. In 1968 the subject 'Acetylene or ethylene as feedstocks for the chemical industry' received particularly close attention at a joint conference of the Society of Chemical Industry (London) and the DECHEMA (Frankfurt) [11.1].

The few hopeful aspects from the acetylene viewpoint, such as BASF's contribution of the Submerged Flame Process or Hoechst's crude oil cracking HTP, could not disguise a clear trend towards ethylene as a basic chemical. With the first oil price explosion in 1973, direct crude-cracking processes too suffered a setback and even newer processes, like the Kureha process, or the very latest developments, such as the UCC–Kureha–Chiyoda ACR process or Dow's PCC process, seem to raise few hopes of a comeback for acetylene chemistry.

But the decline in the importance of acetylene as a basic petrochemical does seem to be slowing down, and this could be attributed to the low capital repayment on 'old' plants, to the exploitation of regional price advantages of petrochemical feedstocks, or to improvements that are making existing processes more energy-efficient.

## ACETYLENE PRODUCTION

The following two figures show the decline in acetylene production for chemical purposes over the last ten to twenty years [11.2].

Figure 11.1 shows the drastic fall in acetylene production in the early 1970s in the United States, duplicated a few years later in Germany. In both countries the losses were principally in carbide-derived acetylene; in fact Germany has been producing acetylene for chemical purposes almost exclusively from petrochemical sources for years now.

Figure 11.2 shows the chemical use of acetylene in the United States by

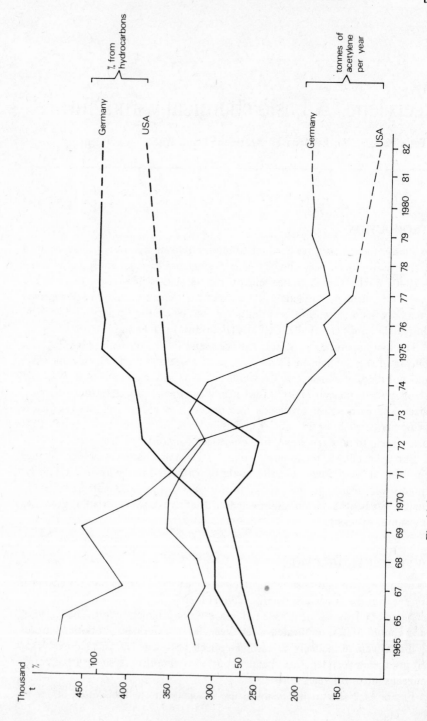

Fig. 11.1 – Acetylene for chemical use: acetylene from hydrocarbons.

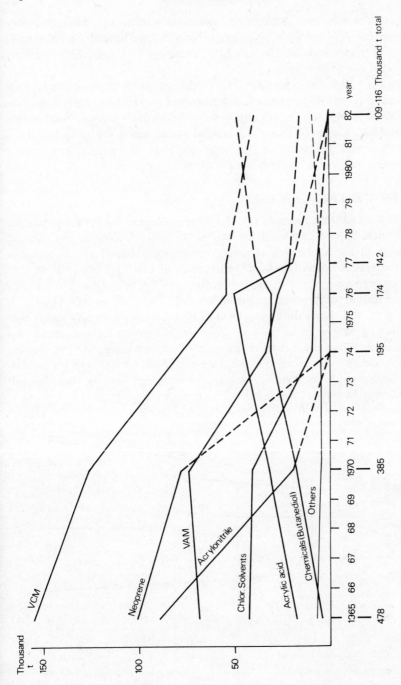

Fig. 11.2 – Acetylene for chemical use in the United States.

product. There is no mistaking the sweeping victories for olefin chemistry (ethylene for VCM and VAM, propylene for acrylonitrile, butadiene for neoprene) in the sixties and seventies. The only growth has been for chemicals, in particular butanediol and its derivatives.

This is why in the immediate future, acetylene production facilities should be sited only where petrochemical feedstocks, in particular natural gas, are available in surplus, that is, cheaply, or where prices of these are limited by state intervention. 'Crude oil parity' for natural gas as aimed for by OPEC would certainly leave little chance of growth for acetylene in countries like Germany that have to rely almost exclusively on imported oil and gas.

## THE FUTURE ENERGY SCENE

The latest oil industry forecasts [11.3] appear to discount further sharp increases in oil prices between now and the end of the century. Reasons given are the development and application of energy conservation technology and economy consciousness in homes following the last price hike in 1979/80. In 1980, oil's share of primary energy consumption in Germany fell below the 50% mark for the first time in twelve years and should fall to 29% by the year 2000 [11.4].

Figure 11.3 shows the development of West Germany's primary energy consumption by source up to the year 2000. In contrast to earlier forecasts this shows an anticipated reduction in the share of oil, in both relative and absolute terms, matched by an increase in the share of coal and nuclear energy. Considerable diversification is therefore anticipated in energy sources, and this will undoubtedly be true in other industrial countries too.

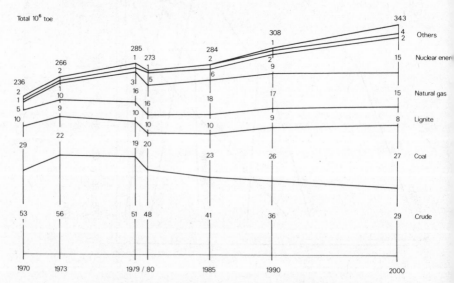

Fig. 11.3 – Consumption of primary energy in West Germany.

This would result in petrochemical feedstocks being available to the chemical industry over a longer term than earlier predictions had suggested. A desirable development would be for heat and power to be generated from other sources with gas and oil reserved for petrochemical processing.

## THE COMPETITIVE POSITION OF ACETYLENE
### Disadvantages of acetylene technology

Acetylene's chances of a speedy comeback as an economically viable chemical feedstock cannot be rated high precisely because of its weak competitive position *vis-a-vis* oil-based ethylene/olefin chemistry.

The cause of the weak competitive position is the high price of acetylene compared with olefins, since acetylene manufacturing processes require large amounts either of raw materials or of electric power. A high proportion of input energy is generally lost, that is, the processes have low thermal efficiency. This lies in the thermodynamically necessary high temperature reaction and in the way the reaction is carried out, also thermodynamically conditioned. Since the acetylene-containing reaction gases have to be quenched from 1500°C to 300°C in a few milliseconds, the energy regained is only suitable for low-level uses.

A further disadvantage is the need, for yield reasons, to work at low pressures, which means additional power input for subsequent compression. Another disadvantage of acetylene technology is the acetylene molecule's roughly four-fold energy of formation compared with the ethylene molecule. As a result syntheses with acetylene are strongly exothermic and take place at relatively low temperatures, so that the best use cannot be made of reaction heat.

Olefins by contrast are produced at lower energy-input levels. Syntheses with them have a lower reaction heat which is often given off at higher temperatures and is therefore more utilisable. A final disadvantage of acetylene manufacture is that because of the lower capacity of acetylene plants compared with olefin plants, fixed costs (investment, labour, etc.) are about twice as high.

As a result of all this, higher capital costs and lower yields have been accepted where conversion from acetylene- to ethylene-based synthesis is economic overall.

Despite the rather sombre outlook for an early revival of acetylene, now, with considerable effort going into coal gasification and liquefaction and the possibilities opened up by these, is precisely the moment to demonstrate why and under what conditions acetylene can regain its former more significant position.

### Advantages of acetylene technology

In comparison with ethylene, acetylene can be converted to a greater number of chemicals and chemical intermediates with greater ease, under milder reaction conditions and with high selectivity.

The basic reactions of acetylene can be classified as shown in Fig. 11.4. Four main products are still produced by addition, vinylation, carbonylation and ethynylation; they are vinylchloride, vinylacetate, acrylic acid and 1,4 butanediol. But as shown in Fig. 11.5 a much larger number of important chemicals can be readily manufactured from acetylene using known and well-developed technology.

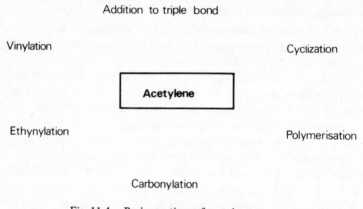

Fig. 11.4 — Basic reactions of acetylene.

Unlike ethylene, acetylene can be manufactured from practically any petrochemical feedstock, including coal, *direct,* that is, in one process step (see Fig. 11.6). Natural gas is particularly suitable for acetylene production since it is only the thermodynamic conditions of acetylene synthesis that can convert methane direct into other hydrocarbons. Natural gas should continue to be available until well beyond the year 2000 with a restructuring of energy technology from gas to coal, particularly in the United States; with the use of coal-derived synthesis gas for chemical syntheses ($NH_3$, methanol, etc.); and increased utilisation of currently flared-off associated gas. Thus, acetylene can be a thoroughly sound proposition, as an example will later demonstrate. On the other hand, ethylene can only be obtained from natural gas via intermediate transformation stages [11.5] and can therefore at present hardly compete with naphtha-based ethylene.

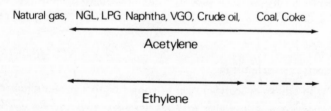

Fig. 11.6 — Feedstocks for acetylene and ethylene.

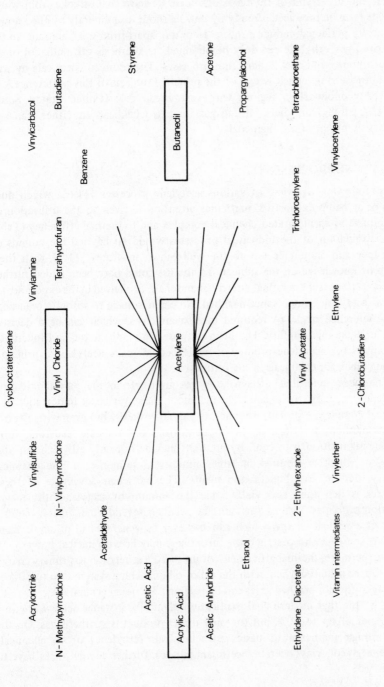

Fig. 11.5 – Chemicals available from acetylene.

At the other end of the feedstock scale we have coal or coke from which acetylene can be produced directly by new methods, or indirectly by the known, technically perfected carbide process. Here too, apart from small amounts in the coke oven gas, ethylene can only be obtained via synthesis gas/methanol or via hydrocarbons from coal liquefaction processes. The route to chemicals by way of coal processing, which is one of the principal subjects of this conference, can only be economical in regions with very cheap coal (United States, South Africa, Australia) or where oil and gas for petrochemicals are either scarce or sufficiently higher priced than coal.

## ACETYLENE PROCESSES

In the following overview of various acetylene processes [11.6], which does not aim at being exhaustive, particular attention is given to the utilisation of by-products in an integrated chemical complex and to thermal efficiency (Table 11.1). Evaluation of the individual processes would go beyond the bounds of this paper and should be left to the well-known institutes [11.7] with their wealth of knowledge on the subject. To the list could have been added further processes either still operating, already abandoned, or shelved in the experimental stage as acetylene's importance declined. No claim is made to absolute precision, nor is reference made to required investment costs, which can be a decisive factor in economic viability. The main purpose is to indicate the multiplicity of possibilities for acetylene production which could make acetylene a principal building block for chemicals in the future too.

The huge quantities of synthesis gas given off in the partial-oxidation acetylene processes, for example, could more than ever be of importance in a chemical complex, especially for the methanol synthesis. But even with electro-thermal methods acetylene can be obtained from any available feedstock, from methane right through to coal. By-products include not only ethylene but also hydrogen, which is required in large amounts in refineries, in the chemical industry, and in coal liquefaction plants. The particular advantage of these processes is that they also yield twice the amount of acetylene/ethylene as oxidation processes. Against this must be set their several times greater require-ments of electrical energy, which can however be generated in principle from whatever source is cheapest, and the future here may lie with nuclear power.

Furthermore, thermal efficiencies of up to 70% in relation to primary energy input can be achieved when, with the aid of oil-quenching systems, up to 80% of the heat of the hot cracked gases can be regained for steam generation.

It is true that modern coal gasification plants for instance achieve thermal efficiencies of up to 80%, but the only direct product is synthesis gas. On the way through methanol to basic petrochemicals (ethylene) or to chemicals (ethylene glycol, vinyl acetate, acetic anhydride), further energy losses have to be incurred.

**Table 11.1**

Acetylene processes

| Process | Company | Feedstocks | Main products | By-products | Input of primary energy per tonne of acetylene† of which | | | Thermal‡ efficiency |
|---|---|---|---|---|---|---|---|---|
| | | | | | Total LHV (10³ kWh) | % Feedstock | % Electricity | |
| Arc | Hüls | Natural gas, LPG | Acetylene Ethylene | Hydrogen BTX, Fuels | 76 | 41 | 59 | 47 |
| Plasma | Hüls | Crude oil (LPG, Naphtha) | Acetylene Ethylene | Hydrogen Hydrocarbons Fuels | 80 | 54 | 46 | 66 |
| WLP | Hoechst | Naphtha Oil for quench | Acetylene Ethylene | Hydrogen | 76 | 49 | 51 | 61 |
| HTP | Hoechst | Naphtha (Oxygen) | Acetylene Ethylene | – | 71 | 82 | 18 | 69 |
| Partial oxidation Waterquench | BASF | Natural gas, LPG Naphtha Raffinat (Oxygen) | Acetylene | Synthesis gas | 74 | 79 | 21 | 58 |
| Oilquench | BASF | Natural gas, LPG Naphtha Raffinat Aromatic oils (Residues) (Oxygen) | Acetylene | Synthesis gas BTX Aromatic oil Metallurgical coke | 86 | 82 | 18 | 70 |
| Submerged flame | BASF | Crude oil (Oxygen) | Acetylene Ethylene | Synthesis gas Propylene | 119 | 81 | 19 | 73 |
| Carbide autothermic | BASF | Coke (Oxygen) Quicklime | Acetylene | Carbon monoxide | 69 | 89 | 11 | 65 |
| Carbide electrothermic | | Coke Quicklime | Acetylene | Fuel gas | 54 | 41 | 59 | 31 |
| Arc waterquench | AVCO | Coal | Acetylene | Char/Carbon-black, Fuels | 68 | 35 | 65 | 39 |
| Arc hydrocarbonquench | AVCO | Coal Propane | Acetylene Ethylene | Char/Carbon-black, Fuels | 64 | 56 | 44 | 58 |

† Thermal efficiency of electricity production assumed as 33%
‡ Thermal efficiency = heating value of the products as a percentage of primary enery input

## A COMPETITIVE GAS-BASED ACETYLENE PROCESS

After the description of the advantages and disadvantages of acetylene technology, it will be demonstrated that in spite of present energy consumption and price patterns the acetylene route can be highly competitive.

Thus the last oil price rise of 1979/80 made naphtha, the main feedstock for olefins in Europe, considerably more expensive and one result of this is that syntheses with acetylene can under certain circumstances once more be competitive. This will be shown with reference first to a gas-based acetylene-complex known to the writer and then to coal-based processes taken from literature.

Figure 11.7 shows the flow sheet of a 100 000 t/year gas-based partial-oxidation acetylene plant in a chemical complex. The required oxygen facility and an acetylene offgas (AOG) based methanol plant are included. The main products of the complex are acetylene and methanol. As can be seen from the use of aromatic oil, such as residue oil from steamcrackers, the acetylene process is operated with improved quenching technology allowing a high proportion of energy to be regained in the form of steam. At the same time the aromatic residue is rearranged into high-purity coke and light aromatics. Electric power supply and other utilities have been assumed.

The production cost estimate in Table 11.2 is based on power, feedstock

### Table 11.2
Production costs of acetylene

| Estimated unit cost: first quarter 1982 | | |
|---|---:|---|
| natural gas | 5.8 | $/GJ |
| naphtha | 333 | $/t |
| methanol | 220 | $/t |
| power | 34 | $/10³ kWh |
| Acetylene | 100 000 | t/yr plant (Fig. 11.7) |
|   Raw materials | 1 546 | $/t |
|   By-products | −1 485 | $/t |
|   Utilities | 240 | $/t |
|   Operating costs | 90 | $/t |
|   Others | 260 | $/t |
|   Net production cost | 651 | $/t |
|   ROI 25% of TFC | 373 | $/t |
|   Product value | 1 024 | $/t |
| Ethylene | 300 000 | t/yr plant |
|   Net production cost | 600 | $/t |
|   ROI 25% of TFC | 178 | $/t |
|   Product value | 778 | $/t |

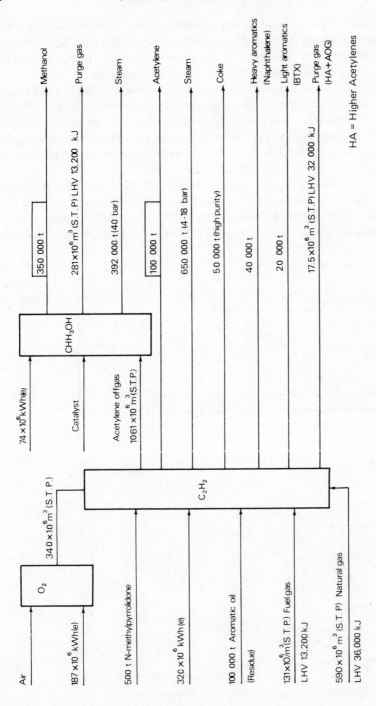

Fig. 11.7 – Acetylene-methanol complex.

and product prices roughly corresponding to present market prices. Natural gas and purge gases are set at 5.8$/GJ, methanol at 220$/t and electric power at 3.4 ¢/kWh, to name only the most important cost items. 'Other costs' comprise depreciation (10%), interest (5%), overheads and pollution control costs. The calculation shows that low production costs for acetylene can be reached under the conditions described despite the relatively high natural gas price (5.8$/GJ) which in heat-value terms comes very close to that of crude oil (6.2$/GJ). The difference from production costs for ethylene manufactured from naphtha (contract price $333/t) is so slight that acetylene from new plants can once again compete with ethylene for certain syntheses.

Figure 11.8 illustrates the profitability of the VCM synthesis via acetylene. This shows for any given ethylene price the corresponding acetylene price for an equal VCM selling value. The competitive position of acetylene from the acetylene-methanol complex is clearly recognisable. Even if some production cost factors

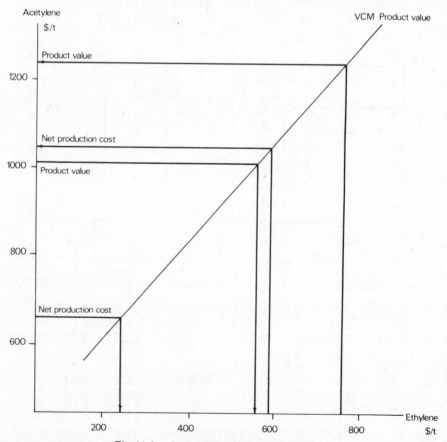

Fig. 11.8 – Competitive position of acetylene.

have been set too optimistically, there still remains a considerable margin before the acetylene route becomes uneconomical. Based on net production costs acetylene could thus be around 370$/t more expensive, or even with 25% ROI included, still around 220$/t more. An even more favourable result appears from a consideration of the VAM synthesis from acetylene compared with ethylene, while with acrylic acid acetylene's margin is not quite as wide.

Viewed as a whole, direct transformation of natural gas into the basic chemical acetylene should show that under particular conditions acetylene chemistry can be advantageous not only today but perhaps even more so in the future, provided that the difference between gas and naphtha prices, at present 1.9$/GJ, increases with any rise in these prices.

This is precisely the central problem in estimating the viability of chemical and energy technology in general. The enormous efforts being made today to use coal more widely than before as an energy source and feedstock for the chemical industry, spring primarily from an expected high price differential between coal and petrochemical raw materials, which is linked naturally to availability of supplies.

## COMPETITIVENESS OF COAL-BASED ACETYLENE PROCESSES

For acetylene too, coal, once the only source of this basic chemical, is being considered again (Table 11.3). As mentioned above, acetylene, unlike ethylene, can be obtained direct from coal, too. The indirect way passes through the electrothermal carbide process, which has been known for eighty years and is still in operation worldwide. In the United States and most Western European countries, however, acetylene for chemical processes is now, for reasons of cost, only produced from carbide on a very small scale. In the mid-fifties attempts were made to replace the high amounts of electric power required for the electrothermal carbide process with thermal energy, by burning part of the coke used with pure oxygen, which also produced carbon monoxide. Trials reached the pilot project stage before being broken off after petrochemical acetylene had made its breakthrough at this time. New trials would have to be conducted if this method of coal-processing should appear promising.

Since the 1960s the AVCO Corporation in particular has run trials aimed at direct transformation of coal to acetylene, that is, of the volatile matter in the coal in an arc or plasma reactor. This direct route to acetylene met only academic interest at first, but increasingly with the two oil crises, trials were intensified until 1980 [11.8] when AVCO published a new process and economic data which have aroused keen interest in the literature [11.9]. From these the AVCO process seems an attractive alternative as a way to acetylene and as a coal-processing method. Here too the break-even point of the process depends not only on future price movements, but also on the cost effectiveness of other coal-processing methods.

Table 11.3
Acetylene from coke and coal

| Process | Carbide electrothermic | Carbide autothermic | AVCO-Arc waterquench | AVCO-Arc hydrocarbonquench |
|---|---|---|---|---|
| Raw materials | Coke   1.65 t<br>CaO   2.40 t | Coke   6.5 t<br>CaO   3.4 t<br>$O_2$   3 720 m³(STP)<br>Electricity   2 640 kWh(e)<br>CO (96%)   3 370 m³(STP) | Coal   3 t<br>Electricity (Arc)   13 125 kWh(e)<br>Electricity   2 000 kWh(e)<br>Fuel   1 290 m³(STP)<br>Char/Carbon-black   1.19 t | Coal   1.70 t<br>Propane   1.77 t<br>Electricity (Arc)   7 560 kWh(e)<br>Electricity   2 000 kWh(e)<br>Ethylene   0.69 t<br>Fuel   2 280 m³(STP)<br>Char/Carbon-black   0.83 t |
| Energy | Electricity   10 570 kWh(e) | | | |
| By-products<br>Consumption/t Acetylene | Fuel (net)   750 m³ (STP) | | | |
| Net cost of materials and energy | ~770 $/t | ~595 $/t<br>(CO = 1.4 × Fuel) | ~660 $/t | ~375 $/t |
| Investment for 100 000 t Ac/year | $110 000 000 | $145 000 000 | $180 000 000 | $180 000 000 |
| Operating costs† Overhead expenses | 280 $/t | 350 $/t | 440 $/t | 440 $/t |
| Production cost | 1050 $/t | 945 $/t | 1100 $/t | 815 $/t |
| Input of primary energy | 54 × 10³ kWh† | 69 × 10³ kWh | 68 × 10³ kWh | 64 × 10³ kWh |
| Thermal efficiency | ~30% | ~65% | ~40% | ~60% |

†Excluding coke and CaO production

It can be seen that an evaluation of these processes under central European conditions shows up substantially higher production costs for acetylene than the process described above based on natural gas. This hardly makes them competitive with the olefin route, particularly when a 25% ROI is taken into account. Yet the AVCO process in particular with its hydrocarbon quenching system comes within range of being competitive through its relatively high energy utilisation. Its sensitivity to raw material price rises is lower, because it yields a second valuable product, than the water quench process.

Without going into further detail, it can be said that prospects for coal-based acetylene processes will improve with the anticipated steeper price rises in petrochemical feedstocks compared with coal and electric power. Low efficiency in relation to primary energy input is principally due to the low efficiency of the thermal/electrical transformation, which has here been taken to be 33%. Here too the use of nuclear power would help to reduce the use of carbon fuels substantially.

## CONCLUSIONS

These are simplified observations on the future prospects for the once-dominant basic chemical acetylene. Its chances of making a comeback, or at least of consolidating its position in the chemical industry, lie in the variety of uses to which it can be put with known technology and high yields, and in the fact that it can be obtained directly from natural gas and coal. Optimum integration in a chemical complex and improvements in processes at present operating or being developed can, even in countries less favoured in terms of raw materials, make acetylene chemistry economically attractive compared with olefin chemistry, or even the only alternative.

Just as the steep rise in oil prices is expected to lead, as shown earlier, to wide diversification of primary energy sources, one should also expect a reduction in the dominant role of crude oil based olefin chemistry in favour of acetylene, based at first on natural gas and in the longer term on coal. In this light a future discussion on acetylene might be entitled: 'Chemicals from coal-based acetylene or from the synthesis gas route?'

## REFERENCES

[11.1] 'Acetylene or Ethylene as feedstocks for the Chemical Industry', Conference proceedings, DECHEMA and Society of Chemical Industry, London, March 1968.

[11.2] Mitteilung des VIC, Frankfurt am Main.

[11.3] 'Shell: Energieverbrauch wächst noch langsamer' Frankfurter Rundschau, 17 Dec. 1981; 'Deutsche Shell AG: Bis zum Jahr 2000 kein neuer Ölpreisschub' Düsseldorfer Handelsblatt, 17 Dec. 1981.

[11.4]  'Der Energiemarkt von morgen' OEL-Zeitschrift für die Mineralölwirt-schaft, July 1981, 182.

[11.5]  Sherwin, M. B., 'Chemicals from Methanol' *Hydrocarbon Processing* March 1981; 'Natural gas-based ethylene process set for commercial use' *European Chemical News* February 1982.

[11.6]  'Acetylen' *Ullmanns Encyklopädie der technischen Chemie* 7, 1973, 43–71.

[11.7]  *Acetylene Report* No. 76-2, Chem. Systems Inc., November 1976; *Acetylene* Stanford Research Institute, PEP No. 16, September 1966; No. 109, September 1976; and No. 16A, November 1981.

[11.8]  'Plasma-Arc Coal Chemicals' technical paper for *Feedstock Alternatives* Session AIChE Meeting, Philadelphia, PA, June 1980.

[11.9]  Bittner, D., Baumann, M., Peuckert, C., Klein, J., and Jüntgen, M., 'Direktumwandlung von Kohle zu Acetylen' *Erdöl und Kohle-Erdgas-Petrochemie vereinigt mit Brennstoff-Chemie* 34, No. 6, June 1981; 'An electric way to make feedstocks *Chemical Week,* 4 February 1981; 'AVCO's new arc coal process may bring back acetylene as a chemical feedstock' *European Chemical News,* 2 February 1981.

# 12

# Economics of the production of synthesis gas from carbon fuels

**R. B. Moore**, Senior Engineering Associate, Air Products and Chemicals Inc., USA

## INTRODUCTION

Synthesis gas, an important feedstock and energy component in the European chemical industry, has historically been generated from different carbon fuels, primarily natural gas, oil, and coal. The choice of the fuel and the related process has been determined mainly on the availability of the feedstock and its relative price. During the last decade the interruptions in oil supply, changes in availability of natural gas and relative changes in gas, oil, and coal pricing have made significant changes in synthesis gas manufacturing technologies used in Europe as well as throughout the industrial world. In Europe there has been a shift towards the use of natural gas for synthesis gas production. With the new Russian pipeline providing additional supplies, further utilisation of natural gas for hydrogen-rich synthesis gas is being produced increasingly from heavy petroleum liquids and natural gas is being deregulated and is rising in cost relative to oil and coal, synthesis gas is being produced increasingly from heavy petroleium liquids and coal. This is particularly true for large-scale facilities where the benefits of economies of scale for partial oxidation of oil and coal gasification are significant. Applications for synthesis gas are broadening with the development of new markets, new products, and new technology routes. The complexities of synthesis gas production, and the magnitude of the capital investment required for its manufacture are both increasing dramatically. This paper surveys the major routes to synthesis gas production and treatment; discusses some of the new markets, products, and application technologies which are developing; and reviews some of the key issues involved in making investment decisions for the production of synthesis gas. The economics presented in this paper can be used as a guide to the relative viability of other 'new' processes presented in other chapters.

## Uses

There are many basic chemicals produced in whole or in part from synthesis gas. These include ammonia, methanol, hydrogen, acetic acid, synthetic natural gas, oxo alcohols, synthesis gasoline, etc. Recently there has also been an active interest in using synthesis gas from coal as a clean fuel for boilers, furnaces and in gas turbines for combined cycle plants.

The need for clean fuels in Europe depends on legislation being adopted, particularly by EEC countries. United States experience in the production of synthesis gas from coal and its use in boilers, furnaces and combined cycle plants will provide valuable experience for the future uses in Europe.

## Definition

The term synthesis gas is generally defined by industry as any gas mixture containing hydrogen and carbon monoxide. Typically synthesis gas contains other components such as carbon dioxide, water, methane, nitrogen, hydrogen sulphide, etc.

## METHODS OF PRODUCTION

Synthesis gas is most commonly produced by:
(1)  steam reforming of natural gas and light hydrocarbons;
(2)  partial oxidation of heavy oils;
(3)  gasification of coals.

### Steam reforming

The steam reforming of methane is typified by the following reactions:

$$CH_4 + H_2O + heat \rightleftharpoons CO + 3H_2$$

and

$$CO + H_2O \rightleftharpoons CO_2 + H_2 + heat$$

A typical analysis of a synthesis gas leaving a steam–methane reformer is given in Table 12.1.

### Table 12.1
Synthesis gas composition, steam–methane reformer

| Component | Reformed natural gas 791 kN/m² (100 psig) % by volume |
|---|---|
| $H_2$ | 74 |
| CO | 18 |
| $CO_2$ | 6 |
| $CH_4$ | 2 |

The steam and light hydrocarbons are catalytically reacted inside a chrome-nickel alloy tube with the heat being supplied by radiant combustion of fuel gas or oil. The metallurgy of the tubes usually limits the reaction temperature to 760-925°C (1400-1700°F). Low pressure favours the reforming of the hydrocarbon and permits minimum usage of steam. High pressure, however, has certain economic advantages in that it reduces subsequent purification, heat recovery and compression costs. Recycled and/or imported carbon dioxide is used as a supplemental feedstock when high $CO:H_2$ ratios are desired.

**Partial oxidation**
Partial oxidation is normally used when the feedstock is a heavy hydrocarbon which can not readily be reacted over a catalyst. Partial oxidation of natural gas is also economically attractive for high $CO:H_2$ ratios and high synthesis gas pressures. The hydrocarbon-steam reactions are similar to those shown for steam reforming. The heat of reaction is supplied by burning additional feedstock with oxygen or, in limited cases, air. The partial oxidation reactions are typified as follows:

$$C_n H_m + (n/2)O_2 \rightleftharpoons nCO + (m/2)H_2 + \text{heat}$$

$$CO + H_2O \rightleftharpoons CO_2 + H_2 + \text{heat}$$

Typical anlyses of synthesis gas leaving a partial oxidation reactor are given in Table 12.2.

**Table 12.2**
Synthesis gas composition, partial oxidation

|  | Heavy oil 6170 kN/m² (880 psig) % by volume | Natural gas 6170 kN/m² (880 psig) % by volume |
|---|---|---|
| $H_2$ | 46 | 60 |
| CO | 46 | 34 |
| $CO_2$ | 6 | 4 |
| $CH_4$ | 1 | 1 |
| $N_2$ + Ar | 1 | 1 |

To achieve the desired reaction the amount of oxygen admitted is controlled to produce an exit temperature of 1 150-1 300°C (2 100-2 400°F). Because of this high exit temperature it is possible to reach favourable equilibrium conditions at much higher pressures than with a reformer. The partial oxidation reactor is much less expensive tha the steam reformer. However, the capital cost of the oxygen supply more than off-sets this advantage.

Combined steam reforming and partial oxidation is occasionally practised on light hydrocarbons and can provide an economic advantage over either by itself. The combination permits the less expensive steam reforming of the bulk of the gas and the advantages of using a partial oxidation reactor for the final feedstock conversion at elevated temperatures and pressures. This combined technology is widely used in the ammonia industry where a primary reformer is used for about two-thirds of the reforming duty. Then air is introduced into a secondary (catalytic) reformer to produce a $3:1$ $H_2:N_2$ synthesis gas stream with a minimum of residual methane. This technology, using oxygen instead of air, is also used in the production of carbon monoxide, methanol and to produce a synthesis gas stream rich in carbon monoxide relative to carbon dioxide and hydrogen. This combination is particularly attractive when oxygen can be obtained at average cost from a large-scale air separation plant.

### Coal gasification

The third method of producing synthesis gas is the gasification of coal. The technology is very similar to partial oxidation of heavy oils. The required heat for reactions (12.2) and (12.3) is provided by oxygen or air by reaction (12.1).

$$C + O_2 \rightleftharpoons CO_2 + \text{heat} \tag{12.1}$$

$$C + CO_2 + \text{heat} \rightleftharpoons 2\,CO \tag{12.2}$$

$$C + 2H_2O + \text{heat} \rightleftharpoons CO_2 + 2H_2 \tag{12.3}$$

Typical analyses of synthesis gases leaving coal gasifiers are given in Table 12.3.

### Table 12.3
Synthesis gas composition, coal gasification

|  | *Lurgi* Type low temperature 2860 kN/m² (400 psig) % by volume | *Koppers–Totzek* Type high temperature 100 kN/m² (0 psig) % by volume | *Texaco Type* High H₂O temperature 4240 kN/m² (600 psig) % by volume |
|---|---|---|---|
| $H_2$ | 39 | 29 | 34 |
| CO | 21 | 60 | 48 |
| $CO_2$ | 30 | 10 | 17 |
| $CH_4$ | 9 | – | – |
| Ar + $N_2$ | 1 | 1 | 1 |

The exit reaction temperature varies from 370–1480°C (700–2700°F). The low temperature reactor typically uses the least amount of oxygen and achieves the highest thermal efficiency. Much of the coal leaves the low temperature reactor as unreformed molecules such as methane and tars. In the high temperature reactors the coal molecules are totally reformed to $H_2$, CO and $CO_2$. The feedstock coal is usually much lower in cost for coal gasification than are the oils or light hydrocarbons for the competing processes. The combined cost of the coal handling facilities, gasifiers, and other facilities result in a significantly higher capital cost.

## METHODS OF PROCESSING SYNTHESIS GAS

After the raw synthesis gas has been produced it is usually necessary to further treat it prior to synthesising it to final products. There are many different treatment steps, but they can generally be classified as follows:
(1)  acid gas removal,
(2)  shift conversion,
(3)  cryogenic, chemical or physical separation of hydrogen and carbon monoxide,
(4)  selective oxidation of residual carbon monoxide or methanation of residual carbon oxides,
(5)  compression of synthesis gas or separated products.

### Acid gas removal
The major competing acid gas removal processes are: chemical absorption, physical absorption and physical adsorption of carbon dioxide and hydrogen sulphide. The quantity, partial pressure, and types of the impurities, along with the required synthesis gas purity, creates a complex analysis of the most economical process. Chemical absorption processes, such as MEA (monoethanolamine) and DEA (diethanolamine), are usually low capital processes, but high utility consumers. They are typically used for low concentrations of acid gas, particularly at low operating pressure (below 70 kN/m$^2$ (10 psi) partial pressure). Physical absorption processes such as Rectisol, Sulfinol, and Selexol, on the other hand, are most economical at high acid gas partial pressures (above 700 kN/m$^2$ (100 psi) partial pressure) and in large-scale plants where the low utility cost offsets the higher capital cost. Between these limits are a number of chemical/physical absorption processes such as Benfield and other hot potassium carbonate processes which utilise some of the advantages of each system. Another process which is used alone or following the above is mol sieve physical adsorption for gas drying and removal of trace acid gas impurities.

### Shift conversion
Additional hydrogen can be produced from synthesis gas by shift conversion, which is the reaction of $CO + H_2O \rightleftharpoons H_2 + CO_2 + heat$. There are several catalysts which promote this reaction both at high temperature 340–480°C (650–900°F) and at low temperature 200–260°C (400–500°F). Recently a sulphur-tolerant catalyst has been developed which operates in the temperature range of 200–430°C (400–800°F). Typically it is more economical to shift as much as practical using high temperature catalysts because these catalysts are

cheaper per unit hourly volume of gas converted. The low temperature catalysts are used for the final shifting which is possible at the lower (more favourable) equilbrium temperature.

## Separation

There are a great number of techniques for separating or enriching the synthesis gas components. The most commonly used are the cryogenic methods. They can be used separately or sometimes with others in a combined process. The most common forms of cryogenic separation of hydrogen and carbon monoxide are: partial condensation of carbon monoxide; methane wash; and nitrogen wash as used in the ammonia industry. Cryogenic processes normally are best for large-scale facilities with concentrated carbon monoxide feed streams. The advantage of dry/$CO_2$-free product streams can also be attractive.

Separation of carbon monoxide from hydrogen is also accomplished by chemically complexing the carbon monoxide molecule as is done in the Cosorb process or older copper liquor process. The carbon monoxide is recovered by heating the chemical solution. This technique is particularly effective compared to cryogenic separation when there is a need to separate the carbon monoxide from a nitrogen-containing synthesis gas. This is because the boiling points of carbon monoxide and nitrogen are too close for good cryogenic separation. Chemical complexing is also economically more attractive where the carbon monoxide concentration is relatively low.

Mol sieve adsorption is used effectively for hydrogen purification/separation where product hydrogen is needed at pressure and the impurities are consumed at low pressure. In small plants mol sieve purification can effectively serve the role of acid gas removal, alternate to shift conversion and final hydrogen purification. Another physical separation technique which has been recently developed is the Prism permeable membrane [12.2]. This is an economical technique for separating out hydrogen from a synthesis gas stream. It is useful for ratio adjustment (increasing CO concentration) of a synthesis gas stream. This technology appears to be most cost-effective when the synthesis gas stream is available at a high pressure and particularly effective where the separated hydrogen is only required at a low pressure.

## Final hydrogen purification

Final purification of a hydrogen stream can be achieved by selective oxidation of any residual carbon monoxide prior to acid gas removal [12.3]. Also competing with this process is methanation of any residual carbon oxides. Mol sieve purification can yield a high purity hydrogen product but is not practical when high recovery is required or other desired components are present in the synthesis gas stream, such as nitrogen as in the case of an ammonia plant.

## Compression

A further consideration in the production and processing of synthesis gas is the desired pressure of the products. Economics normally favours producing the synthesis gas under sufficient pressure as to provide the treated synthesis gas at the desired final pressure without further compression. However, certain practical equipment limits frequently set an optimum pressure. In the case of steam-methane reforming, pressure and temperature are limited by the creep strength of available reformer tube alloys. Also high pressure at a set temperature increases costly methane leakage. Practical pressure limits range from 1480-2100 kN/m² (200-300 psig). In the case where secondary reformers are incorporated the optimum pressure will be higher, 2100-3550 kN/m² (300-500 psig). In the case of partial oxidation facilities, pressures of 5600-10450 kN/m² (800-1500 psig) are frequently found to be optimum. The technology for coal gasification at atmospheric pressure through 2860 kN/m² (400 psig) is well proven. The economics of higher pressure units is well known and there is much effort currently underway to utilise high pressure 2860-8380 kN/m² (400-1200 psig) units. When it is necessary to compress the synthesis gas following its production, there are certain economic consideration that influence plant design and compressor selection. Because of low capital cost, higher reliability and easier operation, a centrifugal compressor is usually preferred over a reciprocating compressor. The combination of gas volume, discharge pressure, and molecular weight of the synthesis gas determine if a centrifugal compressor can be used. In some cases the synthesis gas has been compressed prior to carbon dioxide removal to increase the gas volume and molecular weight so that a centrifugal compressor can be used. The power cost for compressing the extra gas can be recovered through a less expensive acid gas removal unit.

## THERMAL EFFICIENCY

As can be imagined there are many different technology combinations which are possible for producing a treated synthesis gas stream. Each selection will obviously have its own overall thermal efficiency.

However, each of the above processes has a characteristic thermal efficiency depending upon the desired final $H_2$:CO ratio (see Appendix. Table 12A.1). Figure 12.1 is a composite of several different process combinations for reforming natural gas and then further processing the synthesis gas to a desired CO concentration. In Fig. 12.1 it can be seen that the maximum thermal efficiency results when the natural ratio† of about 5:1 $H_2$:CO is produced by steam-reforming of natural gas. The reduction of thermal efficiency to the left of the peak results from shifting CO to $H_2$ and removing the resulting $CO_2$ or rejecting and burning of CO as fuel. The reduction in thermal efficiency to the right of the peak results from the separation and rejection of excess hydrogen to fuel or 'back-shifting' with recycled and imported $CO_2$.

---

† This includes the practical restriction that there is excess steam present to prevent carbon formation at reaction temperatures.

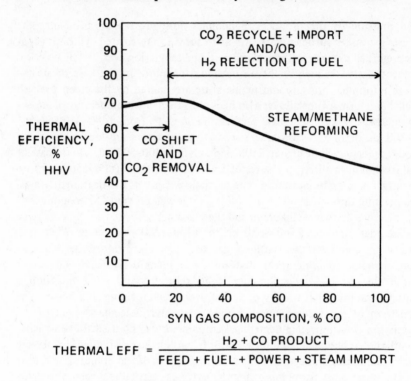

Fig. 12.1 — Thermal efficiency vs syn gas composition.

Figure 12.2 shows the typical efficiencies achieved with the alternate processes of steam reforming, partial oxidation, and coal gasification. The coal gasification efficiency curve shown is typical of a wet feed gasifier such as a Texaco gasifier where the peak efficiency is about 50–60% CO. In the case of a dry feed K–T gasifier, the peak efficiency is closer to 66% CO. A Lurgi reactor will reach a peak efficiency at about 35% CO.

Also shown are the most common synthesis gas products and the required $H_2$:CO ratio. Obviously there are thermal efficiency advantages of picking a reactor that most closely matches the $H_2$:CO ratio of the product slate.

## FEEDSTOCK COST

The cost of feedstock compounded by thermal efficiency is the single most important consideration for the reforming of natural gas and relatively un-important for coal gasification. Typically, feedstocks are priced in the order shown in Table 12.4 (Europe) and Table 12.5 (United States).

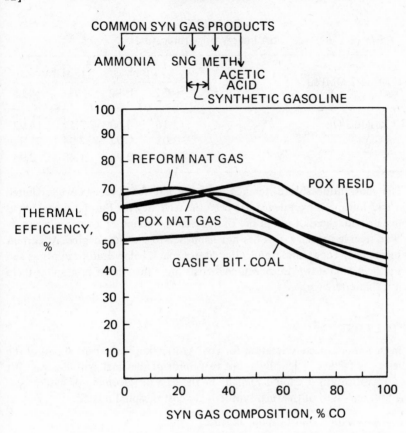

Fig. 12.2 – Thermal efficiency vs syn gas composition.

**Table 12.4**
European energy price forecasts

|  | 1982 | | 1992 | |
|---|---|---|---|---|
|  | $/toe | $/MMBtu† | $/toe | $/MMBtu |
| (1) Heavy residual oil (most expensive) | 253 | 5.86 | 448 | 10.38 |
| (2) Natural gas | 242 | 5.60 | 453 | 10.47 |
| (3) Coal (least expensive) | 137 | 3.17 | 281 | 6.49 |

† IMM Btu = $10^6$ Btu = 1.055GJ

Table 12.5
USA Energy price forecasts

| Costs in $/MMBtu | 1978 forecast | | 1981 forecast | |
|---|---|---|---|---|
| | 1982 | 1992 | 1982 | 1992 |
| Heavy Residual Oil | 2.16 | 4.60 | 5.52 | 14.92 |
| Natural Gas | 3.05 | 7.92 | 3.54 | 21.82 |
| Coal | 1.48 | 2.96 | 1.48 | 2.96 |

The most noticeable difference over the last few years between United States and European experience has been the tendency for European coal prices to move up with those of oil. In the United States, because of lower mining costs and large reserve capacity, this has not happened to the same degree. At current coal prices in Europe, imported United States coal is below European prices and this will help hold down European coal costs more than in the past unless there is political interference.

**CAPITAL COST**

The most important consideration for coal gasification is the capital cost of the synthesis gas facility. Typically about two-thirds of the final synthesis gas cost for coal gasification is capital related. In the case of reforming of natural gas, only about one-third of the final synthesis gas cost is capital related:

(1)  coal gasification (most capital expense)
(2)  partial oxidation of heavy residual oil
(3)  reforming of natural gas (least capital expense).

**OPTIMISATION**

As a result of variable process thermal efficiency, different feedstock costs, and different capital costs, the optimum feedstock cost and process can be expected to change with: (1) plant size; (2) desired end product, and (3) relative energy costs.

The following economic camparisons are based on an elaborate computer program used to represent the capital, utility and operation costs of all the major production and processing steps in producing synthesis gas. This program has proved to be very effective as a screening tool for selecting the best competing processes for a number of different conditions. It is possible to rapidly develop a synthesis gas cost for different feedstocks and feedstock costs. The change in this synthesis gas cost can be studied over a period as feedstock prices vary with time and change in relationship to each other. This same program can be used to examine the economy of scale of different processes and particularly the synergism of co-production of several synthesis gas products. Mathematical analysis of alternative depreciation periods, effects of leveraged financing and differing company profit requirements are easily accomplished.

An example of the results of the computer calculation is shown in Fig. 12.3. It shows typical synthesis gas cost as a function of % CO in the synthesis gas, the balance being hydrogen. Figure 12.3 compares the 'year one' cost of this synthesis

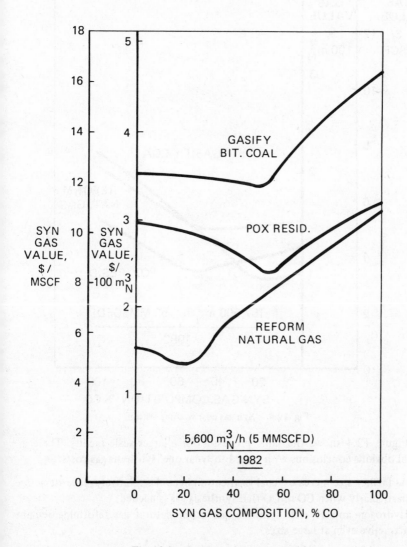

Fig. 12.3 – Syn gas value vs composition.

gas produced by steam reforming, partial oxidation and coal gasification all at a production level of $5600\ m_N^3/h$ (5 MMSCFD†).

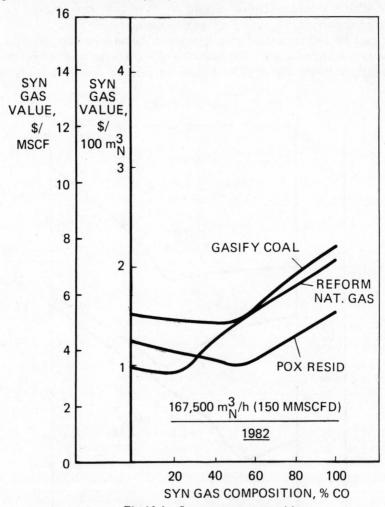

Fig. 12.4 – Syn gas cost vs composition.

Figure 12.4 shows the same processes for a larger scale facility. There are several obvious conclusions with regard to 'year one' synthesis gas costs:

(1) As facility size increases, coal gasification becomes relatively more attractive, particularly when CO or CO-rich synthesis gas is needed.
(2) Hydrogen and $H_2$-rich synthesis gas from natural gas reforming remains attractive even at large sizes.

† $10^6$ standard ft³/day at 60°F 760 mm absolute pressure.

(3) At large sizes, the cost of feedstock is more significant, therefore relative changes in feedstock cost can easily change the choice of process.
(4) High inflation rates, investment tax incentives, long plant life, and leveraged financing will enhance the selection of the most capital intensive process, that is, coal gasification.
Cost and economic data are summarised in the Appendix.

## PRODUCT COST VERSUS TIME

All of the previous comparisons have been at one point in time, 1982. It is also important to examine the resulting synthesis gas costs over a period of time to facilitate the selection of the best technology. Figure 12.5 shows the typical cost

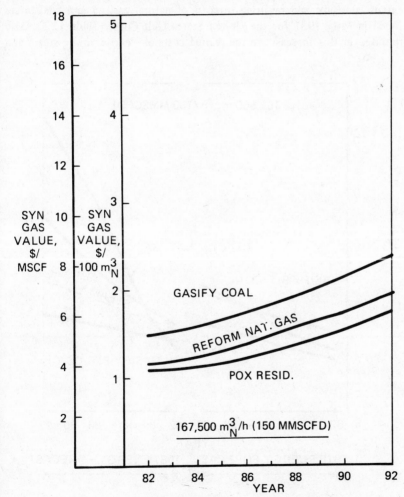

Fig. 12.5 − 2:1 $H_2$:CO syn gas value vs time (Western Europe).

of producing a 2:1 $H_2$:CO synthesis gas for the period 1982 to 1992. The size of the facility was set at 167 500 $m_N^3$/h (150 MMSCFD), about the size of a world scale 1360 tonne/day (1500 t/day) methanol plant. The available energy forecast for Europe (Table 12.4) shows an orderly change in the price of synthesis gas from the three different processes. The conclusion reached from Figure 12.5 is that partial oxidation of residual oil will most likely be the process selected for producing methanol in Europe, that is, if methanol from low cost natural gas sources such as Saudi Arabia does not supply the entire European market.

To dramatise the effect of changing energy forecasts, Fig. 12.6 has been included to show the same size synthesis gas facility located in the United States. The right-hand section of this graph shows the synthesis gas cost based upon an 'extensive' study and resulting forecast of natural gas, oil and coal prices developed in early 1981 for the United States Gulf Coast (Table 12.5). Also incorporated in this forecast are the related costs of electric power, steam and

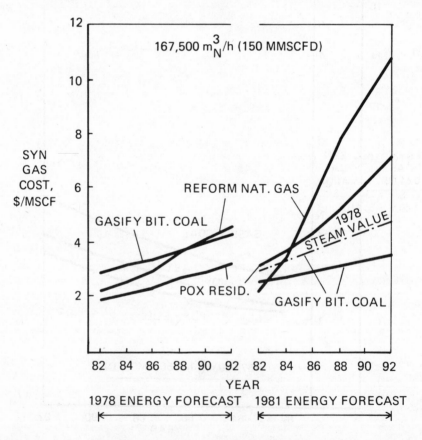

Fig. 12.6 – 2:1 $H_2$:CO syn gas cost vs time (United States).

general inflation projected for this energy forecast. The left-hand side of this graph shows the synthesis gas cost based upon a similar study conducted three years earlier. The obvious message from these comparisons is that in the three years it would take to build this complex, the relationship of the alternatives has changed significantly. These relationships demonstrate the importance of accurately projecting feedstock costs when making a feedstock/technology selection. This is particularly true for natural gas reforming and oil partial oxidation, both of which are energy intensive.

## CO-PRODUCTION

Considerable economies can be achieved by co-production of products. It is possible to: (1) improve the overall thermal efficiency; (2) achieve economy of scale of the synthesis gas facilities; and (3) select a cheaper or more favourable feedstock. In industry, this is observed in several product combinations:

(1) $NH_3$ and MEOH from natural gas is better than either alone. Some plants transfer CO (as $CO_2$) from the ammonia plant to the methanol plant. Others produce ammonia from the excess $H_2$ in the methanol plant.
(2) Methanol and CO from partial oxidation of oil or gasification of coal is more efficient and cost effective than either separately. The resulting methanol and CO produced is close to the desired ratio for acetic acid.
(3) Air Products is in the merchant $H_2$ and CO business at Rotterdam, Netherlands and at Houston, Texas. $H_2$ and CO produced from reforming of natural gas is sold through pipelines to customers with different needs. The economy of scale and the $H_2$:CO balance makes this an attractive arrangement both for the customers and the producer.
(4) There are other co-products which are less obvious, but with significant impact on the process selection. These are co-products such as waste steam, electric power, gaseous nitrogen, and carbon dioxide. On Fig. 12.6 a significant amount of co-produced steam is credited to the coal gasification process. The increased 'value' of this steam in 1981 causes a reduction in the cost of synthesis gas production.

## OPERATING/ENGINEERING EXPERIENCE

The economics and technologies discussed in this paper are based on the experience that Air Products has acquired over the years as an engineering/operating company. The following experience is listed to provide an indication of the 'data base' available for this paper.

As an engineering company, the corporation designs and constructs cryogenic equipment for the separation of hydrogen, carbon monoxide, carbon dioxide and light hydrocarbons. This equipment is used in its own operating plants and is sold competitively. Other synthesis gas processes and equipment for acid gas removal, shift conversion, chemical or physical separation of synthesis gas and final purification are routinely purchased and operated.

As an operating company the corporation has experience in operation of over twenty synthesis gas facilities for production of ammonia, methanol, liquid and gaseous hydrogen, carbon monoxide, and annealing gases. These facilities include steam–methane reformers designed by Chemico, Foster Wheeler, Girdler, Howe Baker, Kellogg, KTI, Selas and Schultz. Air Products has operated Texaco partial oxidation reactors both on crude oil and on natural gas. We have licensed and studied extensively the Shell partial oxidation technology. As part of our solvent-refined coal efforts, many detailed studies have been conducted in the competing coal gasification technologies.

## CONCLUSION

As discussed, there are many sources of synthesis gas and many products commercially made from synthesis gas. Matching up the right source of synthesis gas with the desired end products requires a complex analysis balancing efficiency, energy cost and capital cost. Site specific conditions such as economic availability of raw materials gas, oil, coal, and $CO_2$, and the potential market for co-products such as $N_2$, coal fines, medium Btu gas, $CO_2$, steam, and power, can easily dictate the selection of the optimum process.

## REFERENCES

[12.1] Heck, J. L. and Johansen, T. 'Process improves large scale hydrogen production' *Hydrocarbon Processes* Jan. 1978 (Pressure Swing Adsorption).

[12.2] MacLean, D. L. and Graham, T. E. 'Hollow fibres recover hydrogen' *Chemical Engineering* 25 Feb. 1980.

[12.3] Bonacci, J. C. and Otchy, T. 'Selective CO oxidation process can increase ammonia plant yield' AICHE Symposium, Denver, Colorado, Aug. 1977.

[12.4] Reed, C. L. and Kuhre, C. J. 'Production of synthesis gas by partial oxidation of hydrocarbons' 86th National Meeting, AICHE Houston, Texas, April 1974 (Shell).

[12.5] Eastman, D. 'Synthesis gas by partial oxidation' *Industrial and Engineering Chemistry* July 1956 (Texaco).

[12.6] Stokes, K. J. 'The economics of $CO_2$ removal in ammonia plants' 72nd Annual Meeting AICHE, San Francisco, California, Nov. 1979.

[12.7] Strelzoff 'Partial oxidation for syngas and fuel' *Hydrocarbon Processing* Dec. 1974 (Shell & Texaco).

[12.8] Gregory, D. P., Tsaros, C. L., Arora, J. L. and Nevrekar, P. 'The Economics of Hydrogen' *Chemtech* July 1981.

**APPENDIX**

**Process and Economic data, synthesis gas plant**

Table 12A.1
Energy efficiency of synthesis gas plant

| Steam-methane reforming | | Partial oxidation | | Coal gasification | |
|---|---|---|---|---|---|
| $H_2$ % | Efficiency HHV % | $H_2$ % | Efficiency HHV % | $H_2$ % | Efficiency HHV % |
| 100 | 69 | 100 | 64 | 100 | 51 |
| 90 | 70 | 75 | 68 | 67 | 53 |
| 83 | 70 | 65 | 70 | 49 | 54 |
| 75 | 70 | 45 | 74 | 25 | 43 |
| 67 | 67 | 40 | 71 | 0 | 35 |
| 50 | 60 | 25 | 63 | | |
| 25 | 51 | 0 | 54 | | |
| 0 | 44 | | | | |

$$\text{Efficiency} = \frac{[H_2 + CO]}{[\text{Feedstock} + \text{Fuel} + \text{Steam Import} + \text{Power}]}$$

$H_2$, CO, Feedstock + Fuel, in Btu HHV.
Steam, in Btu consumed or exported.
Power, in 10 000 Btu/kWh.
Gas mixture at least 99 + % $H_2$ + CO at 200 psig.

Table 12A.2
Capital costs of synthesis gas plant

| Product | Steam-methane reforming | Partial oxidation | Coal gasification |
|---|---|---|---|
| 100% $H_2$ | 100 | 171 | 358 |
| 67% $H_2$ | 150 | 154 | 344 |
| Methanol | 230 | 234 | 424 |

Capital costs in 1982 million $ for 150 MMSCFD plant
Greenfield site: includes utilities, $O_2$ supply, site work, feedstock storage, storage of disposal of waste; raw water; electricity and natural gas purchased at plant battery limits.

Table 12A.3
Composition of capital costs; 150 MMSCFD plant producing 2:1 $H_2$:CO

|  | Steam–methane reforming % | Partial oxidation % | Coal gasification % |
|---|---|---|---|
| Oxygen | – | 38 | 26 |
| Reformer | 49 | – | – |
| Gasifier | – | 49 | 66 |
| Shift/Clean-up | 14 | 13 | 8 |
| Hydrogen Rejection | 37 | – | – |

Table 12A.4
Economic factors

| | |
|---|---|
| Income taxes | 48% |
| Long-term interest | 15% |
| Short-term interest | 15% |
| Local taxes and insurance | 2% |
| Return on investment after tax | 6% |
| Straight line taxes | – |
| Grants | 10% |
| Depreciation | 15 years |
| Labour costs | $50 000/year (1982) |

## Table 12A.5
Operating costs; 150 MMSCFD plant producing 2:1 $H_2$:CO (1982)

|  | Product Cost $/MSCF |
| --- | --- |
| *Reforming* | |
| Natural gas at $5.60/MMBtu | 2.81 |
| Operating costs, labour, maintenance, power, etc. | 0.31 |
| Capital charges at $\sim$ 31% of investment (3 yr build) | 0.88 |
| Total | 4.00 |
| *Partial oxidation* | |
| Feedstock at $5.86/MMBtu | 2.36 |
| Operating costs, labour, maintenance, power, etc. | 0.49 |
| Capital charges at $\sim$ 32% of investment (4 yr build) | 0.94 |
| Total | 3.79 |
| *Gasification* | |
| Coal at $3.17/MMBtu | 1.75 |
| Operating costs, labour, maintenance, power, etc. | 0.77 |
| Capital charges at $\sim$ 34% of investment (5 yr build) | 2.54 |
| Total | 5.06 |

# 13

# The Toscoal process: Production of petroleum feedstocks from utility steam coals

Douglas H. Cortez, Director, Development Engineering, Tosco Corporation, USA

## INTRODUCTION

During the past three years, the energy industry has been swept through another economic cycle from shortage and crisis to surplus and complacence. Following the 1979 Iranian revolution, many United States Government and industry leaders called for a massive national synthetic fuels programme. At one point, the Carter Administration proposed a 10-year crash programme providing $88 billion in Federal government assistance. A $22 billion programme was passed by Congress in 1980 and the United States Department of Energy was flooded with proposals seeking financial assistance.

Since the Reagan Administration has taken office, the oil industry has been deregulated and the national synthetic fuels programme has been scaled back considerably. The worldwide business recession, energy conservation and conversion to alternative fuels have caused oil demand and prices to fall. Now every week it seems, we read about a synthetic fuels project that is being shelved or indefinitely postponed. Many of the same people who rushed to develop synthetic fuels in 1979-1980 are now declaring that synfuels plants are no longer needed. At Tosco Corporation, the current recession in the synfuels industry is a replay of the many downturns we have witnessed during our twenty-five years in the business.

Tosco is an independent energy company in the forefront of the commercial oil shale industry. From its beginning in the 1950s, Tosco has unswervingly pursued commercial production of shale oil in the United States. That dedication has been and continues to be based on the belief that shale oil can be profitably produced at a price competitive with the current long-term crude oil price. Temporary excursions in the world oil price above and below the long-term trend have not deterred us from that goal. That goal is now close to reality.

Today, the Tosco-organised Colony Shale Oil Project stands alone as the only commercial-scale synthetic liquids plant under construction in the United States.

In addition to proving up Tosco's technology for shale oil production, the Colony project may have significant implications for coal processing. The Tosco II process for oil shale retorting may be modified to pyrolyse coal to recover liquids, gases and by-product char (the Toscoal Process). If the economics of shale oil production are competitive, then it is reasonable to expect that retorting coal containing similar amounts of liquids should also be economically attractive.

The international steam coal trade has grown significantly in recent years as coal has become a viable alternative to fuel oil. Most of this trade is in high quality steam coal which should be an attractive feedstock to a Toscoal plant. Every large power plant represents an opportunity to produce synthetic crude oil from existing coal supplies. It is that opportunity which is discussed in this paper.

## THE COLONY SHALE OIL PROJECT

Before presenting information on the Toscoal process and its potential for use in Europe, I would like to describe our current commercialisation efforts on oil shale in Colorado.

Tosco organised the Colony shale oil venture in 1964. Prior to its organisation, we successfully demonstrated the Tosco II retorting process in a 25 ton†/day pilot plant at our research centre near Denver. In 1964, the technology was ready to be tested at a demonstration plant scale. The partners in Colony constructed a 1000 ton/day semi-works plant at our oil shale reserve in Western Colorado. This plant was operated as a development facility between 1965 and 1972. About 180 000 barrels of shale oil were produced during the semi-works plant programme.

At the same time, a commercial-scale test mine was opened to obtain commercial mine design information. A room-and-pillar mine was operated to produce about 1.5 million tons of ore. The rooms in the test mine are about 20 m high and 20 m wide.

In 1972, there was sufficient information in hand to design a commercial plant. An engineering contractor was retained to prepare a definitive design and cost estimate for a 66 000 ton/day plant. The Tosco II retort was scaled-up to a commercial module each designed to process 11 000 short tons/day of raw shale.

In 1974, the detailed design of a Colony plant was completed. The plant includes six Tosco II retorts and refining units to upgrade raw shale oil to a light, low sulphur crude oil product. The plant will produce 43 000 barrels per calendar day of product or about 2 million annual metric tons.

---

† Short tons are used throughout this paper.

In 1974, the economics of the Colony project were favourable and field construction was started. Unfortunately, several developments beyond our control forced the Colony partners to suspend construction. The key factors deterring construction were runaway escalation of construction costs, uncertain and confusing environmental regulations, complex federal rules regulating the pricing and allocation of crude oil and petroleum product prices, and uncertain government policy toward synthetic fuels development. Under these conditions, it was impossible to determine what the plant would cost, to whom and for what price the product could be sold.

Considering the sharp increase in real oil prices since 1974, it is now apparent that the Colony project would have been enormously successful if it had been constructed and placed in production in 1978. However, the economics of the Colony project have never relied on rising real oil prices as do so many other synfuels projects.

In June 1980, the Colony partners (Tosco and Atlantic Richfield Co.) recommenced construction of the project. In August 1980, Exxon acquired ARCO's interest in the project and assumed the role of Project Operator. Although the change in operator has affected project cost and schedule, Exxon's impact on the project has been highly favourable. Exxon has committed its most experienced project management and technical staff to the Colony project. After a detailed review of the plant design, the basic process configuration established in 1973 remains unchanged.

Construction of the plant is now proceeding at full speed. In 1981, plant access roads were constructed and the materials handling and process plant sites were cleared and readied for construction. The commercial mine bench was constructed. A new community designed to house 20 000 people was started and the first residents have moved into their new homes.

This summer, the foundations for the pyrolysis units will be installed. Our schedule calls for completion and start-up of the plant in the first half of 1986. Detailed technical and economic information on the Colony project and Tosco's technology has been published elsewhere, [13.1] and [13.2].

### Implications of the Colony project for coal processing
The Colony project will include the largest single train pyrolysis plants constructed in the world. With suitable modifications, the Tosco retorting technology can be applied to the pyrolysis of coal to produce hydrocarbon gases and liquids and a solid char.

Coal pyrolysis is quite different from coal liquefaction and gasification. The major emphasis in recent years has been directed at producing liquids by direct or indirect coal liquefaction. Direct coal liquefaction requires the use of expensive hydrogenation catalysts or hydrogen donor solvents to add hydrogen to coal at extremely high pressures. Indirect liquefaction processes start by completely gasifying the coal at very high temperatures by adding hydrogen in the form of expensive steam. The gasified coal is then reacted by Fischer–Tropsch synthesis to produce low yields of liquids.

When coal is pyrolysed, it is heated at low temperatures and pressures. Instead of adding hydrogen, carbon is rejected in the form of char which can be burned in power stations. The pyrolysis vapours are rich in hydrocarbons which can be treated to produce synthetic liquids. In a typical coal pyrolysis complex no supplemental hydrogen source, such as methane reforming or coal gasification, is needed to produce an upgraded liquid product.

## THE ROLE OF COAL PYROLYSIS IN MEETING EUROPEAN ENERGY DEMAND

The use of coal pyrolysis to recover hydrocarbon liquids and gases is not new. Prior to and during World War II, about twenty-five coal pyrolysis plants and twelve coal tar hydrogenation plants were operated in Germany [13.3]. At its peak, about 2.8 million annual tons or over 50 000 barrels per day of coal liquids were produced. The small-scale shaft furnaces used at that time would not be economic today. However, according to Tosco studies, the concept of coal pyrolysis applied on a large scale in efficient retorts should be economically attractive today.

A large quantity of by-product char is produced by coal pyrolysis. In order to utilise this char it will be necessary to site the pyrolysis plant near a large coal-burning power station. Fortunately, the char produced by the Toscoal process is a potentially attractive boiler fuel in that it has burning characteristics similar to the raw coal.

The opportunities to apply commercial coal pyrolysis to utility steam coals is growing in the United States, Europe and the Far East. The developments which point to this favourable situation are:

(1) continuing need by most industrial nations to develop domestic sources of oil and reduce oil imports from unstable sources;
(2) an increase in the use of coal, domestic and imported, for power generation to reduce dependence on oil and nuclear fuels;
(3) new construction of large, efficient and clean coal-fired power plants in excess of 500 MW capacity;
(4) the unfavourable economic outlook for producing synthetic liquids from coal using direct and indirect coal liquefaction technologies.

These trends are as noteworthy in most European countries as they are in the United States and the Far East. In Western Europe, about 150 million tons of hard coal are now being consumed for electric power generation. About 50 million tons are imported. Most of this coal is consumed in the United Kingdom, Germany and France. However, significant steam coal is used for power generation in Belgium, Denmark, Italy, Netherlands and Spain. Also, in Germany, about 130 million tons of brown coal is used.

Table 13.1 summarises the potential utility market for coal pyrolysis in the major European countries utilising coal for power generation. Because coal pyrolysis is most economic practiced on a large scale, we restricted the market to power stations larger than about 400-500 MW.

**Table 13.1**

Large European coal-fired power stations†

| Country | Existing Stations | | Planned by 1985 | | Total | |
|---------|------|------|------|------|------|------|
| | No. | MW | No. | MW | No. | MW |
| Denmark | 15 | 6 000 | 3 | 2 900 | 18 | 8 900 |
| France | 5 | 3 000 | 2 | 2 400 | 7 | 5 500 |
| Italy | 4 | 3 000 | 2 | 1 920 | 6 | 5 300 |
| Netherlands | 2 | 2 600 | 3 | 1 500 | 5 | 4 100 |
| West Germany | 22 | 20 000 | 9 | 4 000 | 31 | 24 000 |
| Spain | 1 | 500 | 4 | 2 220 | 5 | 2 700 |
| United Kingdom | 26 | 35 000 | N/A | N/A | 26 | 35 000 |
| Total | 75 | 70 100 | 23 | 14 900 | 98 | 85 000 |

† Appropriate estimates for power stations with capacities over 400-500 MW.

Table 13.1 shows that there are currently about seventy-five power stations with a total capacity of about 70 000 MW burning coal in Western Europe. In addition, many new plants are under construction. If we include the plants expected to be on stream by 1985, the total number of plants swells to about ninety-eight with a total capacity of about 85 000 MW.

The quantity of syncrude that could be produced in these large power stations is difficult to estimate. It depends on many factors such as coal quality, plant utilisation rates, and upgrading requirements. Assuming medium and high volatile coals are used in these plants, the total potential syncrude production can be roughly estimated at 400 000 to 600 000 barrels per day. Due to site restrictions, coal quality, regulatory and economic factors, this potential will not be fully achieved. However, these estimates show that the long-term potential is great and worth pursuing.

## SELECTION OF COALS FOR TOSCOAL PROCESSING

In general, the coals with the highest liquid yield are the most attractive feedstocks for a Toscoal plant. Unfortunately, it is not possible to accurately predict oil yield based on the conventional proximate and ultimate analyses. Most high volatile steam coals are potential candidates for pyrolysis. A coal with a high hydrogen to carbon ratio and low oxygen content is usually an attractive pyrolysis feedstock.

Table 13.2 contains typical analysis for foreign coals of interest to European

**Table 13.2**

Bituminous steam coals imported to Europe

| | United States | | Australia Newcastle & NSW | Poland Upper Silesia | South Africa North Transvaal | Columbia El Cerrejon |
| | Eastern | Utah | | | | |
|---|---|---|---|---|---|---|
| Proximate analysis (% by weight) | | | | | | |
| Moisture | 8 | 9 | 3 | 6-10 | 7-8 | 9 |
| Ash | 7 | 8 | 11-15 | 12-16 | 12-15 | 9 |
| Volatile matter | 31 | 41 | 31-33 | 29-34 | 22-30 | 34 |
| Fixed carbon | 54 | 42 | 51-53 | 50 | 50-56 | 48 |
| Sulphur | 0.7-2.8 | 0.6 | 0.4-0.8 | 0.7-1.0 | 0.6-2.0 | 0.6 |
| Approximate 1980 Imports (million tons) | 12.4 | – | 4.2 | 21.1 | 19.9 | – |

## Table 13.3

Tosco material balance, assay retort yields and product properties for various coals

| | Units | Utah bituminous as received | Utah bituminous ash free moisture | Texas lignite as received | Texas lignite ash free moisture | Wyoming subbituminous as received | Wyoming subbituminous ash free moisture | Illinois bituminous as received | Illinois bituminous ash free moisture | Australian Waloon as received | Australian Waloon ash free moisture |
|---|---|---|---|---|---|---|---|---|---|---|---|
| *Coal Properties* | | | | | | | | | | | |
| Gross Heating Value | Btu/lb | 12 173 | 14 418 | 7 447 | 12 800 | 8 419 | 12 570 | 12 235 | 14 194 | 9 891 | 14 420 |
| Free Swelling Index | | 1.5 | | 0.0 | | 0.0 | | 4.5 | | 0.0 | |
| Proximate Analysis | % by weight | | | | | | | | | | |
| Moisture | | 7.63 | — | 33.34 | — | 27.55 | — | 4.20 | — | 6.75 | — |
| Volatile Matter | | 41.22 | 48.82 | 30.80 | 52.94 | 31.83 | 47.52 | 34.20 | 39.68 | 39.59 | 54.26 |
| Fixed Carbon | | 43.21 | 51.18 | 27.38 | 47.06 | 35.15 | 52.48 | 52.00 | 60.32 | 33.38 | 45.74 |
| Ash | | 7.94 | — | 8.48 | — | 5.47 | — | 9.60 | — | 20.28 | — |
| Ultimate Analysis | % by weight | | | | | | | | | | |
| Carbon | | 67.58 | 80.05 | 41.64 | 71.57 | 46.64 | 69.88 | 66.60 | 77.26 | 55.37 | 80.04 |
| Hydrogen | | 5.21 | 6.17 | 2.85 | 4.90 | 3.31 | 4.96 | 5.10 | 5.92 | 4.45 | 6.16 |
| Nitrogen | | 1.47 | 1.74 | 0.71 | 1.21 | 0.71 | 1.06 | 1.20 | 1.39 | 0.83 | 1.74 |
| Oxygen | | 9.59 | 11.35 | 12.27 | 21.09 | 15.27 | 22.88 | 9.60 | 11.14 | 12.04 | 11.35 |
| Sulphur | | 0.58 | 0.69 | 0.71 | 1.23 | 0.31 | 0.46 | 3.70 | 4.29 | 0.23 | 0.69 |
| Ash | | 7.94 | — | 8.48 | — | 5.47 | — | 9.60 | — | 20.28 | — |
| Moisture | | 7.63 | — | 33.34 | — | 27.55 | — | 4.20 | — | 6.75 | — |

| Parameter | Units | 1 | 2 | 3 | 4 | 5 | 6 | 7 | 8 | 9 | 10 |
|---|---|---|---|---|---|---|---|---|---|---|---|
| **Retort Yields** | lbs/ton | | | | | | | | | | |
| Char | | 1 257.0 | 1 420.2 | 1 487.1 | 1 473.9 | 1 332.4 | 1 001.8 | 1 283.3 | 916.2 | 1 338.2 | 1 288.7 |
| Oil | | 411.8 | 246.5 | 238.7 | 205.8 | 235.6 | 157.8 | 270.3 | 157.3 | 420.7 | 355.2 |
| Gas | | 183.4 | 109.8 | 165.1 | 142.3 | 212.1 | 142.1 | 235.8 | 137.2 | 123.1 | 103.9 |
| Water | | 147.8 | 223.5 | 109.1 | 178.0 | 219.9 | 698.3 | 210.6 | 789.3 | 118.0 | 252.2 |
| | | 2 000.0 | 2 000.0 | 2 000.0 | 2 000.0 | 2 000.0 | 2 000.0 | 2 000.0 | 2 000.0 | 2 000.0 | 2 000.0 |
| **Liquid Yields** | gals/ton | 50.1 | 30.0 | 27.0 | 23.2 | 28.6 | 19.1 | 33.3 | 19.4 | 50.7 | 42.8 |
| **Product Properties** | | | | | | | | | | | |
| *Char* | | | | | | | | | | | |
| Volatile Matter | % by weight | 23.0 | | 15.0 | | 18.77 | | 20.9 | | 16.2 | |
| Heating Value† | Btu/lb | 9 771 | | 13 020 | | 12 595 | | 12 110 | | 12 342 | |
| Sulphur | % by weight | 0.19 | | 3.12 | | 0.36 | | 0.93 | | 0.64 | |
| Ash | % by weight | 47.0 | | 13.0 | | 10.9 | | 18.5 | | 12.3 | |
| *Oil* | | | | | | | | | | | |
| Gravity | °API | 11.8 | | 1.6 | | 11.4 | | 14.0 | | 10.9 | |
| Heating Value† | Btu/lb | 16 821 | | 16 060 | | 16 520 | | 17 249 | | 17 323 | |
| Sulphur | % by weight | 0.17 | | 2.06 | | 0.19 | | 0.62 | | 0.24 | |
| *Gas* | | | | | | | | | | | |
| Mole wt. | lbs/mol | 27.5 | | 28.0 | | 33.4 | | 33.0 | | 24.9 | |
| Heating Value‡ | Btu/scf | 874.7 | | 970.6 | | 1 068.9 | | 899.0 | | 1 059.7 | |

† Gross Heating Value
‡ Gross Heating Value $H_2S$ and $CO_2$ free basis

utilities. Of the coals now being imported to Europe, the United States eastern, Australian and Polish bituminous coals are attractive because of their high volatile matter and low moisture, sulphur and ash contents. Typical medium volatile South African coals may not be as attractive unless data on oil yield indicates an unusually high oil content. The Utah and Columbian bituminous coals are being considered for export markets. Our data indicates these are very attractive coals for Toscoal processing.

In order to obtain more reliable estimates of pyrolysis yields, Tosco utilises a laboratory method referred to as the Tosco Material Balance Assay (TMBA). The TMBA test provides a good indication of the potential oil, gas and char yields from a Toscoal plant. Extensive pilot plant runs are required to obtain reliable data for commercial plant design.

The data in Table 13.3 contains TMBA results for several United States coals and a coal from Australia now being investigated by the State Electricity Board of Queensland. These data show that the Utah and Australian coals are attractive because they yield 20-21% oil by weight. The Wyoming sub-bituminous coal and Texas lignite are less attractive because they yield only 12-14% oil by weight. However, the heating value of the char from these coals is greatly enhanced and may improve boiler efficiency. These coals also produce more gas which is valuable as a clean fuel or for hydrogen manufacture. The Illinois coal produces a lower yield of liquids due to the pre-oxidation required to reduce swelling properties. These data illustrate the importance of testing a coal sample prior to judging its potential for coal pyrolysis.

## TOSCOAL PROCESS DESCRIPTION

Figure 13.1 is a schematic flow diagram of the Toscoal process for non-swelling coals. The Toscoal process is essentially the same configuration as the Tosco II process, crushed dry coal is preheated in lift pipes using hot flue gas from the ball heater. Preheating of coal in this manner increases the thermal efficiency of the process. Preheated coal is fed to a rotating pyrolysis drum where it is con-tacted with hot ceramic balls. The ceramic balls are heated in a ball heater fired by clean, process-derived fuel gas. In the pyrolysis drum, coal is decomposed at about 900°F to produce coal char and hydrocarbon vapours. The coal char is separated from the balls by a trommel screen and withdrawn from the char hopper. The balls are elevated, reheated and returned to the pyrolysis drum. Hydrocarbon vapours are cooled and condensed to recover light hydrocarbon gases and coal liquids.

For coals with a swelling index above about 1.0 to 1.5, it is necessary to preheat the coal. For some coals, this is best accomplished by substituting a mild oxidation reactor for the lift pipe preheater.

The Toscoal process has several features which are advantageous for integrat-ing with a coal-fired power plant:

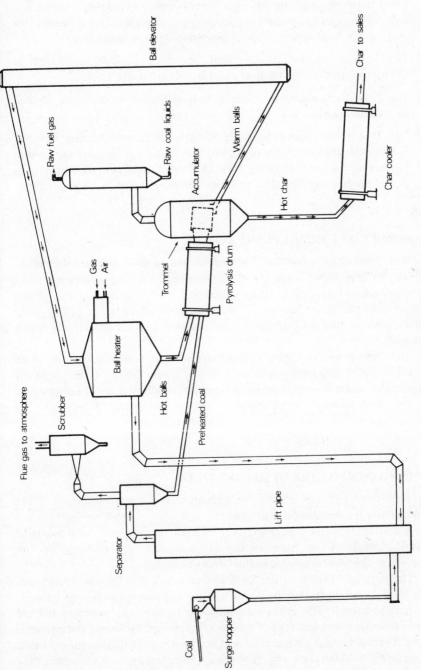

Fig. 13.1 — Toscoal process.

(1) The Toscoal char contains sufficient volatile matter to permit combustion at high percentages in conventional existing power plants. This is because the char has not been subjected to temperatures higher than about 900°F.

(2) The Toscoal process produces high Btu retort gases for conversion to hydrogen for upgrading the heavy coal liquids at the plant site.

(3) The Toscoal process can be scaled-up to large single-train capacities required to minimise capital costs.

(4) The Toscoal process will be ready for commercial application in the mid-1980s. Since the key commercial components of the process are expected to be demonstrated in the Colony Shale Oil Project, small pilot plant operations should provide sufficient data to design commercial coal processing retorts.

## ECONOMICS OF TOSCOAL PLANT IN UTAH

In 1980, when the Colony Shale Oil Project was reactivated, Tosco decided to increase its coal research programme. Preliminary screening studies examined the economics of processing Utah, Illinois and Wyoming coals. Although the economics of processing Illinois and Wyoming coals are attractive [13.4], these studies indicated that the Utah bituminous coal was the most attractive pyrolysis feedstock.

The Utah Power & Light Co. expressed interest in a Toscoal plant to be located at one of their large power complexes in Emery County, Utah. With their cooperation, Tosco has completed a site specific feasibility study of a commercial Toscoal plant feeding a 1650 MW power plant. The results of our study are presented below.

## DESIGN BASES FOR THE UTAH CASE STUDY

A typical Utah coal was selected for evaluation as a Toscoal feedstock. Table 13.4 contains the feedstock design basis.

A conceptual plant design and cost estimate was prepared for a two-train Toscoal plant. Each train has a nominal design capacity of 11 000 tons/day, the same size as the retorts included in the Colony project.

The pyrolysis yield structure for Utah coal was derived from laboratory and pilot plant data and adjusted to develop a consistent material and energy balance. The hydrotreated liquids yields were obtained from process licensors and are based on recent pilot plant tests of coal liquids provided by Tosco. The estimates of the Toscoal Process performance and economics presented herein are based on currently available data and may change significantly as more refined data become available.

**Table 13.4**
Utah coal analysis

|  | As received<br>% by weight |
|---|---|
| *Proximate analysis* | |
| Moisture | 9.00 |
| Ash | 8.38 |
| Volatile Matter | 41.03 |
| Fixed Carbon | 41.59 |
| Total | 100.00 |
| *Ultimate analysis* | |
| Moisture | 9.00 |
| Carbon | 65.56 |
| Hydrogen | 5.19 |
| Oxygen | 9.97 |
| Nitrogen | 1.35 |
| Sulphur | 0.55 |
| Ash | 8.38 |
| Total | 100.00 |
| Gross Heating Value, Btu/lb | 11 735 |

## PLANT DESCRIPTION

A block flow diagram for the conceptual Utah Toscoal plant is shown in Fig. 13.2. The plant is located at the power plant site and operates in parallel with the existing coal delivery system. The plant design capacity is 21 250 tons per stream day.

Coal from an existing storage pile is conveyed to the Toscoal plant. The coal is crushed to 100% minus 3/8 inch and transported to process feed bins. Crushed coal is fed to parallel entrained flow dryers. The dried coal is preheated to about 550°F and fed to the Toscoal retort. The hot char from the retort is cooled and transported to char bins ahead of the existing pulverisers feeding the power plant.

The vaporised hydrocarbons are collected, cooled and the liquid and gaseous hydrocarbons are separated. The gases are processed in conventional refinery equipment to remove carbon dioxide and hydrogen sulphide. The clean retort gas is used for ball heater and other plant fuel, and feedstock for the hydrogen plant.

The raw retort liquids consist principally of a heavy coal liquid high in nitrogen and oxygen which cannot be upgraded in existing petroleum refineries.

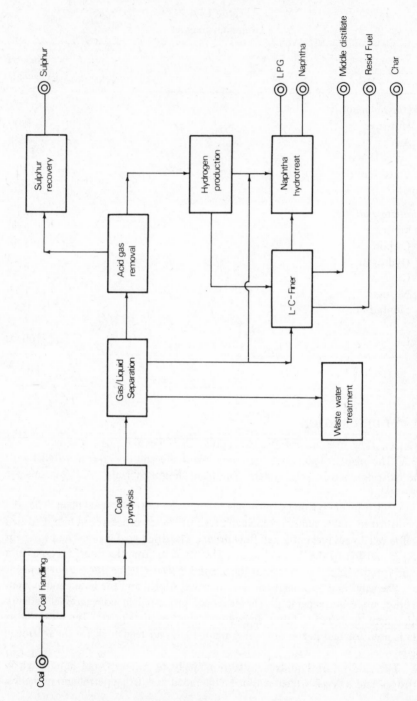

Fig. 13.2 – Toscoal process flow diagram, Utah coal.

An important advantage of the Toscoal process is the production of a high Btu gas by-product which can be used to manufacture the hydrogen needed to upgrade the raw liquid product.

In order to produce a marketable syncrude, an ebulated-bed catalytic hydro-treater has been included in the plant design. For this study, an LC-Fining unit licensed by CE-Lummus was used to upgrade the heavy portion of the raw liquids. Lummus has successfully hydrotreated Utah coal pyrolysis liquids supplied by Tosco. The naphtha boiling range liquids from the retort and LC-Finer are hydrotreated in a fixed-bed catalytic reactor to further reduce hetero-atom impurities to meet reformer feedstock specifications. The unconverted residual oil from the LC-Finer is consumed in the power plant.

The remaining units in the conceptual plant consist of a Claus sulphur recovery plant with tail gas clean-up, a sour water stripper, waste-water treatment and other conventional processes and off-site facilities.

**Table 13.5**
Toscoal process plant feed and product summary; Utah coal

|  | Power plant | Power plant with Toscoal |
|---|---|---|
| *Plant capacities* | | |
| Power Rating, MW | 1 650 | 1 650 |
| Plant Factor, % | 75 | 75 |
| Fuel Rate: | | |
| Coal TPSD† | 14 000 | |
| Char TPSD | | 13 060 |
| *Plant feed* | | |
| Coal, TPSD | 14 000 | 21 250 |
| MMTPY‡ | 4.60 | 6.98 |
| *Plant production* | | |
| Power, billion kWh/yr | 10.82 | 10.82 |
| Syncrude: | | |
| LPG, BBL/SD§ | | 820 |
| Naphtha, BBL/SD | | 7 770 |
| Middle Distillate, BBL/SD | | 13 110 |
| Total | | 21 700 |
| *Incremental Yield:* | | |
| Coal Required, TPSD: | | 7 250 |
| Syncrude/ton, BBL/ton: | | 3.0 |

† tons per stream day        ‡ million tons per year        § barrels per stream day

## PLANT MATERIAL BALANCE

The plant feed and product summary is shown in Table 13.5. The Toscoal plant is located adjacent to a 1650 MW power plant capable of burning all of the char product. Assuming 75% availability, the power plant without a Toscoal unit consumes 14 000 tons per stream day of as-mined coal, or 4.6 million tons per year. The Toscoal plant, processing 21 250 tons per stream day of coal and operating at a 90% availability, produces 4.29 million tons per year of char. This amount of char should be sufficient to fuel the same power plant at a 75% load factor.

The net yield from the Toscoal plant is projected to be 21 700 barrels per stream day of blended syncrude liquids consisting of 13 100 barrels/day of middle distillates, 7 770 barrels/day of naphtha and 820 barrels/day of LPG. The yield of refined liquids per incremental ton of coal mined to support the Toscoal unit is 3.0 barrels per ton.

## PROPERTIES OF SYNCRUDE PRODUCTS

The properties of the syncrude products are shown in Table 13.6. The naphtha is low in sulphur, oxygen and nitrogen and is expected to be a premium reformer or petrochemical feedstock. The middle distillate is a syncrude which will require further refining to produce transportation fuels or petrochemical feedstocks, or can be sold directly as a low-sulphur fuel oil product.

In general, the quality of coal liquids is lower than natural crude oils. However, existing refining technology can be used to upgrade coal liquids at some cost penalty. The relative quality of various petroleum liquids can be observed from the data in Fig. 13.3. This figure, published by Bridge et al, [13.5] shows the hydrogen content of hydrocarbon liquids versus boiling range and molecular weight, expressed as carbon number. Data for the raw and hydro-treated Toscoal liquids have been added to the figure for purposes of comparision. These data show that the hydrogen content of raw Toscoal liquids is significantly higher than most coal liquids and only slightly lower than raw shale oil. The hydrotreated Toscoal syncrude is a bottomless 900°F end-point liquid with a hydrogen content slightly below the distillate material in Arab heavy crude oil.

Unlike direct coal liquefaction processes which add hydrogen to achieve maximum conversion to low grade liquids, the Toscoal process rejects carbon in char for boiler fuel and recovers hydrogen-rich liquids and gases which can be upgraded to petroleum products. Although the precise costs of refining the Toscoal liquids to finished products are not yet available, Fig. 13.3 indicates that those costs can be expected to be substantially less than the direct coal liquefaction processes.

**Table 13.6**
Production specifications

*Syncrude Products*
LPG:

| | | |
|---|---|---|
| | Gravity, °API | 107 |
| | Density, lbs/BBL | 207.3 |
| | $C_3H_8$, % by weight | 49.0 |
| | $C_4H_{10}$, % by weight | 51.0 |
| | Higher Heating Value, Btu/lb | 21 420 |

Naphtha:

| | | |
|---|---|---|
| | Gravity, °API | 55.0 |
| | Density, lbs/BBL | 265.5 |
| | Boiling Range | 500–400°F |
| | Sulphur, ppm | 1.0 |
| | Nitrogen, ppm | 6.0 |
| | Oxygen, ppm | 200 |
| | Higher Heating Value, Btu/lb | 20 140 |

Middle Distillate:

| | | |
|---|---|---|
| | Gravity, °API | 23.0 |
| | Density, lbs/BBL | 320 |
| | Boiling Range | 400–900°F |
| | Sulphur, % by weight | 0.02 |
| | Nitrogen, % by weight | 0.35 |
| | Oxygen, % by weight | 1.2 |
| | Higher Heating Value, Btu/lb | 19 040 |

*Char Product*

| | | *% by weight* |
|---|---|---|
| Proximate analysis | | |
| | Moisture | 0.0 |
| | Ash | 13.40 |
| | Volatile matter | 16.10 |
| | Fixed carbon | 70.50 |
| Total | | 100.00 |
| Ultimate analysis | | |
| | Moisture | 0.0 |
| | Carbon | 75.84 |
| | Hydrogen | 3.25 |
| | Oxygen | 4.94 |
| | Nitrogen | 1.97 |
| | Sulphur | 0.60 |
| | Ash | 13.40 |
| Total | | 100.00 |
| Higher Heating Value, Btu/lb | | 12 600 |

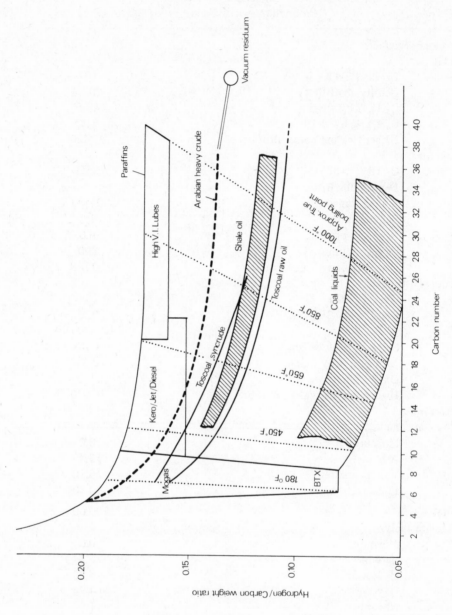

Fig. 13.3 — Hydrogen/carbon ratios in fuels and raw materials.

## PETROCHEMICAL FEEDSTOCK POTENTIAL

The partially refined syncrude produced from a Toscoal plant may have potential as a petrochemical feedstock. Although the liquids described in Table 13.6 are specific to Utah coal, the liquids from most bituminous steam coals are expected to be similar after the initial hydrotreating steps.

The three petroleum fractions produced in the Utah Toscoal plant should be suitable for petrochemical production. The propane and butane in the LPG are excellent steam cracking feedstocks for ethylene manufacture. Propylene would also be produced as a by-product.

The hydrotreated naphtha is rich in naphthenic and aromatic compounds. This product can be separated into feedstocks for aromatics and ethylene production. The naphtha is first catalyticly reformed to increase aromatics content. The aromatics are extracted and separated into benzene, toluene and xylene. The raffinate from the extraction step is paraffinic and suitable for ethylene plant feedstock. The heavy reformate which contains $C_9$ aromatics is best suited for gasoline blending.

The raw Toscoal naphtha may also be a source of phenols and benzene. Most low-temperature coal tars are rich in alkylphenols and cresylic acids. The phenolic compounds can be separated by extraction and hydro-dealkylated to produce large yields of phenol with benzene and toluene as by-products [13.6].

The middle distillate Toscoal liquids contain multiple ring aromatic compounds and are poor petrochemical feedstocks. However, this material can be hydrocracked to produce lighter oils which can be extracted to produce aromatics and ethylene plant feedstock.

The economics of petrochemicals production from Toscoal liquids has not been studied. However, extensive technical and economic investigations of other coal liquids similar to the Toscoal liquids have been completed [13.7]. These studies confirm that petrochemicals, principally aromatics and olefins, can be produced at prices competitive to crude oil derived from petrochemicals.

## PROPERTIES OF THE CHAR PRODUCT

The char product specifications are shown in Table 13.6. The char contains 16.1% volatile matter. The intermediate level volatile matter char is considered an important and unique advantage of the Toscoal process which results directly from the use of inert heat carriers.

The preferred level of volatile matter in the char product can only be determined after optimising the integrated pyrolysis, up grading and power plant system. The Toscoal retort operating conditions can be varied to produce chars over the range of volatile matter believed to contain the optimum. In general, as temperature increases, the quality of the char and liquid products deteriorates. For the study presented here, a 900°F pyrolysis temperature was selected. More pilot plant data and engineering work are planned to determine the optimum plant design.

## CAPITAL AND OPERATING COSTS

The capital cost estimate for the conceptual Toscoal plant in 3rd Quater 1981 dollars is presented in Table 13.7. The plant cost is estimated at $363.5 million. Including indirect charges, a 15% contingency and working capital, the total capital requirement is estimated to be $510.3 million. Table 13.8 lists the major accounts typically included in the installed plant cost.

The estimated plant operating costs are shown in Table 13.9. Excluding the coal cost, the direct operating cost is projected to be $53.0 million per year or about $7.59 per ton of coal feed or $7.43 per barrel of syncrude product.

## ECONOMIC EVALUATION

The economics of the Toscoal plant have been analysed assuming private industry financing. The basis for economic analysis for a privately financed Toscoal plant selling the char to a utility and syncrude to a refinery are summarised in Table 13.10. The discounted cash flow rate of return (DCF-ROI) has been calculated assuming a 10% escalation rate for 100% equity financing, and 50:50 debt/equity financing. Five year depreciation allowed by the Economic Recovery Act of 1981 is included.

**Table 13.7**
Toscoal process Utah coal, capital estimate (3rd Quarter 1981)

|  |  | $ thousand |
|---|---|---|
| Installed plant cost |  | 363 500 |
| Project contingency (15% installed plant cost) |  | 54 500 |
| Total plant investment |  | 418 000 |
| Plant indirects (other capital requirements) |  |  |
|     Process licence fees | 5 000 |  |
|     Initial catalyst & chemicals | 3 000 |  |
|     Spare parts† | 1 800 |  |
|     Owners expense‡ | 54 500 |  |
|  |  | 64 300 |
| Total plant & indirect costs |  | 482 300 |
|     Working capital§ |  | 28 000 |
| Total capital requirement |  | 510 300 |

† Spare parts      – 0.5% installed plant cost
‡ Owners expense – 15% installed plant cost
§ Working capital – one month's operating expense plus 3% of installed plant cost as
      allowance for accounts receivables

**Table 13.8**
Installed plant cost account

---

(1) Major processing units
(2) Support facilities
    Relief and blowdown systems
    Tankage
    Interconnecting piping
    Chemical and catalyst storage
    Electrical distribution
    Cooling and service water systems
    Fire protection
    Sewer and effluent treatment
    Site preparation
    General buildings
    Mobile equipment
(3) Indirects (field distributables, home office and fees)
    Staff support
    Construction equipment
    Construction facilities
    Tools and supplies
    Job insurance
    Travel communications
    Workmans comp and FICA
    Loose materials
    Construction management
    Sales tax
    Construction insurance
    Project management
    Taxes during construction
    Engineering and design

---

The two principal economic determinants are the value of the syncrude and char products. Table 13.11 presents the DCF-ROI as a function of char value assuming a syncrude value of $40 per barrel. If the char is priced at coal value, the hypothetical Toscoal plant owner would achieve a 28.3% return on 100% equity or a 38.8% return if the project is leveraged with 50% debt. If the char is discounted to the utility by 50%, the DCF-ROI is reduced to 24.4% (100% equity) and 33% for the 50% leveraged case.

Table 13.12 shows the effect of syncrude price on project economics with char priced at coal value. Over a range of $36 to $42 per barrel, the return varies only slightly.

**Table 13.9**
Toscoal process Utah coal, direct operating cost estimate (Stream Factor 90%)
Basis: 3rd Quarter 1981

|  | TOSCO Estimate $ thousand/yr |
|---|---|
| *Direct labour and overhead* | |
| Operating labour† | 2 900 |
| Operating labour supervision‡ | 600 |
| Maintenance labour§ | 7 500 |
| Maintenance supervision¶ | 1 700 |
| Administration & general overhead∥ | 8 900 |
| Subtotal | 21 600 |
| | |
| *Supplies* | |
| Catalyst & chemical†† | 3 400 |
| Operating supplies‡‡ | 900 |
| Maintenance supplies§§ | 5 000 |
| Subtotal | 9 300 |
| | |
| *Raw materials* | |
| Coal¶¶ | 151 800 |
| Water∥∥ | 600 |
| Electricity††† | 10 200 |
| Subtotal | 162 600 |
| | |
| *Fixed costs* | |
| Taxes and Insurance‡‡‡ | 11 300 |
| TOTAL | 204 800 |

† Operating labour: 120 men at $24 167/yr.
‡ Operating labour supervision at 22% operating labour.
§ Maintenance labour at 1.8% total plant investment including contingency.
¶ Maintenance supervision at 22% maintenance labour.
∥ Administration and general overhead at 70% total labour (27% payroll burden and 43% overhead).
†† Annual catalyst usage at 90% stream factor.
‡‡ Operating supplies at 30% operating labour.
§§ Maintenance supplies at 1.2% total plant investment.
¶¶ Coal at $22.00/ton at 21 000 TPSD ($0.9374/MMBTU).
∥∥ Raw water at $0.50/Mgals.
††† Electricity at $0.04/kWh.
‡‡‡ Taxes and insurance 2.7% total plant investment.

### Table 13.10
Utah coal; basis of economic analysis (base: 3rd quarter 1981)

---

I *Materials*

| | |
|---|---|
| Fresh water makeup | $0.50/Mgal |
| Utah coal | $0.94/MMBtu |
| Operating supplies | 30% of operating labour |
| Maintenance supplies | 1.2% of plant investment |

II *By-product credits*

| | |
|---|---|
| Sulphur | $25.00/long ton |

III *Labour costs*

| | |
|---|---|
| Operating labour | $24 100/man-year |
| Operating labour supervision | 22% operating labour |
| Maintenance labour | 1.8% of plant investment |
| Maintenance supervision | 22% maintenance labour |

IV *Capital related costs*

| | |
|---|---|
| Insurance and taxes | 2.7% of plant investment |
| Project contingency | 15% of plant installed cost |
| Project life | 25 years |
| Operating factor | (90%) 329 days/year |
| Federal taxes | 46% |
| State taxes | 6.0% |
| Working capital | 1 month plus operating expense plus allowance for accounts receivables |
| Investment tax credit | 10% |
| Owner's equity | 100% and 50% |
| Debt financing | 14% for 10 years |
| Depreciation | 5 years |
| Construction period | 4 years |

Capital spending during construction:

| year | % |
|---|---|
| 1 | 12 |
| 2 | 23 |
| 3 | 38 |
| 4 | 27 |

V *Depreciation*

| year | % |
|---|---|
| 1 | 20 |
| 2 | 32 |
| 3 | 24 |
| 4 | 16 |
| 5 | 8 |
| | 100 |

---

**Table 13.11**
Discounted cash flow rate of return analysis

*Char value*

| $/MM BTU | Percent of Coal Value | 100% Equity | 50% Equity |
|---|---|---|---|
| 0.94 | 100% | 28.3 | 38.8 |
| 0.70 | 75% | 26.5 | 36.1 |
| 0.47 | 50% | 24.4 | 33.0 |

| | |
|---|---|
| Plant capacity | 21 700 BBL/SD |
| Capital investment | $510 MM |
| Base | 3rd Quarter 1981 |
| Oil value | $40/BBL |
| Resid value | $0.94/MM Btu |
| Coal cost | $0.94/MM Btu |
| Escalation | 10%/year on all costs and product values |

**Table 13.12**
Toscoal process Utah coal, discounted cash flow rate of return analysis

*Syncrude value*

| $/BBL | 100% Equity | 50% Equity |
|---|---|---|
| 36.00 | 26.2 | 35.7 |
| 38.00 | 27.3 | 37.3 |
| 40.00 | 28.3 | 38.8 |
| 42.00 | 29.3 | 40.2 |

| | |
|---|---|
| Plant capacity | 21 700 BBL/SD |
| Capital investment | $510 MM |
| Base | 3rd Quarter 1981 |
| Char value | $0.94/MM Btu |
| Resid value | $0.94/MM Btu |
| Coal cost | $0.94/MM Btu |
| Escalation | 10%/year on all costs and product values |

## CONCLUSIONS

Preliminary feasibility studies of a large Toscoal plant operating on Utah bituminous coal indicate that synthetic crude oil can be produced at economically attractive prices. The use of the Toscoal process in conjunction with large coal-fired power plants appears to be a more economic source of liquids than other more complex coal conversion technologies. The European steam coal market is now large and expected to grow. There appears to be many opportunities to site Toscoal units at large coal-burning power plants in Europe.

## FUTURE PLANS

Encouraged by the results of recent work, Tosco is continuing to develop the Toscoal process at the pilot plant scale. Discussions with utilities are now underway to operate a pilot plant programme on several promising steam coals.

The programme will include testing coals to establish yields and operability and to produce representative samples of oil and char for product upgrading and testing. The objective of this programme is to collect sufficient information to begin construction of commercial plants after the Colony Shale Oil Project is operating in 1986.

## REFERENCES

[13.1] Nutter, J. F. and Waitman, C. S. 'Oil shale economics update' 14th Annual Meeting, AIChE, Anaheim, California, April 1981.

[13.2] Vawter, R. G. and Waitman, C. S. 'Scale-up of the TOSCO II process. Energy Technology Conference, ASME, Houston, Texas, Sept. 1977.

[13.3] Rammler, R. W. 'Lurgi coal pyrolysis for production of synthetic fuels' AIChE Winter Meeting, Orlando, Florida, 3 March 1982.

[13.4] Cortez, D. H. and LaDelfa, C. J. 'Coal pyrolysis looks good' *Hydrocarbon Processing* Feb. 1981, 111–117.

[13.5] Bridge, A. G. and Gould, G. D. 'Chevron Hydroprocesses for up grading petroleum residue' Japan Petroleum Institutes's Symposium, Tokyo, Japan, Oct. 1980.

[13.6] Duddy, J. E. *et al.* 'Dynaphen process; phenol and benzene from coal liquids' 2nd World Congress of Chemical Engineering, Montreal, Canada, Oct. 1981.

[13.7] Harris, G. A. *et al.* 'Chemicals from Coal' *Chemtech* Dec. 1981, 764.

# 14

# Coal, crops, chemicals, costs and competition

**Dr S. P. S. Andrew**, Process Technology Section Manager, ICI Agricultural Division

## INTRODUCTION

Carbon, as atmospheric $CO_2$, and energy at high potential, as sunlight, are available at no cost throughout the world. Water is also very widely distributed as are tidal energy, wind energy, wave energy and geothermal energy. Energy from atomic fission adds to the picture of abundance. Coal, oil and gas are also widely distributed as are agricultural products. Over many years chemists have produced schemes for the synthesis or extraction of desirable organic compounds using every one of these carbon feedstocks. Indeed, the present flood of papers on this subject is so great that it is legitimate to conclude that there will be no significant technical problem associated with finding alternatives to crude oil when its supply diminishes. There remains, however, the commercial situation to be considered and this is the subject of the rest of this paper. The theme can however be set down very simply (see Table 14.1). Of the industrialised regions, Japan

**Table 14.1**

| Population | Country | Population Density |
|---|---|---|
| millions | | person/km$^2$ |
| 260 | USSR | 12 |
| 220 | USA | 23 |
| 54 | France | 100 |
| 57 | Italy | 190 |
| 56 | UK | 230 |
| 61 | W. Germany | 250 |
| 115 | Japan | 310 |

and Europe are much more densely populated than the United States and the Soviet Union. In consequence in Japan and Europe there is a much restricted choice of sites for gathering of feedstocks which consequently can only be obtained with greater difficulty, and at higher cost than in the United States and Soviet Union. Table 14.2 illustrates this clearly with respect to coal, where the United States obtains coal significantly more cheaply than the United Kingdom and Germany. In the even less densely populated Australia and South Africa, coal is even cheaper. Similar arguments can be applied to agricultural products used as basic feedstocks, as will be seen later, though in this case the geographical situation of much of the Soviet Union probably outweighs the advantage of its low population density.

**Table 14.2**

| Coal producing country | Coal price $ (1981)/tonne dry ash free |
|---|---|
| Germany | 150 |
| UK | 90 |
| USA | 60 |
| Australia | 50 |
| South Africa | 30 |

Firstly, however, it is valuable to obtain a rough quantitative picture of the manner in which various feedstocks could compete with crude oil for the production of a few basic petrochemicals.

## ROUTES TO BASIC PETROCHEMICALS

Major existing and proposed routes, not based on crude oil, to paraffins, olefins (particularly ethylene) and benzene, toluene and xylenes are sketched in Fig. 14.1. Of those shown, the route to paraffins, olefines and BTX from coal via synthesis gas (a mixture of $CO$, $CO_2$ and $H_2$) using the Fischer–Tropsch reaction is operated on a very large scale in South Africa, and that from biomass (sugar cane) to ethylene via fermentation ethanol in India, and ethanol itself is produced as an automative fuel by fermentation on a large scale in Brazil. The oil shale or tar sands pyrolysis route to ethylene could be very important as a by-product in future large liquid fuel producing plants using these feedstocks. It is significant that major oil companies appear to favour this means of augmenting the supply of liquid fuels compared with either of the two routes based on coal.

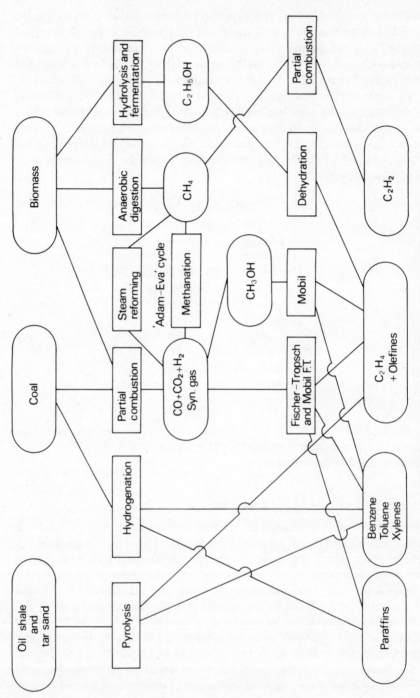

Fig. 14.1 – Major existing and proposed routes to paraffins, olefines, and benzene.

These two routes are coal hydrogenation, the updated Bergius process, and the route through partial oxidation of coal feed to make synthesis gas, followed by build-up of the desired products by selective catalytic hydrogenation of the carbon oxides.

The catalytic schemes for the synthesis gas route are shown diagrammatically in Fig. 14.2. Catalysts fall into two main classes, the metal 'oxide' type, chiefly copper and zinc, and the 'carbide' type such as iron, nickel, cobalt and ruthenium. In the former, the C–O bond remains intact and synthesis produces alcohols. In the latter the C–O bond is disrupted and synthesis produces paraffins and olefines. Both these main classes may be modified by the addition of acidic or basic catalytic species. The addition of acidic species, for instance a hydrogen zeolite, results in dehydration of alcohols to form ethers and, with strong

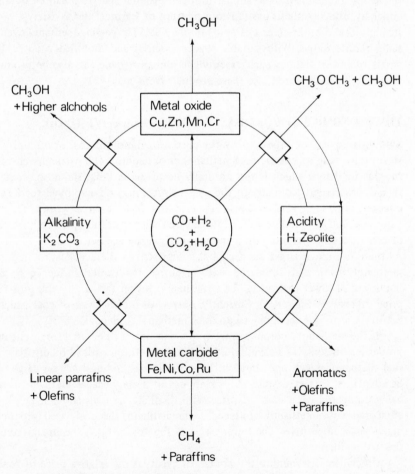

Fig. 14.2 – Catalytic schemes for syn gas route.

acidity, the dehydration of dimethyl ether to form ethylene and the isomeration of paraffins and olefines to form cyclic compounds. The addition of basic species, for instance potassium carbonate, results in the formation of longer chain alcohols or longer chain paraffins. Commercial processes for producing methanol therefore utilise zinc or copper catalysts with no alkaline or acid additives. Similarly no additives are used when producing methane using a nickel or ruthenium catalyst. When higher alcohol production is required alkali is added to, for instance, a zinc based catalyst. When long chain paraffins are required an alkali is added to an iron, cobalt or ruthenium catalyst (the Fischer–Tropsch synthesis). Addition of weak acid component to a methanol synthesis catalyst produces copious quantities of dimethyl ether. A strong acid added to an iron, cobalt or ruthenium catalyst produces ring closure and aromatics (one form of the Mobil process). A very similar ultimate product spectrum can be obtained either by using synthesis gas fed to a mixture of 'carbide' metal catalyst plus a strongly acidic zeolite (for example Mobil's ZSM5) or passing methanol over the same zeolite alone. With suitable space velocities and modified zeolites, high yields of lower olefines such as ethylene and propylene can also be produced thus offering another route to these products from coal.

## THE ECONOMICS OF COAL-BASED ROUTES RELATIVE TO OIL

Synthesis gas can be produced by the partial oxidation of coal or oil, or by the steam reforming of natural gas (methane) or of naphtha. If a particular chemical such as methanol which is not naturally occurring in bulk, should be required then synthesis gas from any of the above routes may be employed for its production. For a large methanol plant the economic break-even between production from coal and production from naphtha varies with the ratio of the relative price of coal to that of naphtha on a thermal basis in the manner shown in Fig. 14.3. As there is a much larger plant capital component in the cost of production of methanol from coal compared with naphtha, the capital influence is most important at lower real prices for heat than at higher (shown by the term 'real price of coal' in Fig. 14.3). Currently, only with the cheapest of coals, such as South African, is it economic to produce methanol from coal.

If the methanol so produced from the plant is not to be used for its chemical properties but only as a clean liquid fuel in competition with an oil distillate — a fuel methanol which one could envisage would be produced in very large coal-based plants — then despite the economics of scale, because methanol is now being compared on a basis of price with distillate whereas above it was being compared with methanol produced from distillate, the coal-based product is much less competitive. The lower curve in Fig. 14.3, shows the break-even for the fuel methanol.

Crucial in determining the course of events is the relative price of coal to that of oil. As was seen earlier, the price of coal at present is very dependent on

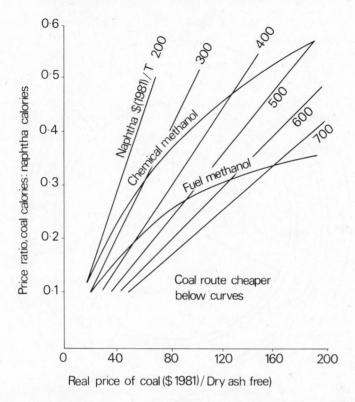

Fig. 14.3 – Competition between coal-based methanol used as a chemical and naphtha-based methanol, and between coal-based methanol as a fuel and naphtha as a fuel.

its location and there is no reason to believe that this feature will change in the future. Certainly, the cost of production of deep-mined European coal will remain much greater than that of strip-mined South African, Australian or American coal. In Western Europe the chemical industry, in so far as it looks towards a coal-based future, must speculate on what will be its competitive position relative to the United States or to industries in, for instance, Australia, perhaps owned by the Japanese using cheaply mined Australian fossil energy. Europe and Japan currently import 60% of their primary energy almost all in the form of oil or gas. Most of this energy is used for the generation of heat (see Fig. 14.4). The international fuel market thus determines, at present, the price of oil as a feedstock. There is no reason to believe that in the future when coal increases its share significantly that the international fuel market will not continue to determine the price of coal. What the relation of this price will be to that of oil will be determined by whether supply leads demand or demand leads supply. Two contrary views may be held. One assumes that an orderly transition will

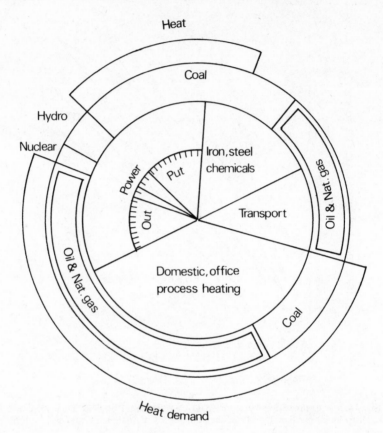

Fig. 14.4 – Energy uses and sources in Europe and Japan.

occur from a predominantly oil- and gas-based international fuel market to one based on coal, a transition dictated by geological factors acting through increased oil extraction costs causing an opening up of new mines. The other assumes a crisis situation resulting from politico-economic changes in the oil market leading to a sudden hunger for fuel in the fuel-importing countries. The two views are laid out in Table 14.3, together with their likely consequences relative to the price of coal. The writer believes that the increasing politico-commercial weakness of Europe makes the crisis situation more likely, with coal sold in a seller's market at up to 80% of the price of crude oil. Under these circumstances the creation of a large chemical industry in Europe based on the hope of cheap imported coal would seem most imprudent, as it would then find itself in competition with United States and Canada with cheaper home supplies of fuel, no doubt at a price controlled to restrict excessive profits for the coal producers, and under circumstances where coal exports are probably politically discouraged.

Table 14.3
Alternative consequences of diminution of world oil and gas supplies.

| | Caused by geological factors | Caused by politico-commercial action |
|---|---|---|
| Rate of diminution of supply | Relatively low (time $\mathrm{cons}^T > 50$ years) | High (time $\mathrm{cons}^T < 10$ years) |
| Predictability of time of onset | Moderate | Very poor |
| *Coal* | | |
| Supply | Gradual increase led by moderate demand | Demand increasing more rapidly than supply |
| Investment in mines | Steady increase backed by adequate financing | Recession in energy-importing countries leading to inadequate financing of overseas mines. |
| Price | Slow increase related to cost of production through market competition | Rapid increase until related to oil and gas price by customer's willingness to pay |
| Determining factor | Commercial and political strength of energy-importing countries relative to exporters | |

History suggests that for the EEC the right-hand possibility is the more probable.

Competition from feedstocks produced as by-products from tar sands and oil fuel industries in the United States and Canada, as well as possibly by Japanese-stimulated ventures in Australia, Brazil, etc., also appears likely. This matter will be returned to in our conclusion after considering biochemically produced feedstocks.

## BIO SOURCES OF CHEMICALS

A large, though circumscribed, range of chemicals may be obtained from microbial and plant life, mostly as intracellular products primarily harvested from inside the cells, but some extracellular that are produced inside the cells and transferred by the cells to the surrounding medium. In addition, all life forms may be used as fuel or as carbon-containing feedstocks. The range of chemicals is indicated in the by no means exhaustive Table 14.4. Microbial extracellular metabolic waste products, such as the various degradation products of sugar — very notably

Table 14.4

| | Plant | Fungal and microbial | | |
|---|---|---|---|---|
| | Differentiated | Intracellular | Extracellular Overflow & other | Metabolic wastes |
| **Carbohydrates** | | | | |
| Structural | Cellulose | Glucan Glucosamine | | |
| Energy stores | Starch Sugar | Glycogen Starch Mannitol P.H.B. Arabitol | Pullulan Xanthan gum Dextran Alginate | |
| Lipids and phospholipids | Oil, fats Glycerol Lecithin | Oil, fats Glycerol Cephalin | | |
| Hydrocarbons and long chain alcohols, acids and terpenes | Waxes Tall oil Rubbers Terpenes | | | |
| Phenols, Polyphenols Heterocycles | Tannin, Indigo, Lignin, Furan, Alkaloids, Resins | Alkaloids Giberillin Quinones | | |
| Amino acids | | Lysine Methionine | Glutamic acid | |
| Proteins | Human food | Human and animal food | | |
| Enzymes | Papain Amylases Peroxidase Facin | Amylases Glucosidase Isomerase | Proteases | |
| Defensive agents | Pyrethrins Rotenoids Nicotine Caffeine Spices | Antibiotics Lysergic acid | | |

Table 14.4 — *continued*

| | Plant | | Fungal and microbial | |
| --- | --- | --- | --- | --- |
| | | | Extracellular | |
| | Differentiated | Intracellular | Overflow & other | Metabolic wastes |
| Vitamins | Carotenoides | Riboflavin B12 Carotenoides | | |
| Acids, alcohols, ketones, esters | Essential oils (perfumes) | | | Ethanol Butanol Citric acid Tartaric acid Gluconic acid Acetone, etc. |

ethanol — are probably the easiest simple chemicals to harvest, as they can be arranged to be produced in relatively high concentrations at a relatively high rate. Microbial extracellular products such as polysaccharides can also be produced in adequate concentrations by subjecting species to live under conditions where these products overflow in superfluity into the surrounding medium. Multicellular organisms, such as plants, are constructed of cells and organs which are differentiated so as to perform specialised functions particularly well. These therefore contain special chemicals in increased concentrations. Using appropriate harvesting techniques, use is made of this preconcentration by only harvesting those parts which contain high concentrations of the desired component. Thus seeds are a good source of starch or oil, the food reserve for the new plant, the bark may contain a potent poison for repelling attacking organisms (quinine). The leaves may have a thick layer of paraffin wax to reduce water loss. The stem is well supplied with lignin and cellulose to give mechanical strength. Microbial cells, as can be seen from Table 14.4 contain, to a greater or lesser extent, virtually all the same range of chemicals but usually there is no equivalent of partial harvesting and the whole organism must be treated so as to obtain the desired chemical. Even so, it may be found to be concentrated in either the nucleus, the cytoplasm or the cell wall and by disrupting the cell, initial fractionation can then be accomplished.

## ORGANISM GROWTH RATE

The productivity of biological chemical synthesis process equipment is dependent on organism growth rates. There is a rough general relation between maximum growth rates and organism size shown in Fig. 14.5, in which the maximum growth rate expressed as its doubling time is plotted against the organism dry weight. Roughly, the maximum organism doubling rate appears to be proportional to the one-sixth power of the mass. (Were all organisms geometrically similar, the index would have been one-third. However organisms have increasingly convoluted surfaces as their mass increases; my readers may not have realised that each person has a surface area of nearly 400 m$^2$, virtually all fortunately out of sight!)

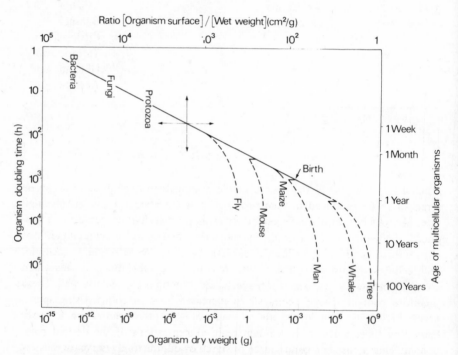

Fig. 14.5 – General relation between maximum growth rate and organism size.

The consequence of this law is that if the desired product is just fuel, biomass can be grown most rapidly when in the form of a large number of small units provided these can be maintained in a steady state of growth and harvesting. The continuously operating fermenter for biomass production is a typical example. Short rotation (2–5 years) forestry in which closely planted trees are grown as fuel is another. As multicellular organisms slow down their growth process markedly and may even stop growing after reaching a certain stage in their life cycle, it is important, when growing for quantity, not to delay harvesting excessively. The optimum cropping frequency for grass is a further example. If, however, a special chemical is required which is only formed in high yields in special organs of the plant, probably as it approaches senescence (e.g. seeds), then the above growth law cannot be utilised. This is the cause of the low productivity of plant cell cultures when used for producing a specialised chemical such as quinine.

Given excellent supplies of nutrients, warmth and moisture, and freedom from infection and pests, typical figures for the production of biomass by the single cell route and by the agricultural route are compared in Table 14.5, from which it will be seen that 1 $m^3$ of medium hold-up in continuous growth at a fairly high cell density produces biomass at about the same rate as 1 hectare of well-tended agricultural land. The moisture held between the soil particles plays the same role as the nutrient medium in the tank and the plant mass transfer surface is the root area compared with the surface area of the single cells. In the case of algal growth depending on photosynthesis as its energy source, the similarities between organism growth in tanks and agriculture are even closer. Nevertheless it must be remembered that in reality agricultural and algal production is normally greatly below the maximum. Though sunlight nearly everywhere permits a growth rate as high as 50 tonnes dry matter/hectare in a

### Table 14.5
A comparison of microbiological and field organism culture

|  | Fermenter culture | Agriculture |
|---|---|---|
| Unit size | 100 $m^3$ capacity tank | 100 hectare (fields) |
| Liquid medium maximum hold-up | 100 $m^3$ | 450 000 $m^3$ (at 15% voids & 3 m depth |
| Number of organisms being cultured | $10^{17}$ to $10^{19}$ | $10^4$ to $10^9$ |
| Weight (wet) at harvesting per organism | $10^{-12}$ to $10^{-10}$ g | 10 to $10^7$ g |
| Harvesting frequency | Continuously | 0.5 to 5 years |
| Maximum density of growth | 10 tonnes per 100 $m^3$ tank | 8000 to 80 000 tonnes per 100 hectares |
| Average density of growth | 10 tonnes per 100 $m^3$ tank | 400 to 4000 tonnes per 100 hectares |
| Average growth doubling time | 1/5 day | 10 to 100 days |
| Average yearly production rate | 15 000 tonne (wet) 5000 tonne (dry) | |

year, the necessity for maintaining the organism at an adequate temperature to promote growth, and in a moist soil, severely restricts the output over most of the earth's surface, see Fig. 14.6. Only in the tropical regions is high sunlight intensity matched by adequate rainfall and an adequate temperature throughout the year. Biomass productivity then tends to be limited by the low level of plant nutrients, especially nitrate in the soil. Of commercial crops, sugar cane comes nearest to maximum output — hence its use for ethanol production. Despite much talk, no commercial whole crops (or at least the 70% above ground) are grown entirely as biomass for fuel. Sugar cane in which the bagasse is used for fuel for operating the sugar refinery and ethanol stills is an approximation to this end. The problem is the low energy density of agricultural products, see Table 14.6, which requires them to be somehow greatly densified at low cost close to where they are grown. The cane to ethanol process, by using high growth rates, cheap harvesting labour and concentration of the sugar fraction and its processing close to the fields, is an optimum solution. A consideration of the utilisation of straw as a fuel or as a feedstock for synthesis gas generation for an industry distant from the farm on which it is produced reveals the problem. The 'energy forests' that are sometimes discussed for Northern latitudes are neither adequately productive, sufficiently extensive, nor commercially viable in the writer's opinion, and they are likely to remain that way.

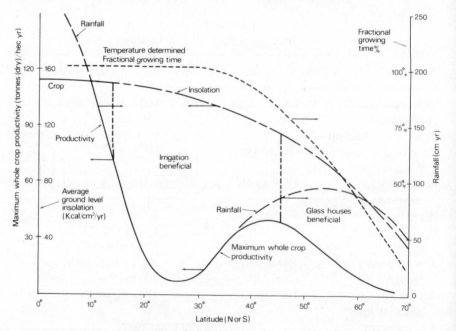

Fig. 14.6 – Simplified diagram showing limitations set by rainfall, insolation, and temperature on maximum crop productivity.

Table 14.6

|  | Relative calories per g | Relative bulk densities | Relative calories per cm$^3$ |
|---|---|---|---|
| Oil | 100 | 100 | 100 |
| Coal | 65 (dry) | 100 | 65 |
| Wood chips | 35 (dry) | 25 | 9 |
| Bamboo | 35 (dry) | 12 | 4 |
| Straw | 35 (dry) | 6 | 2 |

Before leaving the subject of energy from agricultural products mention should be made of the generation of methane by anaerobic digestion, a process which finds favour in sewage plants for disposing of microbial biomass and which can be applied to the effluent of intensive pig farms, etc. Suitable harvesting wastes could also be similarly treated. The methane thus generated could, in theory, be used for producing synthesis gas or cracked to form acetylene as a chemical feedstock. The low density of the feed to such digesters, however, makes this commercial uses, other than as a means of abating a nuisance, of little interest.

The bulk processing of agricultural products, particularly in Northern regions, can never be particularly cheap as the harvesting season is compressed into a small fraction of the year by the climate. The processing plant must be sized for a high throughput over a limited period and then must stand idle over much of the year.

## COSTS OF BIO FEEDSTOCKS

A considerable number of products produced by agricultural and microbiological processes have long been marketed. A general relationship between productivity of either land or fermenter and selling price of product can therefore be drawn up and used for assessing the likely price for new products. This relation is shown in Fig. 14.7. It will be noticed from this figure that whereas 'whole crop' shows the greatest productivity in agriculture, a metabolic waste such as ethanol, has a higher productivity in the microbiological field than does biomass. There are, of course, no such similar products in plant life.

The two bands, it will be seen, infer that the average farmer expects around $600 at very low production rates and $1800 at very high production rates per hectare per year for his efforts. Whilst the average biochemical manufacture requires around $8000 at very low production rates and $30000 at very high production rates per m$^3$ per year for the crop from his fermenter. On this basis

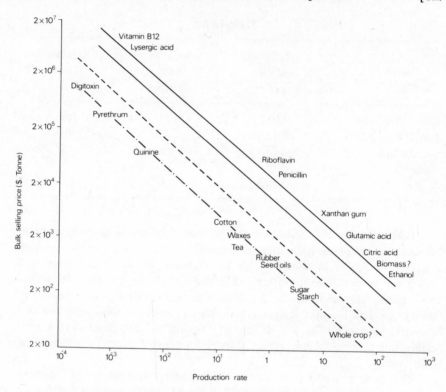

Fig. 14.7 – Price vs production rate: tonnes (dry)/m³/yr or tonnes (dry)/ha/yr.

if, for purposes of illustration, we assume a biomass production rate of 50 tonnes (dry) per m³ year (as in Table 14.5) then, from Fig. 14.7, the biomass selling price should be around $600 per tonne. Clearly this is just a very rough estimate of selling price, as production costs must depend on a large number of factors such as feedstock prices and scale of plant, yield on feedstock, etc. Unless however, a price of this order were received then the biochemical manufacturer would in general feel that he was making a poor use of his assets. Just as the farmer, if he were solely concerned with growing whole crop as a fuel and managed to produce 50 tonnes (dry) per hectare year of which 40 tonne (dry) is harvested, the rest being roots, etc. (productivities requiring a well tended and fertilised tropical environment) would, from Fig. 14.7, expect at least $30 per tonne otherwise he, or someone else, would utilise his land for some other purpose. As his product (let us say sugar cane or bamboo) has only half the calorific value of coal, which could be selling for as low as $30 per tonne in South Africa, and in addition has less than a tenth the energy density, he is most unlikely to find any customers willing to pay that price! Whole crop productivities in the colder regions such as Western Europe are, of course, substantially lower, perhaps about 20 tonne/hectare year at a likely maximum, and less on an average. The corresponding price would be even higher — over \$60 per tonne!

One of the conclusions from Fig. 14.7 is that it is difficult for microbiological methods to compete with agriculture on a world market basis when producing exactly the same product, if this product forms a large fraction of the organism mass in both cases. Competition is only possible with the same product when this forms a relatively small fraction of the organism mass in the plant and consequently there is the possibility that microbial strains can be found which can be grown under condition which result in a many times greater yield of this product than occurs with the plant, so as to offset the cheaper production by suitably located agriculture. With products which are special to either plant or microbe, the question of direct competition with oil- or synthesis gas-based chemicals is, however, most relevant. The case of rubber is an excellent example.

Two further bio-polymers, not yet commercial, are of interest. Pullulan is an extracellular polysaccharide produced by a fungus fed on starch or sucrose. The product can be precipitated, purified and dried and then used for making fibres, films and moulded goods as with other thermoplastics. It has the advantage of being non-toxic, edible and biodegradable. Reported fermenter productivities are, however, low — probably no more than 2 tonnes/m$^3$ year. Figure 14.7 thus suggests that the product is liable to be expensive, around \$10 000/tonne. Polyhydroxybutyrate, an intracellular microbial energy store is a second potentially interesting biodegradable polymer which could be worked up into fibres, film or moulded goods. Little has been published on fermenter productivies, but if its price is to be less than say \$2000/tonne, a typical polymer price, Fig. 14.7 indicates that productivities of over 10 tonnes/m$^3$ year will be required.

## COMPETITION BETWEEN BIO AND SYNTHETIC CHEMICALS

With identical chemicals of a closely similar function and efficacy, the escalation in the real price of oil results in a shift in the relative cost of production of the oil-based product relative to the bio-product. This shift is dependent on the magnitude of the increase in the real cost of oil to the chemical producer and the extent to which the cost of this oil affects the cost of production of his product. Very roughly, the dependence of the real cost increase for the oil-based product relative to the bi-product increases as the real cost per tonne of the product diminishes. Thus a petrochemical product costing \$1000 per tonne is very dependent on the real price of oil, whereas, if it costs \$10 000, it is only slightly dependent. Thus if a factor K is defined such that when the real cost of oil increases (relative to 1981) from 1 to $(1 + X)$, the ratio of real cost of oil-based product to real cost of bio-based product increases to $(1 + KX)$. For products costing in 1981 some \$1000, K could be about 0.7, for those costing \$10 000 K might only be about 0.2, see Fig. 14.8. Thus naphtha-produced ethylene is

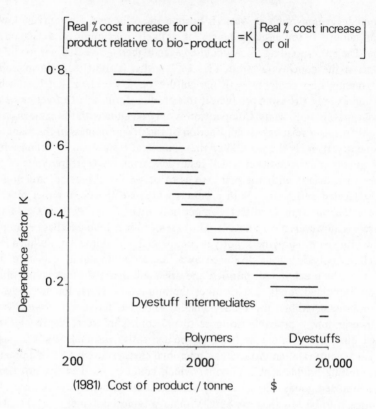

Fig. 14.8 – Ratio of real costs of oil-based to bio-based products.

rapidly rising in cost relative to bio-ethylene and could ultimately become non-competitive. Natural rubber has already severely depressed the production of synthetic rubber ($1400/tonne). Dyestuffs intermediates are, in some cases, somewhat vulnerable $2000-$6000/tonne whereas dyestuffs at $8000+/tonne are, on the whole safe, for many years.

**THE FUTURE FOR EUROPE**

The disadvantageous position of Europe with reference to both the cheap supply of coal, shale oil and tar sands oil, and even with respect to the cheap production of biochemically produced feedstocks such as ethanol, is evident from the above. Japan is in the same, or perhaps in theory, an even worse position. Unlike Europe however, where conferences and research appear to be an acceptable and, what is most important, an inexpensive substitute for action, Japan is wisely attempting to secure cheap future sources of energy and feedstocks on a worldwide basis. Their plan is to set up cooperative ventures involving Japanese chemical and

**Table 14.7**

| | Fossil feedstocks | | Bio-feedstocks | | |
| | Shale and tar sands oil | Coal | Agricultural | Microbiological | |
| | | | Photosynthetic | Photosynthetic | Non-photosynthetic |
|---|---|---|---|---|---|
| **Best technical sources** | USA Canada Brazil Australia | USA, Canada Australia South Africa Indonesia | All tropical regions | Tropical Regions | Industrialised countries |
| **Commercial situation in Europe and Japan** — Price of Imports | Competitive demand from USA, Europe, Japan Sold per calorie at crude oil price | Selling price per calorie rising to crude oil price | Price coupled to world food market Cost of production dependent on labour costs | | Price coupled to world fuel market |
| Products imported or exported | | **Imports** Basic chemicals and intermediates (BTX, $C_2H_4$, $C_3H_6$, VC Monomer, methanol, some fibres, fertilisers | **Imports** Rubber, cellulose and lignin products | **Imports** Glycerol Ethanol | **Imports** Ethanol and other high tonnage metabolic wastes  **Exports** Pharmaceuticals, plant protection products, special effect chemicals |

engineering technology and capital, together with government and commercial organisations in countries with large unexploited deposits of coal, oil shale and tar sands, with the objective of extracting and processing these materials and exporting an agreed fraction to Japan.

The arguments presented above suggest to the writer a future scenario for carbon containing feedstocks in, perhaps, thirty years' time as shown in Table 14.7. In view of the virtual immobility of Europe at present in this matter and our commercially hazardous position relative to the United States, the extent to which Europe participates in this scenario appears to be open to doubt. The extensive and vigorous implementation of a nuclear energy programme — with an emphasis on breeder reactors and fuel reprocessing facilities — is both a commercially competitive and a necessary step towards reducing our dependence on imported oil and gas as energy sources. The current rate of expansion of nuclear power would appear, however, to be, in general, far too low to be effective, bearing in mind both the magnitude of European dependence on imports and on the pace of events which may well be greater than that dictated by geological availability of oil and gas. There are no signs that Europe intends seriously to follow the Japanese example in securing cheap overseas fuels and feedstocks. Perhaps our industrial troops are so demoralised by soft living during the past twenty years that their generals, both political and industrial, consider it imprudent to attempt to engage in the vigorous action required to escape what potentially could be a very damaging situation.

# Part 3

# TECHNOLOGY
# DEVELOPMENTS

*Chairman:* **A. A. L. Challis**, Chief Scientist,
Department of Energy, U.K.

# 15

# The British Gas/Lurgi slagging gasifier

**Dr J. A. Lacey**, Programme Director (SNG) and **H. J. F. Stroud**, Head of Process Studies, British Gas Corporation.

## INTRODUCTION

Whatever the eventual reserves of oil and gas in the North Sea are found to be, the apparently insatiable demand for energy in Western Europe in the long term will prompt every alternative to be evaluated. Imported oil, gas and LNG are immediate options, but as world hydrocarbon reserves become depleted expansion in the use of nuclear energy and coal appears to be inevitable, regardless of the probable valuable contribution from non-conventional or renewable sources. Although the European coal industry has a long history there still remains in the ground vast reserves of indigenous coal. The search for reserves in the other continents is producing enormous finds and the world coal trade is just in its infancy, although bulk coal transport over sea and land is well established. The challenge is, therefore, to develop the technology to enable its use to make high grade fuels, such as fuel or pipeline gases and synthetic liquid hydrocarbons, reducing gases for the metallurgical industries and synthesis gases for the chemical industry. To be successful this technology must combine high thermal efficiency, reasonable capital costs, good operational flexibility, stability, reliability and safety and be demonstrably practical with minimum detrimental effect on the environment. Coal gasification can be made to measure up to these searching requirements.

Large-scale coal gasification has been practised for decades. To meet the anticipated resurgence in the use of coal, numerous new coal gasification processes have been proposed [15.1]. Those relevant to the very large-scale applications mentioned above generally operate at high pressures and ultimately are based on the gasification of coal by steam and oxygen. These processes have been classified according to the way the steam and oxygen reagents are contacted with the coal, the three main contacting modes being entrained flow, fluidised bed and fixed bed. Each type has its attractions and drawbacks, and the only fully commercial high pressure oxygen-blown process is the Lurgi fixed-bed gasifier which has been in widespread use since the 1930s. A development from this already successful process is the British Gas/Lurgi slagging gasifier which combines all the advantages of fixed-bed operation with the benefits that accrue from operating under conditions that allow the mineral matter in the coal to be removed from the gasifier as a molten slag. Briefly the advantages of fixed-bed operation include:

- complete conversion of coal
- stability in operation
- commercially proven design concepts
- high thermal efficiency
- low consumption of steam and oxygen
- no need for coal crushing or pulverisation
- inherently safe operation
- gas cooling and heat recovery is well proven.

The extra advantages of slagging operation briefly include

- much higher throughputs for given reactor size
- much lower steam consumption
- potentially higher efficiency
- much lower production of aqueous effluent
- ability to eliminate net production of tars, oils and phenols
- ability to handle bituminous coals
- ability to handle coals fines.

The development of the process over twenty years, involving 100 ton/day and 350 ton/day gasifiers, has shown that these advantages are achievable. The larger gasifier has consumed over 90 000 tons of numerous coals including over 30 000 tons in one run. The demonstrated reliability and acceptability of the process renders it ready for commercial exploitation.

In this paper the development and application of the British Gas/Lurgi slagging gasifier are described. A description of the process is given with actual performance data.

## PROCESS DESCRIPTION

In concept the slagging gasifier is a simple process [15.2]. A vertical cylindrical reactor is supplied with fuel through a lock hopper and a rotating coal distributor (Fig. 15.1). The fuel moves slowly down the reactor while gases pass up through the bed at a much higher velocity to leave at the top of the gasifier. A mixture of steam and oxygen is injected through nozzles, called 'tuyeres', into the base of the fuel bed at such a velocity as to cause a small zone of turbulence, called a raceway, to be formed. The ratio of steam to oxygen is set at a low value (typically 1.0 to 2.0 molar) and this gives raceway temperatures high enough to cause the ash derived from the coal to melt, giving a mobile slag which drains

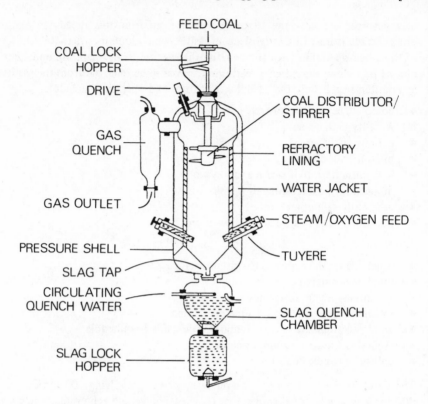

Fig. 15.1 – The British Gas/Lurgi slagging gasifier.

into the hearth that forms the bottom of the reactor. The slag drains from the hearth through a centrally-placed slag tap and is quenched in water to form a glassy frit. The predominant reaction in the raceway is combustion of carbon yielding a very hot gas containing, *inter alia,* steam and carbon dioxide. As this gas enters and ascends the fixed bed, carbon is rapidly gasified by the steam or carbon dioxide. These reactions are strongly endothermic so temperatures drop rapidly, effectively limiting the very high temperature slag liberation zone to a small volume. This is clearly beneficial in reducing both heat losses and potential refractory problems. The carbon gasification reactions proceed up the bed giving progressively lower temperatures and hence lower reaction rates until a point is reached where the reactions effectively stop. Above this point the rate of temperature fall is much less, being caused simply by small heat losses and sensible heat gains by the fuel bed, until a further temperature drop occurs when the ascending gases come in contact with coal near the top of the bed. Here rapid heating of the fresh coal drives off moisture and initiates devolatilisation reactions. These reactions yield tars and oils, significant amounts of methane, some carbon oxides, sulphur compounds, steam and other minor products which are all carried out of the gasifier in the product gas.

This counter-current flow of gases and solids in the fixed bed means that the gaseous products leave the reactor at a modest temperature, so obviating the need for high-grade heat recovery. The small amount of tars in this stream protect the reactor offtake and downstream equipment so that they can be fabricated from inexpensive carbon steels. As explained the counter-current flow does give quite complex temperature profiles in the bed but, paradoxically, this does lead to a system which is very stable in operation. The flow of coal and char is not uniform across the shaft; indeed it appears that there is relatively little flow adjacent to the walls and little or no wear of the gasifier internal surfaces is observed.

The slagging gasifier can handle a large variety of fuels ranging from cokes to highly caking and swelling bituminous coals. Smooth operation with agglomerating coal is achieved by fixing to the coal distributor a stirrer that protrudes into the fixed bed. Rotation of the stirrer mixes fresh coal with non-sticky char and breaks up any agglomerates that form. A wide range of ashes can be handled, although sometimes flux addition with the coal is practised to modify the flow characteristics of the molten slag. By injection at the tuyeres, by-product tar and oils can be gasified completely in the raceway so allowing complete gasification of the carbon in the coal. Pulverised coal can also be added in this way. Normally a better way to handle fine coal (here defined as coal particles less than 6 mm in size) is to feed it with lump coal on to the top of the fixed bed since this is thermodynamically more efficient, but the ability also to inject fine coal in the raceway extends the total amount of fine coal that can be handled so that all the fines in most run-of-mine coals can be accommodated.

## DEVELOPMENT OF THE PROCESS

The development of the slagging gasifier goes back to the 1950s. In 1955 British Gas erected at the Midlands Research Station, Solihull, an experimental gasifier and carried out initial exploratory research [15.3] into slagging gasification, using coke and operating at 75 psig (5 bar).

Encouraging results led to the gasifier being extensively modified for operation on coal at higher pressure to give a design output of 5 million sft$^3$/d of crude gas. Work on this gasifier between 1962–64 [15.4] successfully demonstrated the gasification of coal at pressures of 300 psig (20 bar) and showed that high loadings (1168 lb (daf)/ft$^2$† per hour producing 8 million sft$^3$/d could be achieved. A significant breakthrough was the demonstration of the tapping of slag through a tap in the centre of the hearth, at instantaneous rates of up to 14 000 lb/h, under automatic control.

† Cross sectional area of gasifier.

The next step was to build a larger gasifier, but circumstances in the United Kingdom prevented this becoming a reality until the resurgence in coal gasification in the mid-1970s. In 1974 British Gas set up its Westfield Development Centre (Scotland), the site of a full-scale coal gasification plant based on Lurgi dry-ash gasifiers for making town gas. Here, under the sponsorship of a number of American corporate bodies, a three-year programme to develop a larger gasifier was started.

This Westfield slagging gasifier was built [15.5] by converting one of the Lurgi gasifiers to slagging operation. Its main features are shown diagramatically in Fig. 15.1. The original gasifier was lined to reduce its shaft diameter from 9 ft to 6 ft as the gasifier throughput was limited by the total output of the existing oxygen plant, which had previously served several Lurgi gasifiers. A second gas offtake was added, together with an associated downstream cooling system, to match the greater output. Following the installation of the slag tapping, hearth and tuyere systems, the addition of a slag quench chamber, new instrumentation and control equipment, and the elimination of the grate, the remainder of the Lurgi gasifier continued in use to serve its previous role. The gasifier operates at a maximum pressure of 350 psig (25 bar) and consumes 350 United States tons of coal per day.

In this three-year development, over 20 000 tons of coal were gasified and it culminated in a successful 23-day test which demonstrated the commercial viability of the gasifier and confirmed its excellent performance. This programme marked the beginning of formal cooperation between British Gas and Lurgi.

As a result of this successful project, the United States Government funded a further programme. This included a number of gasifier trials at Westfield in which Pittsburgh 8 and Ohio 9 coals, both having highly caking and swelling characteristics and having high sulphur contents, were successfully gasified.

Since 1978 British Gas has continued development work on the 6 ft gasifier to meet several objectives:

(1) to perfect operating procedures;
(2) to obtain performance data on a wide range of British coals;
(3) to develop systems for the injection of pulverised coal at the tuyeres;
(4) to carry out a long demonstration run;
(5) to demonstrate processing of the crude gas, including its upgrading to SNG using the British Gas HCM process [15.6].

The programme will involve continued operation of the 6 ft diameter shaft gasifier and the construction, commissioning and operation of an 8 ft diameter prototype commercial gasifier. During this British Gas programme, a three-month interruption was made in 1979 to accommodate a three-run project, sponsored by the United States Electric Power Research Institute (EPRI), aimed at demonstrating the potential of the slagging gasifier for electric power generation in a

combined cycle plant using Pittsburgh 8 coal. This showed the gasifier well able
to meet all the load-following and other requirements set by EPRI.

Up to December 1981 nearly 100 000 tons of coal from many United
Kingdom and several United States coal fields have been gasified in 70 runs
covering 7000 hours of operation (see Table 15.1).

**Table 15.1**
Summary of Westfield slagging gasifier projects

| Project | No. of runs | Hours on line | Fuel gasified (US tons) |
|---|---|---|---|
| Sponsors' Programme 1974–1977 | 27 | 1500 | 21 800 |
| DoE Programme 1978 | 15 | 980 | 12 200 |
| EPRI Trials 1979 | 3 | 420 | 4 400 |
| British Gas Programme 1978–1981 | 25 | 4 260 | 58 900 |
| Total | 70 | 7 160 | 97 300 |

Currently a gasifier with an 8 ft diameter shaft is being installed at
Westfield. Operation of this full-sized reactor in 1983 will demonstrate the
commercial status of the process at this larger size.

## PERFORMANCE OF THE SLAGGING GASIFIER

One of the objectives of the Westfield development programme has been to
show the operation of the gasifier on a wide range of coals as shown in Table
15.2. These coals represent not only a wide spectrum of reactivity and caking
characteristics, but also widely different ash compositions. Typical performance
data for British and Eastern United States coals, obtained from the Westfield
gasifier are shown in Table 15.3. The similar performance of the slagging gasifier
for widely different coals is noteworthy. The very low steam demand of the
slagging gasifier with its favourable, and fairly constant oxygen demand, is
clearly evident. Counter-current operation and low steam to oxygen ratio result
in the modest outlet temperatures. The almost complete decomposition of process steam within the gasifier leads to a very small liquor yield and compared
with a dry-ash gasifier of the same diameter, a much greater thermal output.
The low yield of $CO_2$ leads to a crude gas having a high $H_2S/CO_2$ ratio.

These gasification characteristics result in considerable reductions in capital
costs in the process areas of steam raising, oxygen production, gasification,
effluent treatment and desulphurisation, while the high thermal efficiency of the

## Table 15.2
### Coals used in the Westfield slagging gasifier (1975–1981)

| COAL | COMRIE | COTGRAVE | FRANCES | GELDING | HUCKNALL | KILLOCH | LYNEMOUTH | MANTON | MANVERS | MARKHAM | ROSSINGTON | SEAFIELD | BELLE AYR | ILLINOIS No. 5 | OHIO No. 9 | PITTSBURGH No. 8 |
|---|---|---|---|---|---|---|---|---|---|---|---|---|---|---|---|---|
| ORIGIN | Scotland | England | Scotland | England | England | Scotland | England | England | England | England | England | Scotland | USA | USA | USA | USA |
| **PROXIMATE ANALYSIS** (% by weight) | | | | | | | | | | | | | | | | |
| Fixed carbon | 57.0 | 38.9 | 54.0 | 50.7 | 55.6 | 53.7 | 51.4 | 57.1 | 55.5 | 54.3 | 54.7 | 41.8 | 31.3 | 42.3 | 41.4 | 50.2 |
| Volatile matter | 33.2 | 35.1 | 32.9 | 31.3 | 34.1 | 33.7 | 32.0 | 31.5 | 32.6 | 31.4 | 31.2 | 26.5 | 33.0 | 31.1 | 33.6 | 34.1 |
| Moisture | 4.7 | 10.5 | 8.7 | 13.3 | 6.4 | 8.1 | 11.3 | 4.1 | 6.3 | 10.1 | 9.5 | 12.0 | 30.2 | 11.8 | 6.1 | 5.0 |
| Ash | 5.1 | 15.5 | 4.4 | 4.7 | 3.9 | 4.5 | 5.3 | 7.3 | 5.6 | 4.2 | 4.6 | 19.7 | 5.5 | 14.8 | 18.9 | 10.7 |
| CAKING INDEX (GRAY-KING) | F | B | B | C | G | E | E | G6 | F | D | E | A | A | A | G | G6 |
| BS SWELLING No. | 2½ | ½ | 1½ | 1½ | 3½ | 3½ | 3½ | 6½ | 3½ | 1½ | 1½ | 1 | 0 | 0 | 4½ | 7 |

## Table 15.3

### Performance data for British Gas/Lurgi slagging gasifier

| Coal<br>Origin<br>Size, in | Frances<br>Scotland<br>¼–1 | Rossington<br>England<br>¼–1 | Manton<br>England<br>¼–1¼ | Ohio 9<br>USA<br>¼–1 | Pittsburgh 8<br>USA<br>1/8–1 | |
|---|---|---|---|---|---|---|
| *Proximate Analysis* | | | | | | 91 |
| Fixed carbon | 54.0 | 54.7 | 57.1 | 41.4 | 50.2 | 48.3 |
| Volatile matter | 32.9 | 31.2 | 31.5 | 33.6 | 34.1 | 36.1 |
| Moisture | 8.7 | 9.5 | 4.1 | 6.1 | 5.0 | 4.7 |
| Ash | 4.4 | 4.6 | 7.3 | 18.9 | 10.7 | 10.9 |
| *Ultimate analysis* | | | | | | |
| C | 83.0 | 83.5 | 85.1 | 79.6 | 83.7 | 83.7 |
| H | 5.5 | 4.9 | 5.1 | 6.1 | 5.7 | 5.5 |
| O | 9.2 | 7.7 | 5.5 | 7.4 | 6.9 | 7.1 |
| N | 1.4 | 1.7 | 1.6 | 1.2 | 1.6 | 1.7 |
| S | 0.5 | 1.7 | 2.3 | 5.6 | 2.0 | 1.9 |
| Cl | 0.4 | 0.5 | 0.4 | 0.1 | 0.1 | 0.1 |
| Coal calorific value, Btu/lb ar | 12 664 | 12 589 | 13 281 | 10 868 | 10 616 | 10 598 |
| B.S. Swelling No. | 1½ | 1½ | 6½ | 4½ | 7 | 7 |
| Caking Index (Gray–King) | B | E | G6 | G | G6 | G6 |
| *Operating conditions* | | | | | | |
| Pressure, psig | 350 | 350 | 350 | 350 | 335 | 335 |
| Steam to oxygen ratio, vol./vol. | 1.34 | 1.29 | 1.39 | 1.25 | 1.22 | 1.13 |
| Outlet gas temperature, °C | 480 | 480 | 513 | 410 | 516 | 521 |
| *Crude gas composition*<br>(main components), % by vol. | | | | | | |
| $H_2$ | 28.6 | 27.2 | 28.1 | 28.7 | 28.0 | 29.5 |
| CO | 57.5 | 58.1 | 56.8 | 53.2 | 56.4 | 55.8 |
| $CH_4$ | 6.7 | 6.8 | 6.8 | 6.9 | 7.1 | 5.8 |
| $C_2H_6$ | 0.4 | 0.5 | 0.7 | 0.3 | 0.3 | 0.6 |
| $C_2H_4$ | 0.2 | 0.2 | 0.2 | 0.2 | 0.1 | 0.2 |
| $N_2$ | 4.2 | 3.9 | 3.4 | 4.0 | 4.2 | 4.1 |
| $CO_2$ | 2.3 | 2.9 | 3.5 | 5.5 | 3.0 | 3.5 |
| HHV, Btu/scf | 355 | 355 | 357 | 342 | 350 | 347 |
| *By-product yields, lb/US ton coal* | | | | | | |
| Tar and oil | 192 | 138 | 90 | 149 | 122 | |
| Ammonia | 11 | 11 | 6.4 | 8.9 | | 6.2 |
| Phenols | 9.1 | 1.6 | 2.5 | NM | | 1.5 |
| Fatty acids | 14 | 0.21 | 0.64 | NM | | 0.2 |
| Naphtha | 40 | 13 | 29 | 28 | 34 | |
| $H_2S$ | 7.6 | 25 | 34 | 73 | 28 | |
| COS | 1.8 | 3.7 | 4.7 | 14 | | 2.8 |
| $CS_2$ | 0.02 | 0.04 | 0.03 | 0.15 | | 0.02 |
| thiophene | 0.06 | 0.09 | 0.03 | 0.08 | | 0.04 |
| HCN | 0.06 | 0.08 | 0.02 | NM | | 0.02 |
| Coal gasification rate, lb/ft²h § | 852 | 848 | 841 | 664 | 816 | 592 |
| Gas thermal output 10⁶ Btu/ft²/h | 10.6 | 10.6 | 11.0 | 7.8 | 10.0 | 8.0 |
| Steam consumption lb/lb coal § | 0.41 | 0.40 | 0.46 | 0.39 | 0.39 | 0.42 |
| Oxygen consumption lb/lb coal § | 0.54 | 0.56 | 0.61 | 0.57 | 0.57 | 0.64 |
| Liquor production lb/lb coal § | 0.20 | 0.21 | 0.15 | 0.21 | 0.17 | 0.17 |
| Gasifier thermal eff. †% | 94.3 | 93.9 | 95.7 | 91.5 | 97.2 | 97.1 |
| ‡% | 83.4 | 82.1 | 82.7 | 79.7 | 85.1 | 83.7 |

*Notes:*

†   Defined as the total product gas thermal output (based on HHV, including tar, oil, naphtha) divided by corresponding thermal input of coal feedstock.

‡   Defined as total product gas thermal output (based on HHV, including tar, oil, naphtha) divided by corresponding thermal input of coal feedstock and the fuel equivalent of the steam (1700 Btu/lb) and oxygen (2250 Btu/lb) used.

§   Coal expressed as dry, ash-free.

NM  not measured.

¶   For this case alone tar was injected (at rate of 93 lb/US ton of coal feed) through the gasifier tuyeres.

gasifier gives lower operating costs because of the lower coal feed requirements. The slagging gasifier operates with steam to oxygen ratios in the range 1 to 2 (vol/vol) with satisfactory slagging conditions, and over this range the thermal efficiency of the gasifier varies little, but there is some effect upon product gas composition as shown by the summary of experimental data in Table 15.4. Coals with refractory ashes need fluxing, which is achieved by mixing either blast furnace slag or limestone with the coal as it enters the coal lock. The amount of flux added depends on the composition of the ash, the amount of ash in the coal, and the temperature of the molten slag in the gasifier. This temperature is a function of steam to oxygen ratio.

The fixed-bed gasifier generates tar and oil which are carried out from the gasifier with the crude gas. Their presence is beneficial in preventing corrosion of the downstream equipment and washing dust out of the gas. They can be recycled to extinction through the tuyeres of the slagging gasifier with little effect on gasifier efficiency, and produce an increased output of gas per ton of coal gasified. Performance data for Pittsburgh 8 with and without tar injection are shown in Table 15.3. Phenols from liquor clean-up, and, more importantly

**Table 15.4**

Performance of the slagging gasifier at various steam/oxygen ratios for Rossington coal

| Steam to oxygen ratio, vol./vol. | 0.93 | 1.15 | 1.39 | 1.54 | 1.78 |
|---|---|---|---|---|---|
| Outlet gas temp, °C | 529 | 471 | 476 | 468 | 473 |
| Crude gas comp, % by vol. | | | | | |
| $H_2$ | 25.8 | 26.5 | 27.3 | 28.1 | 28.8 |
| CO | 61.2 | 60.7 | 57.2 | 56.3 | 53.8 |
| $CH_4$ | 6.9 | 6.5 | 6.6 | 6.8 | 7.2 |
| $C_2H_4$ | 0.5 | 0.6 | 0.6 | 0.6 | 0.5 |
| $C_2H_4$ | 0.2 | 0.2 | 0.2 | 0.2 | 0.1 |
| $N_2$ | 3.7 | 3.5 | 4.0 | 3.3 | 3.6 |
| $CO_2$ | 1.3 | 1.6 | 3.7 | 4.1 | 5.5 |
| $H_2S$ | 0.4 | 0.4 | 0.4 | 0.6 | 0.5 |
| *Derived data* | | | | | |
| Steam consumption, lb/lb coal | 0.299 | 0.348 | 0.426 | 0.460 | 0.540 |
| Oxygen consumption, lb/lb coal | 0.593 | 0.563 | 0.570 | 0.553 | 0.562 |
| Liquor production, lb/lb coal | 0.16 | 0.18 | 0.19 | 0.21 | 0.27 |

*Note:* Coal expressed as dry, ash-free.

pulverised coal entrained in a suitable carrier gas, can also be injected directly into the reaction zone via the tuyeres. The results of tests in which 15% of the total coal was fed in this way are shown in Table 15.5, and higher amounts are

### Table 15.5
Performance of slagging gasifier with 15% fines injection through tuyeres

| | |
|---|---|
| Coal | Markham |
| Origin | England |
| Size, in | ¼–2 or pulverised |
| *Proximate Analysis* | |
|   Moisture | 7.18 |
|   Ash | 4.40 |
|   Volatile matter | 33.38 |
|   Fixed carbon | 55.04 |
| *Ultimate analysis* | |
|   C | 83.54 |
|   H | 4.61 |
|   O | 7.98 |
|   N | 1.78 |
|   S | 1.62 |
|   Cl | 0.47 |
| BS Swelling No. | 1 |
| Caking Index (Gray–King) | D |
| *Operating conditions* | |
| Steam to oxygen ratio, vol./vol. | 1.18 |
| Outlet gas temperature, °C | 546 |
| % coal feed gasified as fines | 15 |
| Crude gas composition, % by vol. | |
|   $H_2$ | 27.5 |
|   CO | 55.6 |
|   $CH_4$ | 5.7 |
|   $C_2H_6$ | 0.4 |
|   $C_2H_4$ | 0.1 |
|   $N_2$ | 7.2 |
|   $CO_2$ | 3.1 |
|   $H_2S$ | 0.4 |
| *Derived data* | |
| Steam consumption, lb/lb coal | 0.403 |
| Oxygen consumption, lb/lb coal | 0.633 |
| Liquor production, lb/lb coal | 0.22 |

*Note:* Coal expressed as dry, ash-free.

possible. Other ways of handling coal fines yet to be tried are the pumping of coal slurries through the tuyeres and addition of briquettes or extrudates, made from coal fines, into the top of the bed. The gasifier has operated perfectly satisfactorily with Pittsburgh 8 coal containing 25% fines and Manton coal containing 35% fines fed directly to the top of the fixed bed. Thus in most situations the slagging gasifier will be able to consume run-of-mine coal.

A heat balance for a slagging gasifier is shown diagrammatically in Fig. 15.2. This relates to standard operating conditions with complete recycle of tars, oils, naphtha and phenols. The sum of the sensible and latent heat in the product gas is seen to be small owing to both the modest outlet temperature obtained with fixed-bed operation and the relative dryness of the gas. Some of this heat is readily recovered as low pressure steam, while the heat transferred to the gasifier jacket produces steam at high pressure. Most noteworthy is the extremely small loss to cooling water, which includes the heat content of the molten slag, even when the ash plus flux amounts to 16% of the weight of coal used.

The load following performance of the gasifier was well quantified in the 1979 EPRI-sponsored trials [15.7] which were aimed at confirming the suitability of the slagging gasifier for use with combined cycle power generation systems. Following an initial run on Rossington coal, Pittsburgh 8, a high-caking, high sulphur Eastern United States coal, was chosen for the tests, which were particularly orientated towards establishing the ability of the gasifier to respond quickly to load changes and to run steadily at a variety of loads. These objectives were sucessfully achieved, with the gasifier's ability to respond to load changes more than matching the requirements. Table 15.6 highlights the results from over fifty controlled load changes. The gasifier ran stably at all loads between 30% and 110% of standard load and its load could be changed rapidly within this range. There were no significant transients during load changes and the gas composition remained substantially constant for all loadings. The slagging gasifier also has important start up and shut down characteristics. It can be started up,

Table 15.6
Gasifier load following capability

| Normal load change required (%) | Response time required for combined cycle operation (min) | Fastest demonstrated response time at Westfield (min) |
|---|---|---|
| 15 | 3 | 2 |
| 30 | 10 | 2 |
| 50 | 37 | 2 |
| 70 | 100 | 3 |

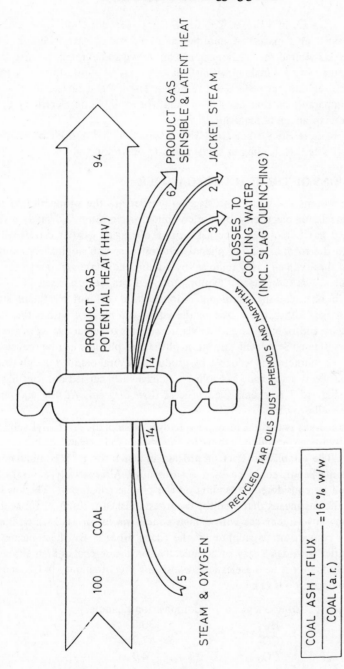

Fig. 15.2 – Heat balance for British Gas/Lurgi slagging gasifier.

from an empty state in 4 hours. Transition from gas production to hot standby can be achieved in a matter of minutes, and held for at least 48 hours. When maintenance is required the gasifier can be shut down and emptied for inspection and then returned to gas making within 7 days. It may be shut down and cooled, kept full of fuel, and yet restarted in 2 hours. The level of management supervision is comparable to that normally found, for example, in electricity utility power stations or an ammonia plant.

These characteristics stem from the presence in the gasifier of relatively large quantities of fuel, leading to stable and safe operation.

## APPLICATIONS OF THE SLAGGING GASIFIER

Among the potential uses of the slagging gasifier are the production of fuel gases, chemicals, or electricity by the combined cycle route. Normally a clean desulphurised gas is required. The gas from the slagging gasifier carries with it impurities in the form of oil, tar, phenols, and nitrogen and sulphur compounds. The bulk of these can be easily removed by cooling the crude gas to separate tar, oil and aqueous liquor. The net liquor produced can be processed in solvent extraction (dephenolation), ammonia stripping, and effluent treatment stages before discharge. Alternatively part of the liquor can be used within the plant, for example by addition to the gasifier via the tuyeres to replace some of the live high pressure steam. Separated tar, oil, naphtha and phenols can be recycled to the gasifier, or marketed as valuable by-products. Several detailed design studies for the whole gasification and gas cooling area have been carried out for full-scale plants and Fig. 15.3 is a simplified process flow diagram. All the equipment employed is well proven.

The cooled gas is treated to remove hydrogen sulphide, carbonyl sulphide, ammonia, hydrogen cyanide, aromatics, naphtha and organic sulphur compounds, etc. The established Rectisol process, in which the gas is contacted with refrigerated methanol, could be used for this duty. Alternatively, several other processes can be considered depending on the precise constraints. The low $CO_2$ content of the gas means that even the use of established techniques for acid gas removal gives an effluent gas with a high concentration of $H_2S$ which is very suitable for conversion to sulphur by the Claus process. By these means the slagging gasifier provides a way of completely converting coal at high efficiency and relatively low cost into a clean gas consisting predominantly of CO and $H_2$. A typical composition is as below:

| Component | Composition (% molar) |
|---|---|
| $H_2$ | 30.8 |
| CO | 57.2 |
| $CO_2$ | 4.9 |
| $CH_4$ | 6.2 |
| $C_2H_6$ | 0.3 |
| $C_2H_4$ | 0.1 |
| $N_2$ | 0.5 |
| | 100.0 |

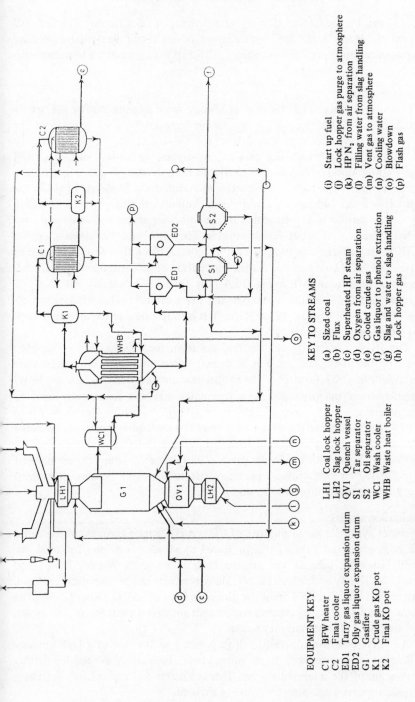

EQUIPMENT KEY

C1    BFW heater                          LH1    Coal lock hopper
C2    Final cooler                        LH2    Slag lock hopper
ED1   Tarry gas liquor expansion drum     QV1    Quench vessel
ED2   Oily gas liquor expansion drum      S1     Tar separator
G1    Gasifier                            S2     Oil separator
K1    Crude gas KO pot                    WC1    Wash cooler
K2    Final KO pot                        WHB    Waste heat boiler

KEY TO STREAMS

(a)   Sized coal                          (i)    Start up fuel
(b)   Flux                                (j)    Lock hopper gas purge to atmosphere
(c)   Superheated HP steam                (k)    HP $N_2$ from air separation
(d)   Oxygen from air separation          (l)    Filling water from slag handling
(e)   Cooled crude gas                    (m)    Vent gas to atmosphere
(f)   Gas liquor to phenol extraction     (n)    Cooling water
(g)   Slag and water to slag handling     (o)    Blowdown
(h)   Lock hopper gas                     (p)    Flash gas

Fig. 15.3 – Simplified process flow diagram for gasification gas cooling and tar-oil  liquor separation stages.

This gas can be used either directly as a fuel gas of calorific value around 350 Btu/sft$^3$ (13 MJ/m$^3$), for SNG manufacture, or for chemicals synthesis. For the latter application it is often necessary to CO-shift the gas to give a hydrogen-rich gas.

## SNG production

SNG can be made at high thermal efficiency from slagging gasifier gas, partly because a substantial proportion of the methane in the SNG is made in the gasifier. The remainder is produced from the CO and $H_2$ in the gas by reaction with steam over a British Gas catalyst. An upgrading route, called the HCM (High Carbon Monoxide) process [15.6] has been developed that is specifically tailored to take advantage of the particular composition of slagging gasifier gas (high CO content and low $CO_2$ and steam contents). In this way the high efficiency of the gasifier is not dissipated in the following stages and an overall coal-to-SNG efficiency of about 70% is possible. The combination of slagging gasifier and HCM also offers a relatively low capital cost. A block flow diagram of one of the possible HCM-based routes for the production of SNG from an Eastern USA Bituminous coal is shown in Fig. 15.4. The final SNG contains less than 3% hydrogen and 0.1% carbon monoxide and has a calorific value without enrichment of about 960 Btu/sft$^3$ (35.8 MJ/m$^3$). It is fully interchangeable with natural gases from many sources.

The HCM process for methanating gas from the slagging gasifier has been proved on the pilot scale and will be demonstrated on a semi-commercial scale at Westfield in 1983. Control of the exothermic methanation reaction will be by recirculation of the product gas, a technique that was successfully used to demonstrate the production of SNG from a Lurgi dry-ash gasifier at Westfield in 1974. In the latter trial a stream of 10 mcfd of purified synthesis gas from the outlet of the Rectisol plant was methanated over a British Gas catalyst to produce a methane-rich gas with a calorific value of 980 Btu/sft$^3$. After LPG enrichment it was distributed locally over a period of two months and, because of its satisfactory interchangeability, it went unnoticed.

### Chemicals synthesis

At present synthesis gas (a mixture of CO and $H_2$) is used throughout the world on a huge scale and most of it is produced by steam reforming of natural gas. Possible chemical uses of the gas has been reviewed by Wender [15.8] and recently by Denny and Whan [15.9]. Many products can be manufactured from synthesis gas and Fig. 15.5 indicates the chemistry of its use. Other possibilities include production of oxygenated compounds such as ethylene glycol, but as yet, these are only in the exploratory stages.

All these processes operate at high pressure so that use of a high pressure gasification process reduces or, in some cases, eliminates the need for further compression of the intermediate gas. This is a major cost and energy-consuming item when atmospheric pressure gasifiers are used.

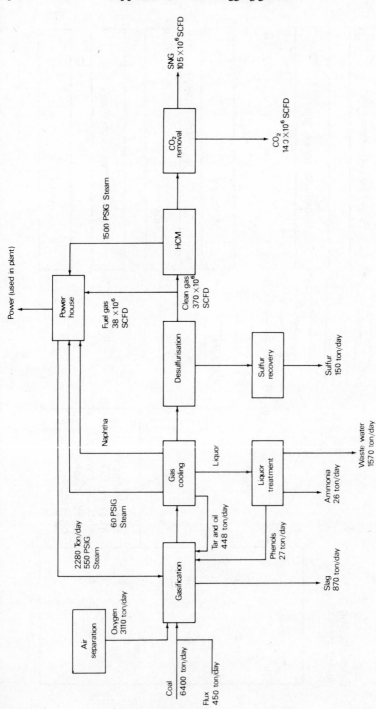

Fig. 15.4 – SNG scheme based on Eastern United States bituminous coal.

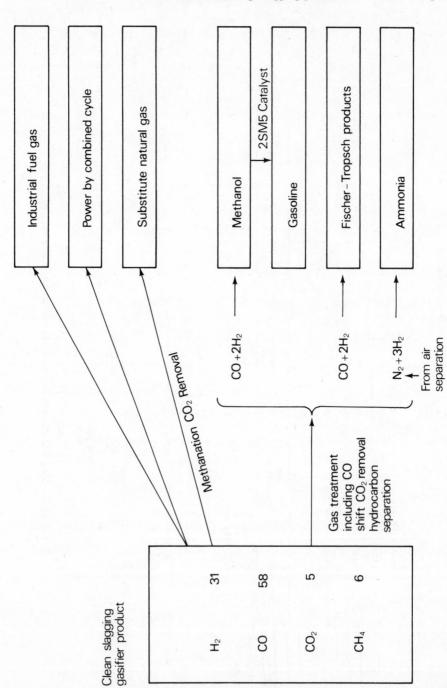

*Ammonia production*
At present the major end-use for synthesis gas is ammonia production. Currently about 70% of the world ammonia production is natural gas based. The salient steps in the route are high temperature methane reforming followed by extensive CO-shift conversion. As purified slagging gasifier gas is somewhat like a partly reformed methane stream, it can directly replace both the natural gas used as process feedstock and that used as fuel (about one-third) [15.10]. Such retro-fitting of existing plant has the minimum risk technically and economically and could be implemented rapidly if desired. A study [15.11] made of five gasifier types for this kind of scheme showed that the slagging gasifier was the most efficient and had nearly the lowest capital cost.

For a new purpose-built coal-based ammonia plant, the purified slagging gasifier gas would be passed directly to a CO-shift unit followed by $CO_2$ removal to adjust the gas composition to give the required hydrogen content.

*Methanol*
Methanol is already an important chemical with many uses. Many have speculated that the methanol market will increase rapidly in future years as new uses develop, these being as a clean liquid fuel in its own right, as an intermediate for gasoline production (as in the Mobil process) and perhaps as a feedstock for propylene plants [15.12]. Chemical uses demand a high product purity (99.5+%) but fuel-grade methanol need be only about 98% pure, the balance being water and higher alcohols, and so can be made with more simple plant.

Methanol synthesis catalysts are highly sensitive to sulphur and so the feed gas must be extensively purified and a chemical guard, say of zinc oxide, may be essential. Basically the synthesis gas should comprise hydrogen and carbon monoxide at a molar ratio of 2, so gas from the slagging gasifier, with its much higher CO content, must be extensively CO-shifted. The purge gas rate from the synthesis loop will have to be sufficient to prevent a build-up of the nitrogen and methane which are present in the purified slagging gasifier gas.

*Fischer–Tropsch*
Although the original Fischer–Tropsch (F–T) process for producing higher hydrocarbons from CO and $H_2$ was first announced over forty-five years ago, it has undergone various stages of development in the United States and especially by Sasol in South Africa where large quantities of vehicle fuels and oxygenated chemicals are produced by this route. Sasol use both fixed-bed and circulating-bed F–T reactor systems each yielding different product distributions.

The $H_2/CO$ ratios used in these reactors at Sasol are 1.7 and 2.8 respectively. As these gases are hydrogen-rich the highly purified gas from the slagging gasifier needs to be subjected to the CO shift reaction.

*Multi-product schemes*
In the above synthesis processes whatever the ultimate source of the feed gas, it

is necessary to purge non-reacting species such as methane and perhaps nitrogen continuously from the synthesis loop. In slagging gasifier-based schemes, the clean gasifier product gas can be processed, say by cryogenic fractionation, to remove from the main stream methane and higher hydrocarbons that can be sold as SNG. In this way the purge gas rate is considerably reduced with associated cost savings. The purge gas can sometimes be used as plant fuel. In ammonia plants the cryogenic separation can be carried out in the liquid nitrogen wash plant which would be present.

The most attractive process configuration for a particular situation will depend on the available market for pipeline or industrial fuel gas. In some circumstances it will be more profitable not to recycle hydrocarbon liquids made in the gasifier but to sell some or all of them as valuable products. There are many ways in which several products can be made simultaneously to give a robust process having increased thermal efficiency and reduced capital cost per unit of total product compared with a single product plant.

**Power generation**

The atmospheric emission of sulphur oxides from conventional coal-fired power stations is becoming less acceptable. Its avoidance by direct coal desulphurisation or stack gas treatment is technically difficult and expensive. It is more attractive to gasify the coal and then to remove sulphur (mainly in the form of hydrogen sulphide) from the crude product gas, particularly if the gas is at a high pressure. Also, by allowing the use of advanced power generation cycles, gaseous fuels can be used more efficiently than the initial feedstock, giving improved thermal efficiency for power generation. Thus electricity generation using combined cycles, that is using an optimised combination of gas and steam turbines to drive alternators (Fig. 15.6) can result in an overall efficiency,

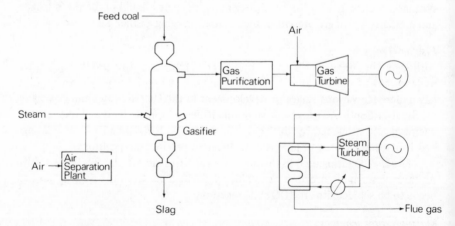

Fig. 15.6 — Coal gasification with combined cycle power generation.

including gasification, of above 40%. This compares with less than 35% for conventional steam cycle power plants fitted with stack gas clean-up devices.

The use of oxygen-blown gasifiers is not essential in this application, but can be economically desirable as it reduces the cost of the compression, gasification and gas purification stages. The low $CO_2$ and steam contents of the slagging gasifier product, are advantages in the combined cycle route. Little $CO_2$ is removed with the $H_2S$ and the capital and operating costs of the sulphur recovery plant are minimised. Processes which produce gases with higher $CO_2$ contents will also lose pressure energy during acid gas removal, which would otherwise be recovered in the gas turbine expander.

An evaluation of combined cycle schemes has been undertaken by EPRI for five different gasifiers [15.13]. Their estimates of both plant capital costs and power generation efficiency are summarised in Table 15.7, which for comparison includes a conventional coal-fired case, employing flue gas desulphurisation. It clearly shows the advantage of the British Gas/Lurgi slagging gasifier.

**Table 15.7**

Capital costs and efficiencies for power generation systems†

| Process | Capital cost requirement ($ million) | Overall efficiency coal to power (%) |
|---|---|---|
| British Gas/Lurgi slagging | 711 | 40.6 |
| Foster-Wheeler (air blown) | 705 | 40.5 |
| Texaco (slurry feed) | 816 | 38.7 |
| Combustion engineering | 931 | 38.1 |
| Lurgi (air blown) | 906 | 35.0 |
| Coal-fired plus stack gas desulphurisation | 838 | 34.4 |

*Basis*
Mid-1976 dollars at 70% operating factor and coal at $1/million Btu.

† From 'Economic studies of coal gasification combined cycle systems for electric power generation, EPRI Report, AF 642, 1978.

When used for power generation in a combined power cycle, a coal gasifier must respond quickly to the fluctuating fuel demands of the gas turbine over the full operating range. The EPRI tests described earlier showed the ability of the slagging gasifier to meet this requirement.

Where the demand for full power output is far from continuous, it can be economic to use the clean fuel gas for chemical sysnthesis. This option is opened by the use of an oxygen-blown gasifier and is made particularly attractive if the purge gases from the synthesis loop can be returned to the gas turbines. In particular methanol made off-peak can be used for supplemental firing of the gas turbines or in other power plant within the system.

## CONCLUSIONS

The British Gas/Lurgi slagging gasifier is at an advanced stage of development and is now ready for commercial exploitation. It is particularly suitable for high volatile, unreactive, bituminous coals and can find wide application in the areas of ammonia, methanol and Fischer–Tropsch synthesis, SNG manufacture, combined cycle power generation and medium calorific value gas production.

The process offers complete gasification of coal at high thermal efficiency. High reactor throughputs are attainable with extremely good operational controllability and flexibility. Investigations have shown the process to be environmentally acceptable [15.14] and indeed it provides now a technology which enables high sulphur coals to be used without atmospheric pollution.

British Gas is involved in a number of design studies for large-scale demonstration of the slagging gasifier and is prepared to provide full commercial and performance guarantees for gasifiers up to 8 ft in shaft diameter. The operation of an 8 ft gasifier at Westfield in 1983 will provide further confidence in operating gasifiers of this and even larger sizes.

## REFERENCES

[15.1] Hebden, D. and Stroud, H. J. F. (1981), 'Coal gasification processes' in Elliot, M. A. (ed.) *Chemistry of coal utilisation* Second supplementary volume, Chapter 24, John Wiley & Sons Inc.

[15.2] Tart, K. R. and Rampling, T. W. A. (1980) 'Fixed bed slagging gasification – a means of producing synthetic fuels and feedstocks' *I. Chem. E.* Symposium Series No. 62.

[15.3] Hebden, D. and Edge, R. F. (1958–59) 'Experiments with a slagging pressure gasifier' *Trans. Inst. Gas. Eng.* **108**, 492–527.

[15.4] Hebden, D., Lacey, J. A. and Horsler, A. G. (1964) 'Further experiments with a slagging pressure gasifier', Gas Council Research Communication, GC 112.

[15.5] Hebden, D. and Brooks, C. T. (1976) 'Westfield – the development of processes for the production of SNG from coal', Institution of Gas Engineers, Communication 988.

[15.6] Tart, K. R. and Rampling, T. W. A. (1981) 'Methanation key to SNG success' *Hydrocarbon Processing,* April, 114.

[15.7] 'US Coal Test Program on BGC-Lurgi Slagging Gasifier' EPRI Report AP-1922, August 1981.

[15.8] Wender, I. (1976) 'Catalytic Synthesis of Chemicals from Coal' *Catal. Rev. Sci. Eng.* **14**, 97.

[15.9] Denny, P. and Whan, D. A. (1978) 'Heterogeneously catalysed hydrogenation of carbon monoxide' in Kemball, C. and Dowden, D. A. (eds.) *Catalysis* **2**, 46, Chemical Society, London.

[15.10] Timmins, C. (1979) 'The future role of gasification processes' Institution of Gas Enginers, Communication 1112.

[15.11] Brown, F. (1977) 'Make ammonia from coal' *Hydrocarbon processing*, Nov., 361.

[15.12] Anthony, R. G. and Singh, B. B. (1981) 'Olefins from Coal via Methanol' *Hydrocarbon Processing* March, 85.

[15.13] Chandra, K. *et al.* (1978) 'Economic studies of coal gasification combined cycle systems for electric power generation' EPRI report, AF-642, January.

[15.14] Sharman, R. B., Lacey, J. A. and Scott, J. E. (1981) 'The British Gas/ Lurgi slagging gasifier' Synfuels International Conference, Frankfurt, Germany, 11-12 May.

# 16

# Electricity and its relationship to feedstocks

**M. R. Cowan**, Director, Electricity Council Research Centre

## INTRODUCTION

During the current century coal and oil, and in the latter years, natural gas have been plentiful and reasonably cheap and as a result have been widely used as fuel and as feedstock. These same fuels have been used by electricity utilities to generate electricity. All three of these fossil fuels have a finite life and oil and natural gas are expected to become scarce early in the next century.

As oil and gas become less plentiful and thus more expensive, electricity generation can, and is already, being gradually transferred to the more plentiful fuels that will be available, for example, to coal or nuclear power. As the reserves of coal and uranium are large it can be suggested that the supply of electricity is virtually unlimited and that it can be considered as being plentiful and will not be threatened significantly by scarcity of oil or gas. If this transfer of fuels used for generation is pursued, increased quantities of oil and gas will be made available for use by the chemical industry as feedstock and by the transport industries for fuel.

Electricity utilities and chemical industries have much in common. Because electricity is traditionally manufactured from coal, gas and oil, electricity utilities are currently competing with the chemical industry for the same feedstocks. Like the chemical industry, the electricity utilities are concerned with forecasting and with the future availability and price of fossil fuels and both industries are concerned to make the best decisions about their future feedstocks to try to ensure that the policy decisions made will endure into the next century.

In Britain the recently adopted strategy of the electricity industry is that of influencing an orderly transition from potentially scarce oil and gas to plentiful electricity. The ability to influence this transition will be affected by many factors including technological advances, the pattern of world trade and the price of each fossil fuel. It will also be influenced by the effectiveness of the salesmanship employed by the electricity industry. A policy of the nature outlined can be considered to be a direct and useful contribution to the conservation of oil and gas resources, provided of course that the utilities do not significantly increase their use of oil and gas to generate electricity.

## COMMON PROBLEMS – ELECTRICITY AND CHEMICAL INDUSTRIES

Electricity utilities and large chemical industries have a great deal in common:

(1) Both require large quantities of fuel and feedstock obtained as cheaply as possible.
(2) Both are traditionally large spenders of capital.
(3) Both do and will probably continue to experience problems in getting large plants built to time.
(4) Both industries are currently suffering from low sales.
(5) Both have historically been rather poor at estimating their future sales.

The two industries are of considerable importance to each other with the chemical industries being large purchasers of electricity. Knowledge of each other's plans is important but planning in both industries is highly dependent on the subject of forecasting.

## FORECASTING

The forward forecasting of sales is of necessity an inexact science and it is interesting to consider the historical performance of forecasting in some of the fuel industries. For this purpose a reference will be made to coal, oil and electricity forecasts made in Britain.

Dealing first with coal, an estimate made in 1956 suggested that coal consumption in Britain would reach 250 Mt by 1970; by 1978 the forecasters were suggesting a take of only 120 Mt by 1985. It is not the purpose of this paper to consider the reasons for the changes that have occurred over the past two decades, but it is reasonable to indicate that the early forecasts for coal included optimistic assumptions about the quantity of coal likely to be used for electricity generation and thus, when electricity sales fail to reach their forecast levels, then the fossil fuel use will also fall short. As an aside it is also worth noting that the use of coal by the chemical industry in Britain fell by 78% between 1971 and 1979 and this was probably not forecast in the 1960s.

The down-turn in oil industry forecasts have all occurred during the past decade. For example, a 1972 forecast for oil demand in Britain indicated likely consumption of 140 Mt in 1980 whereas the actual out turn in that year proved to be only 75 Mt.

The situation in relation to the forecasting accuracy for electricity has taken a slightly different pattern but has considerable similarity. Firm forecasts are made seven years in advance and during the 1970s the forecast figures were consistently higher than out turn. The forecast for 1980 made in 1974 predicted total sales of 290 TWh while the actual out turn for 1980 was only 225 TWh.

These examples for coal, oil and electricity forecasts in Britain are not unique to Britain or to the fuel industries, and other industries including chemicals have experienced a situation where sales have fallen well below forecast.

A general observation can be made arising from the study of forecasting and capital approval decisions. It appears to be fairly common that when sales are good, and this usually means that there is an associated plant shortage, that there is a tendency for forecasts to be over-optimistic. Buoyant markets and optimistic forecasts tend to encourage the approval of capital expenditure plans, and it is more likely that extra capacity will be authorised during periods of buoyant sales than at other times. It is reasonable to say that both the electricity utilities and the chemical industries have approved expenditure at periods of good sales and, after the plants have been built and commissioned, have found that they and their colleagues and competitors have all acted similarly and have created an over-capacity situation and potential crisis. Crises created by over-capacity have an immediate impact on both prices and profitability and may well be more serious than crises created by capacity shortage.

There can be no doubt that forecasting is a vital part of decision-making and that all concerned will have to continue to make forecasts. It is however important that the fallibility of forecasting be recognised.


**THE FUEL SITUATION 2000–2025**

Despite the comments that have been made about the problems of forecasting it remains essential to attempt to look ahead and to make plans well in advance. In practice, long-term plans need to be viewed in a flexible manner and most decisions taken on a comparatively short time scale. Looking ahead for either medium or long-term periods requires consideration of both the current situation and the potential future, and the following comments provide a mixture of both.

Due to the worldwide economic recession, many electrical utilities in the developed world have surplus capacity at the present time. Due to price and the uncertainties about oil many of them are endeavouring to minimise their use of oil and this could work to the advantage of the chemical industry, releasing this fuel for such purposes as chemical feedstock and transport.

Electricity is a potentially plentiful fuel and the main fuels used for generation in the year 2025 will probably be coal and nuclear energy. It also seems likely that in countries where the proportion of nuclear power is high that electricity prices in those countries will be more favourable than in those with a higher percentage of conventional fossil fuel fired plant.

It seems virtually certain that gas will still be available to EEC countries in the next century, but possibly by the year 2025 its future as a fuel may be determined by the price relationship then existing between synthetic gas and electricity. Ignoring the extremes of worldwide war and revolution, it is also probable that oil will still be available by the year 2025 but its use will probably be greatly restricted by its high price at that time.

The role of coal by the year 2025 is extremely difficult to judge. Because reserves are large, suggestions abound for its use in generating electricity, as a feedstock for the chemical industry, as a feedstock for synthetic gas and as the basis for the production of oil and petrol for transport. Groups and organisations that lobby against nuclear energy presumably assume that electricity will be generated from coal. If all these potential uses for coal are favoured then the demand for coal may well exceed production capacity and it is not inconceivable that the price of coal would be raised to levels that became too high to justify burning coal in power stations. Obviously the world's fuel and feedstock problems are in something of a melting pot at the present time but the options should clarify in a reasonably gradual and orderly manner, although it does appear that the pressure on fossil fuel resources will be severe. There thus seems little doubt that the most constructive policy that can be advocated is to steadily increase the share of nuclear generation of electricity to release and conserve scarcer fossil fuels. A gradual transfer of fuel loads currently supplied by gas and oil for electricity would also assist in prolonging the life of gas and oil supplies at reasonable prices.

## ENERGY AND RESOURCE CONSERVATION

Earlier in the paper it has been suggested that it is beneficial for electricity to capture some of the fuel markets now supplied by oil and gas. This could be interpreted as a statement that selling electricity should be a top priority for utilities and the only conservation message involved is that of conserving potentially scarce oil and gas. A policy of this nature is by itself insufficient and would have little or no chance of success. Traditionally, electricity at the point of delivery to a customer's premise is an expensive refined fuel and a customer will only wish to use electricity at its higher prices if its use offers a unique advantage, or if the efficiency of its use is so good compared to a primary fossil fuel that using electricity is genuinely competitive and cost saving. Because of the higher price the electrical designer has always had to concentrate on minimising the energy required to operate an electrical process or a piece of electrical equipment. Thus if the suggested policy is to be effective in increasing the sales of electricity then the electricity utilities have to continue to seek ways of improving the efficiency of electrical processes and products and to pay attention to energy conservation via the minimisation of electrical losses and wastage.

In reviewing the future it appears probable that electricity prices will rise less rapidly than those of oil and gas and it is therefore prudent for all fuel users to consider electrical alternatives when seeking to purchase new plant. Although forward price prediction is difficult, the buyer of new equipment does need to make a judgement on fuel prices and on their rate of escalation.

This section is headed energy and resource conservation and possibly the latter is the more important of the two. If Europe is to maintain its standards of living well into the next century, it is essential that European industries retain their competitiveness. In practical terms being competitive means using minimum resources both human and material. It seems very likely that advancing technology including micro-electronics will increasingly help to reduce the labour required per unit of output and to improve the co-ordination and control of processes to increase efficiency. At the same time inflation is also creating new opportunities to apply technology to the recovery, re-use and recycling of previously discarded resources and this should make a significant contribution to competitiveness in the coming years and might have an even greater effect than energy conservation.

Resource conservation and materials recycling can also reduce energy requirements as the energy required for recycling will usually be less than that involved in the original production of the material.

## TECHNOLOGICAL DEVELOPMENTS

Many of the papers presented in this volume are concerned with advancing technology and the possibilities associated with the development of new processes to produce gas and oil products and substitutes for use as feedstock or even as energy sources. Technical advance in electrical processes and techniques can also have an impact but is more likely to be indirect, that is, it will concentrate on improving the opportunities for sales of electricity and if these represent transfer from oil and gas they will be beneficial to the chemical industry. At the Electricity Council's Research Centre at Capenhurst the main emphasis of the work carried out is on industrial utilisation of electricity. Although the Capenhurst centre has an electrochemical research section its objective is not to invent a new chlorine process or patent the latest drug for the pharmaceutical industry. Its work is less dramatic but is almost entirely in line with the concept already referred to, namely improving the competitiveness of industry via electrical processes. For example, a fluidised electric cell has been developed to extract metals from dilute solutions. This is the Chemelec cell and when coupled to a rinse tank in a conventional copper plating line it acts to recover metal that would otherwise be lost, it maintains the purity of the rinse tank water and virtually eliminates an effluent disposal problem†.

To the buyer, the Chemelec cell represents an additional item of equipment and an addition to electricity consumption. However these additional costs are

† See Appendices for economic summaries.

often recovered in under a year from the value of the metal recovered and the reduction in the process costs. Applied as illustrated to the plating industry it reduces production costs; applied to the recovery of silver from film it allows money to be made from a waste product; and applied to metal winning it offers imporved efficiency for the final stages of the traditional process. It is an example of a device that conserves and recovers valuable resources.

Another Capenhurst development is the Dished Electrode Membrane Cell, the purpose of which is the regeneration of spent oxidising agents such as chromium, cerium and manganese. This cell is intended for use by the chemical industry and like the Chemelec cell is aimed to appeal to customers by saving money.

The topic of electro-kinetic thickening has also been researched at Capenhurst and a process has been developed for removing water from slurries such as clay, PVC or even coal tailings. The solids in the slurry are attracted to an electrode and accurate control of the required moisture content of the separated solids is consistently achieved.

A plant design for drying PVC has a capital cost estimated at half that of a conventional spray dryer and, with estimated running costs lying between £10–13 per dry tonne, a saving on operating costs of over 70% is expected and payback should be achieved in under twelve months.

These three examples of electric cell developments provide an indication of how advancing technology can assist industry via the provision of new tools and concepts. The developments quoted have been deliberately related to electro-chemical work although, of course, work is proceeding in relation to many other industries and electrical applications.

One area of general work is in the field of industrial heat pumps. The principle of heat pumps has been known for many decades but in industrial terms their promise has always been greater than their practicality. They have been adopted for refrigeration and air conditioning where the alternatives are limited, but application of heat pumps within industry has so far been virtually negligible. However heat pumps are now poised for their long-awaited entry into industry, not just because the energy crisis has forced reconsideration of energy savings but because technological development has now caught up with theory. The key to the work that has been done at Capenhurst lies in the field of refrigerants and lubricants rather than any direct and dramatic new developments regarding the actual operation of heat pumps themselves.

In science fiction terms the heat pump is sometimes described as a device that provides heat for nothing but in practical terms it is a device that allows the upgrading of waste or unused heat to a useful temperature. The efficiency of this upgrading is described by the coefficient of performance or COP.

For a heat pump designed for domestic use the COP is about 2, or a little higher, and this means that for the use of one unit of electricity fed to the heat pump compressor, about 2 units of heat can be extracted from the outside air for use at an acceptable temperature inside the house. By definition the COP is inversely proportional to the rise in temperature attempted and, for a house, the difference between the outside air and the inside air temperature is large and thus only relatively low COPs around 2 can be achieved.

However, if one considers industrial drying it is then possible to so design the system that the difference in temperature between the outlet waste air stream and the inlet temperature is kept small and COPs up to 6 or 7 become an attainable target.

Until recently heat pumps for industrial drying were limited to very low temperatures, about 50°C or 60°C but a heat pump for drying timber has now been developed operating in the region of 80°C, and work in the laboratory has indicated that temperatures up to 150°C are feasible. Drying represents a significant proportion of the total process energy requirements of industry. Thus, if over the next two decades a considerable proportion of industry's drying requirements can be transferred to heat pumps, plus new techniques like electro-kinetics together with the older techniques such as infra-red, radio frequency and microwave drying, then the transition away from the use of potentially scarce fossil fuels for drying towards plentiful electricity will be accelerated and will occur in a reasonably orderly and smooth manner.

Whilst there can be little doubt that technology is an essential aspect to the future of industrial development and competitiveness in Europe, there is an ever present danger that investment in new technology will be viewed with excessive caution. There is a need to encourage managers to keep all possible opportunities for technological advance under constant review and to encourage them to put propositions forward for capital authorisation. In a period of economic depression it is all too easy to try to control cash flow by imposing tight budgetary requirements and possibly bureaucratic approval systems for schemes and expenditure. While it is imprudent to argue against these in principle, it can, and does, happen that junior management ceases to make propositions or to fight for them with fervour because they believe the bureaucracy or the rules are too tight and cannot be broken. If technology is the key to remaining competitive it is essential that technical people at the level associated with appraisal of the fundamental suggestions be encouraged to follow up the suggestions they consider to be of merit, almost irrespective of the company's financial control structure.

## SUMMARY

It is impossible to forecast the future with certainty but the future remains encouraging. Fossil fuels and feedstocks are not likely to run out in the immediate future or even to run out quickly. However oil and gas will become scarcer and more expensive as time passes and it is desirable that they be conserved wherever possible by the substitution of electricity as a fuel generated from nuclear sources.

It is extremely difficult to make any reliable forecast for energy use as between the fuels and feedstocks for the year 2025. Many scenarios have been prepared and are available and the most essential action for large businesses is to keep such scenarios and the performance of their own organisation under constant review.

The tendency to over-optimism in times of boom and to underestimate future sales in times of depression may well prove to have been one of European industries greatest weaknesses during the 1960s and 1970s.

One would expect the most successful companies in the year 2025 to be located in the most successful countries and it seems probable that those countries will be the ones which consistently and correctly embrace advanced technology in the intervening years including the increased adoption of nuclear power generation. If the philosophies outlined in this paper are proved viable then oil and gas should still be available in reasonable quantities in the year 2025, albeit that they will be expensive. High prices will accelerate the transfer of the fuel load to electricity. The prediction that electricity prices should rise more slowly than those for oil and gas should be taken into account when making forward decisions on process plant.

A few years ago the electricity supply industry in Britain had an advertising slogan which read 'The Future is Electric'. This slogan is factual but it suffers from the great missing link that worries supply utilities and chemical industries alike, and that is that it does not state the date.

## APPENDIX A

### Chemelec cell used for nickel recovery

The following figures illustrate the general economics relating to the recovery of nickel via a Chemelec cell. They are based on an actual installation. No allowance has been made for the reduction in sludge disposal or water costs incurred on a process line operating without a Chemelec cell. Prices shown are at 1979/80 levels.

Estimated capital cost of cell to recover nickel at the rate of 14 kg/week    £7245

Annual value of nickel recovered at £2.7/kg      $£(700 \times 2.7) = £1890$

Electricity operating costs at £0.3/kg nickel recovered      £210

Annual operating cost saving      $£(1890 - 210) = £1680$

Payback period based on metal recovered only      $7245/1680 = 4.3$ years

Other significant savings to be quantified by individual users are:

(1) savings on existing or alternative effluent treatment plants, reduced sludge disposal costs;

(2) reduced water costs, if dilution is currently needed to meet consent limits;
(3) easing of Water Authority pressures.

For more valuable metals such as silver the payback period as calculated above is often less than one year and the other unquantified savings still accrue.


## APPENDIX B

### Chemelec cell for silver recovery

The following figures illustrate the general economics relating to the recovery of silver via a Chemelec cell. They are based on an actual installation. No allowance has been made for the reduction in sludge disposal or water costs incurred on a process line operating without a Chemelec cell. Prices shown are at 1979/80 levels.

Estimated capital cost of cell to recover silver at the rate of 5 kg/week     £8010

Annual value of silver recovered at £200/kg                 £(200 × 250) = £10 000

Electricity operating costs at £1.5/kg silver recovered     £(1.5 × 250) = £375

Annual operating cost saving                                £(10 000 − 375) = £9625

Payback period based on metal recovered only                8010/9625 = 10 months

Other savings relating to the cost and problems associated with effluent treatment will also accrue, but the process is justified in economic terms on the savings of metal alone.


## APPENDIX C

### Dished Electrode Membrane Cell for sulphuric acid recovery

The following figures are at 1981 prices. The company on whose operations the figures are based is currently operating a pilot plant and the estimated data has been confirmed from the results obtained.

Capital cost of DEM Cell plant to recover 10 tonnes/day sulphuric acid + 8 tonnes/day caustic soda = £393 000

Amortised over 5 years this equates to a cost of £22.5/tonne of sulphuric acid recovered calculated from

$$\frac{£393\ 000}{5 \times 350 \times 10} = £22.5/\text{tonnes}$$

Running cost of operating such a plant at 2.4p/kWh at an operating current
density of 2000 A/m$^2$ is given by

$$£0.024 \times 3780 \text{ kWh/tonne of sulphuric acid} = £91/\text{tonne}$$

Total operating cost (capital and running)/tonne of sulphuric acid

$$= £(22.5 + 91)/\text{tonne} = £113.5/\text{tonne}$$

Savings from operation of DEM Cell/tonne

| | | |
|---|---|---|
| Sulphuric acid | = | £40 |
| Caustic Soda | = | £82 |
| Hydrogen | = | £5 |
| | | £127/tonne |

| | |
|---|---|
| Net savings | = savings − costs |
| | = £127 − £113.5 |
| | = £13.5/tonne of sulphuric acid |
| Savings/day | = £13.5 × 10 = £135/day |
| Savings/annum | = £135 × 350 = £47 250 |

This indicates an annual process saving and should be compared to the operation
of an alternative chemical lime treatment plant of similar capital cost which has
a net cost of chemicals each week combined with a large effluent disposal
problem.

## APPENDIX D

### Electro-kinetic dewatering of PVC emulsion
The figures used in this appraisal of electrokinetic dewatering were prepared in
collaboration with a chemical company as part of a submission to their Board
for authority to proceed with an installation. The figures are in 1980 prices.
Completion of the project coincided with a downturn in PVC production and
the plant was not installed.

The first full-scale commercial installation of electrokinetic dewatering of
slurries is still awaited.

Comparison of Conventional Spray Drying with Dewatering for 10 000
tonne/year plant.

| Capital costs | Spray drying | Dewatering |
|---|---|---|
| Spray dryer | £500 000 | |
| Grinder | 40 000 | |
| Power supply | | £15 000 |
| Electrodes | | 2 000 |
| Cell and handling | | 138 000 |
| Dryer, grinder | | 60 000 |
| | £540 000 | £215 000 |
| Capital cost apportioned over 5 years expressed as cost/tonne | £11/tonne | £4 |
| Running cost/tonne (electricity) | £40–£55/tonne | £6–£9/tonne |
| Total operating cost (range) | £51–£66/dry tonne | £10–£13/dry tonne |

The capital and running costs of dewatering are significantly less than the conventional processing method, labour costs being assumed to be the same in both cases.

## APPENDIX E

### High temperature heat pump (for timber drying)

The first commercial application of the high temperature heat pump developed at the Capenhurst laboratories was used for drying timber. The following figures compare the performance (in 1979/80 figures) of a conventional oil-fired heat and vent kiln with a dehumidifier (heat pump) operating in conjunction with a well insulated kiln.

Energy cost comparison of heat pump dehumidifier drying versus conventional heat and vent drying of Beech from a moisture content of 26.9% to 12.3%.

| | Heat and Vent (Oil Fired) | Dehumidifier |
|---|---|---|
| Energy consumed | 127 gallons oil 615 kWh electricity | 1012 kWh |
| Energy cost | £87.83 | £29.35 |
| Volume of load of timber | 21 m³ | 11.5 m³ |
| Energy cost/m³ | £4.18 | £2.55 |

% saving with dehumidifier drying

$$= \frac{4.18 - 2.55}{4.18} \times 100\%$$

$$= 39\%$$

# 17

# Syngas by underground gasification

**P. Ledent**, Director of the Belgian–German Project for Underground Gasification of Coal

## THE ENERGY SITUATION IN EUROPE

A look at the latest statistics published by the Commission of the European Communities, shows that the consumption of primary energy in Europe is distributed as shown in Table 17.1.

**Table 17.1**
Distribution of energy consumption in the European Economic Community, 1980

|  | % |
|---|---|
| Oil | 51.8 |
| Natural gas | 18.2 |
| Coal and brown coal | 23.8 |
| Electricity generated from hydraulic power | 1.6 |
| Electricity generated in nuclear stations | 4.6 |

*Source:* EEC Statistics Edition 1981, EUR9–1980, Table 53/76.

Liquid fuels represent more than 70% of the total, and the share of nuclear energy is still very small.

If the development of nuclear power is not excessively slowed down by reactions of a political or social nature, the share it could take in Europe's energy supplies could reach 10–15% by the end of this century, but our principal problem is to ensure supplies of gaseous and liquid fuels.

All the forward studies carried out in respect of energy have shown very clearly that world output of oil will fall short of requirements by the beginning of the twenty-first century, and that a growing proportion of the supply of vehicle fuels and liquid fuels must come from the liquefaction of coal [17.1].

   A return to independence in respect of energy within a short space of time
is inconceivable for Europe unless she plays a leading part in the coming transformation, which must necessarily lead to the creation and development of the
industries which will convert coal into liquid and gaseous fuels.

## European coal
It is unlikely that the coal produced in Western Europe will be able to serve as
the basis of a coal chemicals industry of any significant size. The limited scale
of the reserves of coal workable by the traditional methods, the difficulty of
recruiting labour, and the very high cost of production are all obstacles to any
increase in output. It is significant that, despite the crisis, the level of coal
production in the Europe of the Nine has for the last eight years stagnated below
a figure of 250 million tonnes/year.

## Imported coal
The prospects for using imported coal as the basis for a European coal gasification
and liquefaction industry are not much more attractive.
   Imported coal is still appreciably cheaper than the coal produced in Western
Europe, but the price differential is likely to shrink with the passage of time.
Moreover, large-scale imports of coal would necessitate enormous investments in
means of transport by sea and by land, and even now the landed price in Europe
is three to four times that of coal available at the production site (Wyoming,
South Africa or Central Siberia). The price has risen rapidly over the last few
years (see Table 17.2).

**Table 17.2**
Price of imported steam coal, European Economic Community

|                      | Mean value c.i.f. | |
|----------------------|-------|-------|
|                      | $/t   | $/GJ  |
| First quarter 1978   | 38.2  | 1.30  |
| 1979                 | 40.5  | 1.38  |
| 1980                 | 52.0  | 1.77  |
| 1981                 | 68.8  | 2.35  |

   This should lead to the extension of coal liquefaction where the coal is
produced; this view is forcibly expressed in a recent report issued by the Foster-
Wheeler group in the following terms:
'It is difficult to imagine the importation of large quantities of coal to produce
transport fuels. It is more likely that these fuels will be manufactured close to
the source of low-cost coal, and the intermediate or final products shipped to
Europe. This could also be the case for chemicals such as methanol.' [17.2].

## Underground gasification of coal

This paper sets out to show that a return to energy independence for Europe could flow from the development of a new method of underground gasification of the coal deposits found at great depths. Success in developing this new technique would multiply our workable reserves by a factor of ten, and make available to our industry an energy source and a raw material able to compete with gas and liquid fuels produced overseas, on the basis of coal won from opencast workings.

## UNDERGROUND GASIFICATION AT SHALLOW DEPTHS

It was Sir William Siemens and Mendeleyev who, nearly 100 years ago, first proposed that coal be gasified *in situ*. The first trials of the technique go back to the thirties. To date, the process has been tried only at shallow depths (between 50 and 300 metres) without leading to any significant commercial application [17.3].

The chemical reactions which underlie the underground gasification of coal are no different from those which have for over a century been used in conventional surface gasworks.

There are however three fundamental differences between underground gasification *in situ* and surface gasification of the extracted coal. They are:

(1) The rock walls which form the internal surface of the underground gas generator are not necessarily air- and gas-tight.
(2) The coal is present as a compact seam; this limits the total reaction surface between gas and solid.
(3) The circuit in which the underground gas generator is installed may be several kilometres long, and it always offers a not inconsiderable resistance to the passage of the gases.

It is the presence of these three features which explains the failure to achieve industrial development of this process in relatively shallow deposits of coal.

The fact that the underground gas generator is not air- or gas-tight allows infiltrations of water and leaks of gas which cause pollution and reduce the thermal efficiency of the process.

The possible yield in each underground circuit rarely exceeds 2000 $m_N^3$/h in thin seams, with a maximum of the order of 10 000 $m_N^3$/h in very thick seams. Any increase in the yield of gas can be obtained only by adding more circuits and increasing the number of workings. This simultaneously increases the thermal losses by infiltration of water and by the heating of the rock walls.

These handicaps have led to the adoption of an underground gasification technique using multiple faces. In fact, this rather resembles the classical room-and-pillar method; it is illustrated in Figs. 17.1 and 17.2.

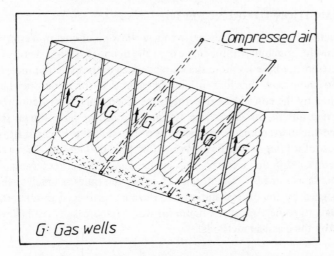

Fig. 17.1 – Multiple-face gasification in a steeply dipping seam.

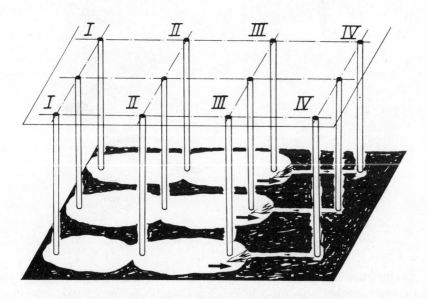

Fig. 17.2 – Multiple-face gasification in a horizontal seam.

On the whole the results obtained have been rather poor: the calorific value of the gas rarely exceeds 4000 kJ/m$_N^3$, the thermal efficiency reaches a ceiling of some 50%, and the useful power produced by a single gas borehole generally lies between 2 and 10 MW, the latter value being attainable only in thick seams.

## THE TRANSITION TO GREAT DEPTH

The attractive feature of the underground gasification of coal in Western Europe stems from the possibility of making use of the deposits at great depth.

At present we have no more than a vague idea of the extent of these deposits, but we do know now that it is infinitely greater than the resources which could be exploited by the conventional methods. As an example, the recent prospecting work carried out in the North Sea to find natural gas have shown that there exists a coalfield stretching from the Ruhr to the Midlands of Great Britain (Fig. 17.3), most of it lying at depths between 1000 and 5000 metres. The preliminary estimates made by the German and Netherlands geologists suggest that this deposit contains some $6 \times 10^{12}$ tonnes of coal – ten times as much as the largest figure assigned by the official statistics to the coal reserves of Europe; this figure corresponds to some 50% of the figure for world resources of coal believed to be workable by the classical methods.

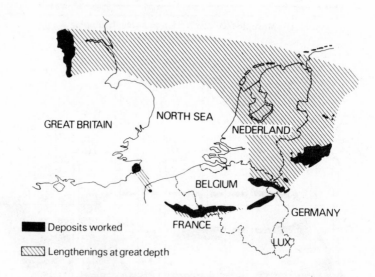

Fig. 17.3 – Coal deposits of Northern Europe.

Investigations into the possibility of developing underground gasification of the deep deposits was begun in Belgium towards the end of 1974. This very modest beginning has grown since 1976, with the signing of a Belgo–German cooperation agreement; the result is the establishment in Belgium, on the Thulin site, of an experimental installation for the underground gasification of coal at great depth – a world first [17.4]. The Belgo–German agreement has also served as a framework for the development of a series of fundamental studies carried out in cooperation between various research institutes and university departments.

In France, the interest in underground gasification of deep-lying coal deposits has led to the setting up of an Underground Gasification Study Group (GEGS), in which Gaz de France, Charbonnages de France, the Institut Francais du Petrole and the Bureau de Recherches Geologiques et Minières all participate.

In Great Britain, investigations are being carried out by the Mining Research and Development Establishment of the National Coal Board, with a view to assessing the prospects of various non-conventional coal-winning methods; underground gasification is in the first rank of the methods which might well be developed in the decades ahead [17.5].

The multiplicity of research projects shows that the solutions which will be selected for the development of the underground gasification of deep deposits could be very different from those which were perfected for the gasification of coal seams at shallow depth.

Already, it seems possible to sketch the more significant characteristics of the processes which will be applied, the choice thereof being dictated by four variables whose role in the economics of the process is crucial:

- the gasification yield;
- the quality of the gas;
- the productivity of the boreholes;
- the area of the deposit which can be gasified from each borehole.


## UNDERGROUND GASIFICATION UNDER HIGH PRESSURE

The transition to great depths should provide us with an air- and gas-tight gas generator which can be utilised under high pressure.

Two arguments can be used to support this statement:

(1) the plastic behaviour of carboniferous shales when subjected to high lithostatic pressures;
(2) the increase in the hydrostatic pressure in the aquiferous rocks and in the fissured zones which cover the deposit, the water in these formations acting as a water seal which prevents any escape of gas.

The creation of gas generators able to function under a pressure of 20–30 bars has been one of the biggest improvements during the last 50 years in the techniques used for gasifying extracted coal.

When we consider the underground gasification of deep-lying deposits, we have good reason for thinking that the transition to high-pressure operation constitutes the essential element for any industrial development of the process.

The first advantage which can be expected is an increase in the flow of gas proportional to the gasification pressure, without any increase in the power consumed to overcome the pressure losses.

In practical terms, this means that an underground gasification working operating under a pressure 20–30 bars will give a gas yield of the order of 50 000 $m_N^3/h$ if the working is located in a thin seam, and of 100 000 $m_N^3/h$ in a seam 2 metres thick.

The second advantage of operating the gas generator under pressure is that it raises the reaction rates; this should in turn lead to a very substantial increase in the rate of advance of the gasification faces. For example, during the underground gasification trials at Djerada, Morocco, carried out at the suggestion of the Centre d'Etudes et de Recherches des Charbonnages de France (CERCHAR), the average rate of advance of the face when gasifying with air at a pressure of 1.3 bars lay between 5 and 6 cm every 24 hours [17.6].

It can be expected that – for a gasification pressure of 30 bars and with no change in the temperature or in the composition of the gasification medium – the rate of advance will be between 1.25 and 0.70 metres/day, varying according to whether the rate of reaction is determined by the chemical kinetics of the process or by diffusion phenomena in the gaseous phase.

The increase in the rate of advance of the gasification face is accompanied by a marked improvement in the thermal efficiency, the amount of heat released by the reactions being directly proportional to the rate of advance, while the losses by conduction of heat through the rocks are proportional to the square root of this rate. In addition, the increase in the rate of advance is the most reliable means of improving the behaviour of the surrounding rock and of ensuring that the gas channel along the gasification face is not prematurely blocked by roof falls.

The third advantage of bringing the underground gas generator under pressure is a question of thermodynamics; it is a consequence of two facts: that the gasification operation is always accompanied by a rise in the number of moles of gas; and that the energy expended on compressing the gasification media is appreciably less than the energy which would have had to be supplied to compress the gas.

## THE CHOICE OF GASIFICATION MEDIA

Over the period of 150 years of industrial development of the gasification of extracted coal, there has been a gradual extension of the range of gasification media used.

Air and steam were the only gasification media used up to the First World War.

During the thirties, the advances in the industrial manufacture of oxygen made it possible to construct gas generators operating on a mixture of oxygen and steam.

In the same period there were developed the first pressurised gas generators, but it is only in the last thirty years that – stemming from the work of Professor

Dent [17.12] — we have seen the advent of the new prototypes of very high pressure generators designed to produce methane by the gasification of coal using a hydrogen-rich gas.

In the first efforts to develop underground gasification at shallow depths, oxygen was used only in the most exceptional cases. The reasons for this were its high cost and the low gasification yields obtained because the underground generators were not gas-tight.

Moreover, the high moisture content of the shallow deposits made it quite unnecessary to inject any steam; consequently, industrial applications of the underground gasification of coal were restricted to the production of a lean gas, using air or — exceptionally in very thin seams — oxygen-enriched air as the gasification medium.

The changeover to gasification at depth and under pressure opens up new prospects for the process of underground gasification. At the start of the Belgo-German studies, three variants were envisaged:

(1) production of a lean gas, to be burned in combined-cycle power stations, using a mixture of air and steam as a gasification medium;
(2) production of a gas of medium calorific value, which could be used as mains gas or as a feedstock for chemical synthesis, using a mixture of oxygen and steam as a gasification medium;
(3) direct production of a gas as a substitute for methane-rich natural gas, by means of high-pressure gasification using a hydrogen-rich gas as a gasification medium.

The present state of our research leads us to conclude that the hydrogen method should be abandoned, in consequence of three unfavourable factors:

• the very small area of the gas–solid contact faces in the underground gas generator;
• the low rate of reaction between the carbon and the hydrogen;
• the fact that it is impossible to increase this reaction by raising the temperature, as this would shift the chemical equilibrium conditions in a direction which would lead to breakdown of the methane.

Gasification using air seems to us to have little more chance of industrial development than the previous method, since it results in the circulation of large quantities of nitrogen in the underground gas generator; since nitrogen is totally inert, its presence has several deleterious effects:

• the rates of advance of the gasification face are reduced, by reason of the fall in the concentration of reactive gases;
• the thermal power produced per borehole is reduced;
• it increases the volume of gas to be circulated in the underground gas generator and the loss of energy in the form of the sensible heat of the gas leaving the gasification zone.

The use of oxygen to produce a gas of medium calorific value seems at present to be the most promising approach for developing underground gasification at depth. However, oxygen cannot be used alone without causing an excessive rise in the temperatures and a falling-off of the gasification yield.

Apart from the cost of the demineralised water and of the thermal energy used to obtain it, the steam has the major disadvantage of needing to be injected at a temperature of the order of 200–250°C; this entails a very heavy increase in the costs of the distribution pipes and of the injection boreholes. It would seem that the use of steam must be replaced by a technique of seam infusion by the injection of water ahead of the gasification face.

This technique would considerably reduce the production costs for the gasifying media, for the energy consumed to inject them, and the cost of the pipes used to do so. It would also offer two other beneficial possibilities:

● getting rid of the phenol-containing water from the gas-scrubbing operation, which could be reinjected into the deposit;
● considerably increasing the area of the gas–solid reaction surface by causing breakdown of the coal seams and promoting crushing of the coal under the action of the lithostatic pressure.

One could also consider using carbon dioxide to dilute the oxygen [17.7]. It would offer no advantage to use carbon dioxide in the surface installations, by reason of two drawbacks:

● the fact that, at the same temperature, the rate of the Boudouard reaction is less than the rate of the water gas reaction;
● the deleterious effect of an increase in the pressure on the rate of dissociation of the $CO_2$, which falls off more rapidly than does the rate of dissociation of the steam [17.8].

There is however, in the special case of underground gasification, an important argument in favour of using $CO_2$. The need to inject water to cool the gasification gases and to protect the grouting of the extraction boreholes means that one has to bring to the surface a mixture of some two parts of dry gas to one of steam. The low temperature of the mixture ($\leqslant 200°$ C) makes the recovery of the heat of condensation of this steam a very uncertain matter.

The use of $CO_2$ as the gasification agent makes it possible to recover this heat. The gas produced, which is simultaneously rich in CO and rich in steam, is – after a brief cleaning operation – treated in a shift reactor which operates at a temperature of the order of 300–350°C, the reaction being as follows:

$$CO + H_2O_{steam} = CO_2 + H_2 + 41.2 \text{ kJ/mol}$$

This reaction brings the hydrogen content of the gas to a level comparable with that which could have been obtained by directly utilising the steam in the gasifying-medium mixture; the reaction also leads to the recovery – in the shift reactor or in heat-recovery boilers – of an amount of thermal energy which is very close to the amount of sensible heat issuing from the gas generator.

Table 17.3 gives predicted limit values for the gasification yields and gas analyses which can be expected in a gasification working operating under a pressure of 30 bars, using as gasification medium a mixture of 50% oxygen and 50% steam, and a mixture of 50% oxygen and 50% $CO_2$.

These figures were obtained by using mathematical models developed at Liege University, the input data being experimental values obtained in pressure gasification trials carried out at the Institute National des Industries Extractives (INIEX).

Comparison of these figures shows that the oxygen/steam mixture is not obviously superior, since the slight gains in gasification yield and in oxygen consumption are due to the increased injection temperature of the gasifying medium.

Lastly, if one chooses the oxygen/$CO_2$ mixture, the basic consequence is to facilitate the distribution of the gasifying agent underground, with a resultant increase in the dimension of the gas treatment and conversion installation.

### Table 17.3

Predicted results from gasifying an anthracitic coal (7% volatile matter), gasification pressure 30 bars

| | 50% $O_2$ + 50% $CO_2$ | 50% $O_2$ + 50% $H_2O_{steam}$ |
|---|---|---|
| Gasifying medium | | |
| Injection temperature | 25°C | 250°C |
| Equilibrium temperature | 1027°C | 1080°C |
| *Composition of the raw gas* | *% by vol.* | *% by vol.* |
| $CO_2$ | 11.1 | 3.7 |
| CO | 74.5 | 58.4 |
| $H_2$ | 10.1 | 32.9 |
| $CH_4$ | 2.4 | 2.0 |
| $H_2O$ | 1.9 | 3.0 |
| LHV $(kJ/m_N^3)$ | 11 360 | 11 640 |
| $n_G = \dfrac{\text{LHV gas}}{\text{LHV coal used}}$ | 88.0 | 89.2 |
| $O_2$ used (moles per mole of gas) | 0.245 | 0.227 |

The main point thrown up by this assessment of the relative merits of the various gasifying media is that the underground gasification process must aim at producing a gas of medium colorific value, using three gasification media: a mixture of $O_2$ and $CO_2$, injected at ambient temperature behind the gasification face, and water, in measured quantities, injected into the seam ahead of the face.

The choice of these reagents is governed by three criteria:

- keeping the cost of the boreholes and the distribution pipes at a minimum, employing gasification media which can be injected at ambient temperature;
- achieving maximum rates of face advance of the gasification faces, by ensuring that the reactive gases are not diluted by the atmospheric nitrogen;
- achieving a maximum 'thermal motor' effect of the underground gas generator by the use as gasification media of liquid water and of gases which yield a considerable increase in the number of moles of gas during the gasification process.

**THE MIXED METHOD**

The use of an oxygen-rich gasification medium and a gasification pressure of 20-20-30 bars make it possible to achieve a considerable concentration of gas production. In a seam one metre thick, each working can yield an hourly output of $50\,000$ $m_N^3$ of gas of medium calorific value, $8000$–$10\,000$ $kJ/m^3$; this represents a thermal power of $125$ MW. If the face is given the form of a double unit, the gas produced from the two circuits will be extracted via a single borehole; this allows of attaining a thermal power of the order of $250$ MW, which is comparable to that provided by the biggest gas generators which can be built today for the treatment of extracted coal.

With an overall gasification efficiency of some 65%, the quantity of coal consumed per day to keep a single borehole supplied would be $1150$ tonnes.

Such a concentration of production makes it necessary to prepare large workings, since the total surface area of coal to be gasified to supply one borehole for three months would be of the order of 9 to 10 hectares.

Figure 17.4 shows, by way of example, the general appearance of an underground gasification working in a seam at great depth. The working comprises two twin rise faces (a double unit), the faces themselves being 400 metres in total length. They are fed with the gasification medium via vertical boreholes drilled along the edges of the panel to be worked.

The gas from the two faces is extracted via the same borehole, which is connected to the faces by boreholes drilled in the thickness of the seam.

Blocking-out such a working involves drilling holes which can be as long as 200-300 metres in the seam. Two methods are theoretically possible for this blocking-out:

(1) a method similar to that used by the oil industry, all the work being carried out from the surface by means of directional boreholes;

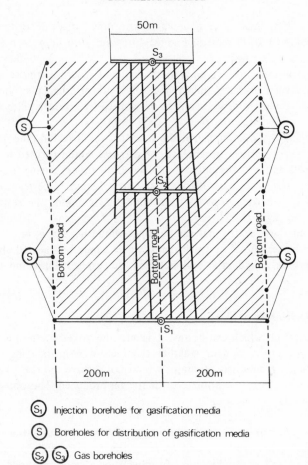

50m

$S_3$

Bottom road

Bottom road

Bottom road

$S_2$

$S_1$

200m 200m

$(S_1)$ Injection borehole for gasification media

$(S)$ Boreholes for distribution of gasification media

$(S_2)$ $(S_3)$ Gas boreholes

Fig. 17.4 – Layout of a gasification working.

(2) a 'mixed method', in which the boreholes which link the underground gas generator to the surface are used solely for injecting the gasification media and for extracting the gas; prospection of the deposit and preparation of the working being carried out from a system of roadways in the virgin ground, driven by conventional mining methods [17.9] [17.10] [17.11].

The technique of drilling directional boreholes was considerably improved when applied to deep-sea drilling; hitherto, however, it has been necessary to steer the drilling in only two directions, North-South and East-West, and there is still no means of creating an 'automated mole', controlled from the surface and capable of maintaining itself within the thickness of the seam and of being steered at will, to drive a passage along a predetermined course.

The mixed method is available here and now for all deposits of depths between 1000 and 2000 metres. Figures 17.5 and 17.6 show such a working, reduced to its simplest form.

The conventional mining operations required would comprise sinking an intake and a return air-shaft, and driving a network of cross-cuts, lying at great depth and separated from the seams to be worked by a safety pillar of at least 50-60 metres thickness.

The preparatory operations for the working would include the drilling of a certain number of vertical boreholes; one of these would be used to bring the gasifying media as far as the bottom roadways, the others serving to extract the gas produced. The lower part of these rise boreholes would be reamed out to a diameter of 2-3 metres, so as to form blind shafts linking the seam and the underlying roadways.

It is from these shafts that the headings would be driven, after which the shafts would be sealed off by thick dams; water would be injected above these dams to serve as a hydraulic seal between the gas generator and the access roadways.

## THE ECONOMIC PROSPECTS

The main objection which can be raised against the mixed method is economic in nature: one might wonder whether the fact of re-introducing miners to prepare the underground gas generator would not involve the risk of entraining an excessive increase in the prime cost of the gas produced. This objection does not resist analysis.

The number of miners required for the preparatory work should not exceed 15-20% of the labour force needed to run a conventional mine; the cost of the system of shafts and roads needed to reach the deposit could be much lower than the cost of underground layout of a mine using classical methods.

Actually, the mixed method allows us to envisage intervals of 200-300 metres between the horizons; the effective diameter of the shafts and access roads could be limited to 3-3.5 metres, because there is no haulage equipment to carry the coal and because the requisite amounts of ventilation air would be very much smaller.

Two other factors augment the economic interest of the mixed method.

(1) It makes possible better prospection of the deposit and constant monitoring of the progress of the gas generators. This correspondingly reduces the risks which are due to incomplete geological information.

(2) The number of boreholes linking the gas generator to the surface is very small – of the order of one borehole to every 10 or 20 hectares – and the distribution of the gasifying media can be carried out by means of a system of small-diameter holes drilled from below upwards, in a virgin deposit, which will not be subjected to ground movements caused by working.

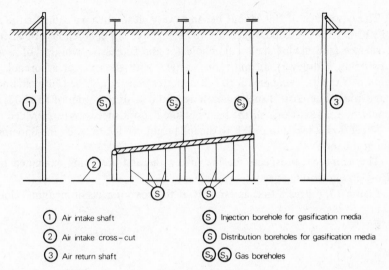

Fig. 17.5 – Layout of a mixed method operation.

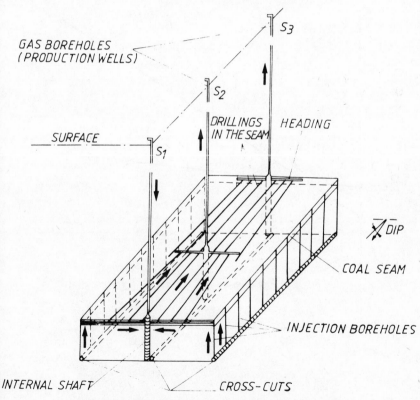

Fig. 17.6 – Perspective view of a gasification working.

The advantage of this layout becomes very clear when we compare the cost of £300 000 to £400 000 which has to be invested for a borehole drilled from the surface (the use of such a borehole for the successive working of several seams being, moreover, an uncertain matter), with the cost of a borehole for injection of the gasifying media 200–300 metres in length, drilled from an underground road. The cost of such a borehole would be of the order of £13 000, and it could be used to work all the seams in one deposit successively, provided that the precaution is taken to drill it to the full height needed to reach the uppermost seam in the series.

The economic prospects for the mixed method are being examined by a Belgo-German working party.

Figure 17.7 gives a first assessment of the cost of a gas of medium calorific

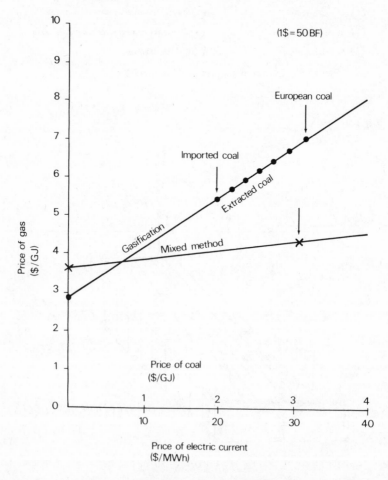

Fig. 17.7 – Cost of synthesis gas.

value, comparing these figures with the cost of gas produced by the conventional method from extracted coal.

In the conventional method, the cost per GJ of gas is made up of two items: the fixed costs of some $2.8 for the operation and depreciation of the plant; and the cost of the coal used, which can vary from $4.2 for European coal to $2.7 for imported coal. The cost per GJ of gas obtained by the mixed method includes a larger proportion of fixed costs — the cost of preparing the underground gas generator being as a rule higher than the saving resulting from the elimination of the surface gas-generator. As against this, the item for coal falls to zero, since the value of the coal *in situ* is deemed to be nil, and the only material consumed and to be charged to the cost of the gas is the electric current used to produce the oxygen employed as the gasifying medium.

The cost of the gas obtained by the mixed method is, all in all, some 20% lower than the cost of an equivalent gas produced from imported coal, and 35% to 40% lower than a gas produced from European coal.

The foregoing facts lead us to believe that it is now the moment for Europe to make a decisive effort in developing the underground gasification of coal industrially.

There is no area in which Europe could enjoy the benefits of a technological advance comparable to that which is available in respect of mining at great depths, and in no other area of the energy sector are there such excellent prospects of economic viability in the short term. Europe's abundance of fresh-water resources is a major positive factor for the industrial exploitation of all the techniques for the production and conversion of gas.

The high price of coal on the European market — which is a major handicap to the extension of gas-making from extracted coal — is an extraordinary advantage for making underground gasification of coal an economic proposition. By gasifying the coal 'on the spot', we can hope to produce synthesis gas at a price comparable with that achieved in South Africa or in Wyoming by gasifying opencast coal, but production in Europe would have the benefit of three incomparable commercial advantages:

(1) the fact that there are absolutely no competing energy products available, at low price;
(2) the presence of an immense energy consumer market right on the doorstep of the production sites;
(3) the extraordinary stimulus which could be given to the entire European economy by the large-scale production of synthesis gas; it would improve the balance of payments of the Community countries and stimulate growth in all the industries which can find their place in the chain of production resulting from the introduction of this new form of energy — ranging from the production of oxygen for supply to the gas generators to the production of methane, methanol, plastics and synthetic petrol downstream.

## THE DEVELOPMENT PROGRAMME

The trial at Thulin was planned in 1978 and the requisite preliminary work began in April 1979.

The site is at the western limit of the Borinage coalfield, in a zone left unworked because of the existence of major tectonic disturbances between the surface and the 800-metre level.

In preparation for the experiment, four 7-inch diameter boreholes were drilled to a depth of 880 metres. Between 860 and 870 metres, they intersected two twin seams with a total thickness ranging from 2.30 to 4.0 metres.

The work programme invisages three successive stages, as detailed in Table 17.4.

### Table 17.4

The Thulin trial

| | |
|---|---|
| *Phase 1* | Erection of surface installations |
| three years | Drilling four boreholes to a depth of 880 metres |
| 1979-1982 | Linking-up boreholes 35 metres apart |
| | Gasification: small faces; air/stream |
| | |
| *Phase 2* | Drilling three new boreholes to a depth of 880 metres |
| two years | Linking-up boreholes 50 to 70 metres apart |
| 1982-1983 | Gasification: longwall; oxygen/steam |
| | |
| *Phase 3* | Deepening six boreholes to a depth of 980 metres |
| two years | Linking-up the boreholes |
| 1984-1985 | Gasification of a second seam: longwall |

The equipment required for the linking-up work was finished in 1981. It consists of 2 five-stage piston compressors with a nominal rating of 350 $m^3/h$, with a delivery pressure which can reach 300 bars.

Installation of the gasification equipment should be finished in April 1982; this comprises an 8.8 tonne/h steam boiler and a five-stage centrifugal compressor with a delivery of 12 500 $m^3/h$ of air at a pressure of 45 bars.

During 1981, studies were made of the permeability of the deposit by injecting water and nitrogen via the various boreholes. This investigation confirmed that the initial permeability of the coal is very low (below 0.1 millidarcy†), but that the use of fluids at pressures equal to or higher than the lithostatic pressure causes expansion of the fissures which in turn leads to a very considerable increase in the permeability, allowing the injection of large quantities — 600 litres of water or 300 $m_N^3$ of nitrogen per hour.

† A rough idea of one darcy is obtained by considering a one foot cube of sand. If the sand has a permeability of one darcy this one foot cube will pass approximately one barrel of oil per day with a one pound pressure drop.

These preliminary studies also demonstrated that by selecting the most favourable directions of flow, it is possible to obtain a recovery rate of 10% via a borehole located 35 metres from the injection borehole. Injection of air was planned for early January 1982 and the first ignitions should be made in February.

The second stage of the Thulin trial may well be the most interesting, since it should make it possible to start the gasification with oxygen, to increase the distances between the boreholes and to make progress towards longwall operations.

The third phase will be no more than a repetition of the second; its essential aim will be to check whether the same system of boreholes can be employed for the successive gasification of several seams.

It is to be hoped that, in parallel with the second and third phases of the Thulin trial, a new organisation will be created to make it possible to extend the study of the potential of the mixed method, and to develop a sufficiently ambitious programme to lead rapidly to the industrial application of underground gasification of deep deposits.

The financial risks which must be faced to perfect this new technology are low, when compared with the risks accepted by the oil companies when they decided to prospect in the North Sea and to develop the deep-sea working techniques. Moreover, the financial justification of the investments to be made could be of a very different order of magnitude, bearing in mind the enormous scale of the energy resources which would thus be placed at our disposal.

## CONCLUSIONS

The Thulin trials could constitute a decisive step in Europe's return to energy independence.

The experimental installation which we have been enabled to construct with the financial assistance of the Belgian Government, the Government of the Federal Republic of Germany and the Commission of the European Communities is intended to demonstrate, for the first time in the world, that the conventional method of mining which was Europe's source of prosperity for more than 150 years is not the only possible method for working the coal deposits of Europe.

Because, after the switch to oil in the sixties and seventies, we closed our pits and enthusiastically adopted those new forms of liquid fuels, cleaner and easier to use, Europe's industry now stands at a cross-roads.

It is inevitable that we shall have to return to coal in the next few years, but the return could assume two diametrically opposed forms.

Europe can return to the past, enlarge coal-handling ports and its system of transport by water, by rail or by pipeline and progressively remodify its plants to use solid fuels, abandoned twenty years ago. By doing so, she would be exchanging one sort of energy dependence for another, and would be giving up the extraordinary ease of transport and utilisation provided by liquid fuels.

Europe can also solve her problems by making the breakthrough towards the coal of the twenty-first century, the coal of the deep deposits which, to this day, have not yet been reckoned among our workable reserves. Towards deposits of coal whose ash and inert material will remain for ever buried in the bowels of the Earth, while the useful components are brought to the surface as high-pressure gas. After appropriate chemical treatment, this gas could be distributed over an immense trans-European system to feed our towns and our industries. It could serve as a feedstock for producing all the organic and inorganic products currently derived from oil, including fertilisers, plastics and methanol. It could feed the plants which will produce the liquid heating and motor fuels of tomorrow.

It is up to us to accept the effort required to accomplish this technological breakthrough which is capable of guaranteeing our independence for the next 100 years, and so ensure that the opening of the twenty-first century will be the dawn of a new era of prosperity for Europe.

## REFERENCES

[17.1] McDonald, A. (1981) 'Energy in a finite world' IIASA Executive Report 4, May.

[17.2] Foster-Wheeler Power Products Ltd. (UK), Foster-Wheeler Synfuels Corporation (USA) and Foster-Wheeler Energy Ltd. (UK) 'Review of coal conversion processes — Coal Technology Europe '81' Cologne, June 1981.

[17.3] Thompson, P. N., Mann, J. R. and Williams, F. (1976) *Underground gasification of coal* a National Coal Board reappraisal, NCB.

[17.4] Ledent, P., Beckervordersandforth, Chr. *et al.* 'Planning and state of the Belgian-German experiment of underground gasification' Seventh UCG Symposium, Lake Tahoe, Sept. 1981.

[17.5] Bailey, A. C., Chapell, R. S. and Blades, M. J. (1981) *Underground Coal Conversion: Feasibility and Potential in the United Kingdom* MRDE Bretly.

[17.6] Loison, M. (1952) 'Essais francais de gazeification souterraine a Djerada (Maroc)' *Annales des Mines de Belgique* Jan.

[17.7] Ledent, P. 'Esquisse d'une nouvelle methode de production de gaz de synthese par gazeification souterraine du charbon' Seminaire International AILg, ULg 25-27, May 1981.

[17.8] Chapell, R. S. (1980) *High pressure gasification of deep coal* NCB.

[17.9] Ledent, P. 'How in situ conversion of coal can enable a move from conventional coal mining to underground gasification techniques' Forward to Coal Conference, London, 12 May 1981.

[17.10] Ledent, P., 'De l'exploitation minière conventionnelle à la gazéification souterraine du charbon' Colloque *L'énergie en 1981* AIM Liege, 18-21 May 1981.

[17.11] Ledent, P. (1981) 'Vom herkommlichen Kohlenbergbau zur Untertage-vergasung von Kohle' Gluckauf Forschungshefte 4.
[17.12] Dent, F. J. 'Hydrogenation of coal with a view to the production of hydrocarbons' International Conference *Gazeification intégrale de la nouille extraite* Liège, 3-8 may 1954.

# 18

# Power system control: Load shaping and energy management

**F. Ledger,** Central Electricity Generating Board

## BACKGROUND

Electricity supply undertakings generally have the objective of satisfying consumer demand at minimum cost consistent with reliability of supply. Consumer demands have daily and seasonal patterns which, when integrated because of diversity of requirments, result in a daily load shape which varies over the year according to the seasons. Examples of the daily variations of the CEGB system demand in the winter and summer are illustrated in Fig. 18.1. The seasonal variations are caused mainly by climatic conditions which create a need for space heating, air conditioning and by lighting requirements. However some utilities' daily load shapes are less variable because of the effect of high load factor industrial consumers and/or diversity of climatic conditions across the area served. This is illustrated in Fig. 18.2.

Notable features of the daily load shape are: (1) the comparatively low levels of demand at night; (2) the rapid increase in demand between about 06.00 hours and 08.30 hours; (3) the sharp, short duration peak demands at the onset of darkness during the winter months; and (4) the smaller fluctuations in demand throughout the day. Overnight plant output regulation requirements are dictated by the magnitude of the daytime 'plateau' and the night minimum demand, the shape of the night trough, the minimum shutdown period characteristic of individual generating plants. The number of generators available, the degree of plant flexibility and plant mix are also important. Daytime plant flexing requirements are dictated by the secondary peaks and troughs, that is, the smoothness of the day 'plateau' demand and its relative magnitude to the daily peak demand. All these plant regulating and flexing requirements are costly, not only in additional fuel requirements but in capital expenditure for a particular level of energy supply.

The extra costs of plant flexing are incurred because units that have been generating are then not required to operate overnight and/or during any long duration daytime troughs. These units lose heat during the shutdown period which has to be replaced in order to restore temperature and pressure conditions when required to generate electricity again. Generators which low-load also incur extra costs since turbo generators consume more heat per unit of electricity generated at partial load than they do at full output. Other costs of plant flexibility arise from steam plant operating over the daily peak for short periods of 1 to 3 hours. Such usage incurs very high start-up heat consumption, extensive operation at part-load and excessive wear and tear. Over the shorter end of the range its generation cost compares with very high cost gas turbine operation using light distillate fuel oil. Additionally, permanent reduction of the daily

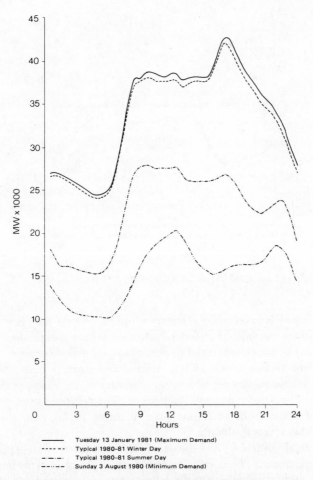

Fig. 18.1 – Summer and winter demands on the CEGB system in 1980–81, including days of maximum and minimum demand.

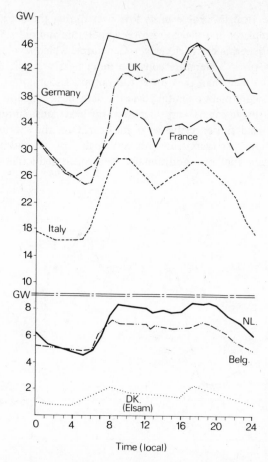

Fig. 18.2 — Load curves for Wednesday, 21 December, 1977.

peak for an equal amount of total energy supplied, reduces the plant capacity requirements with consequent considerable savings in capital expenditure and/or in revenue costs by the possibility of de-commissioning expensive plant.

One of the factors which affects demand and requires generator output flexibility is television viewing and some programmes, for example the Royal Wedding, can cause major demand fluctuations. The demand variations require very rapid response from generating plant in order to maintain the system frequency within acceptable limits.

In the period covered by the theme of this book the generating plant load-following requirement will generally increase in severity because of the probability of a greater proportion of relatively inflexible nuclear capacity in the plant mix, as shown in Table 18.1 for the CEGB. The predicted relatively slow growth

Table 18.1

Plant mix declared net capability of CEGB plant

| Type | 1982 % | 2000 % |
|---|---|---|
| Coal-fired | 71.0 | 67.8–50.8 |
| Oil-fired | 13.8 | 13.5–5.1 |
| (Gas-fired)† | (3.0) | (3.0) |
| Nuclear | 8.3 | 8.5–33.9 |
| Gas turbines | 6.0 | 3.4 |
| Hydro and pumped storage | 0.9 | 3.4 |
| Cross-channel link | Negligible | 3.4 |
| Total | 100.0 | 100.0 |

† Option of coal- or gas-firing (included in coal total)

in demand for electricity in developed countries, the high nuclear plant proportion and very high fossil fuel prices will lead to an even greater need to reduce efficiency losses thus requiring extreme flexibility performance by large fossil-fired units.

There is therefore a very strong incentive to improve the daily load shape in order to reduce costs and to limit the amount of plant flexibility required by endeavouring to reduce the difference between day and night demand level and to establish a constant 'smooth' demand during daytime. The desirable load shape will also permit pre-planned start-up times of units in the morning, followed by acceptable rates of load pick-up and programmed shutdowns in the late evening. Such a 'smoothed' mode of steam plant operation will make a major contribution towards fuel savings and less plant maintenance because of reduced wear and tear.

## HISTORIC APPROACH TO LOAD SHAPING

Striving for financial benefit has always led electricity utilities to seek improvement in daily load shapes. The most common influence towards achieving this improvement has been through tariff cost messages to consumers which have sought to reduce both the costs of production by stimulating a more constant demand for electricity throughout each day, and capital expenditure on plant by constraining the growth of extreme peak demands during the winter months.

An example of this in the United Kingdom is concerned with the tariffs for bulk supplies from the CEGB to the distribution Area Boards. The tariff changes over the past twenty years have been as follows:

1960–61     Capacity charge rebates were offered for the first time for peak lopping, that is, introduction of load management.

1962–63     A night unit rate was introduced to reflect the cheaper cost of producing electricity during this period.

1967–68     A peak period unit rate was introduced for the months of December, January and February to indicate the high cost of meeting peak demands.

A second capacity charge was introduced, that is, peaking in addition to the existing basic.

A generally applied 'potential peak warning' system was introduced, operative for the November to February period as a further contribution towards load management.

1975–76     A new set of running rates was introduced to refine the reflection of production costs at night, peak and the 'plateau'.

1977–78     A 'load management warning' system was introduced to replace the 'potential peak warning' system. It applied to peak half-hours only when the plant/demand position was critical.

1978–79     'Load Management Warning' was subdivided into two categories (A and B) dependent on advance period of notification.

1979–80     Introduction of reformulated night rate to suit the shorter charging period required by modern storage radiators.

1981–82     A third category (C) of load management was introduced enabling load reduction to be implemented by the consumer at as little as 15 minutes notice.

Figures 18.3, 18.4 and 18.5 show the improvements made by the CEGB in daily load shapes since 1960/61 for average seasonal weekdays. It can be seen that the night trough is now less pronounced, the rate of climb from low to high demand has lessened, the peaks in demand during the day are now less pronounced and the daytime troughs have been filled. The annual average daily load factor (the ratio of average to peak demand) has increased from 73% in 1960/61 to 82% in 1980/81. Also the system annual load factor has increased from 49% in 1960/61 to 57% in 1980/81. These improvements have allowed a better utilisation of capital assets and have provided savings in fuel at a time of significant increase in fossil fuel prices.

Complementary to the influence of the tariff messages, utilities have generally been directing major effort at improving the generation pattern of the conventional steam plant by judicious use of pumped storage, low head limited storage and high head high seasonal storage hydro plant. Additionally, high running cost gas turbines that incur negligible stand-by losses and have low maintenance costs have been used for the short, sharp peaks of the day. Trading arrangements with neighbouring utilities have contributed beneficially to improving the use of thermal generation and contracts for purchasing surplus output from auto-producers has aided in load shape improvement.

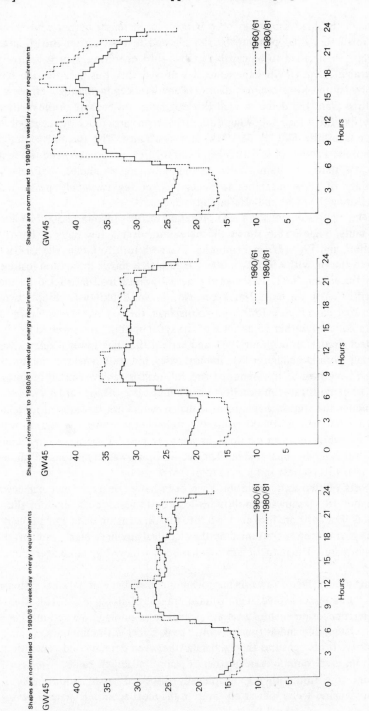

Shapes are normalised to 1980/81 weekday energy requirements

Fig. 18.5 – CEGB winter week day average daily load shape.

Fig. 18.4 – CEGB spring/autumn weekly average daily load shape.

Fig. 18.3 – CEGB summer weekday average daily load shape.

Although the load shape has been improving, there has been a continuing need for fossil fuel fired plant to be flexible and in this respect British large modern plant is specified to be capable of two-shift operation, that is, shutting down overnight. The CEGB experience has proved that this specification has been met by United Kingdom manufacturers and has been increasingly effective. The coal-fired plant has demonstrated this capability but because of the relative fuel prices, extensive load following duties have been carried out by the modern oil-fired units in the 500 to 660 MW unit size range. The flexibility of the oil-fired units at Fawley (4 × 500 MW), Pembroke (4 × 500 MW) and the first two 660 MW units at Grain power stations are good examples, particularly the capability of these machines to generate over the two daily peaks and shutdown during the night and afternoon troughs.

The next figures demonstrate the operating regimes of our oil 500 MW and 660 MW units. Figure 18.6 shows the operation, over a few days at Fawley power station, and Fig. 18.7 at Pembroke. Then we turn to Grain, the CEGB's first oil-fired station with 660 MW units.     Figure 18.8 shows the station loading pattern in December 1981. It shows the actual against load instructed by the Grid Control Centre on units No. 1 and No. 2. No. 2 unit was 'double two-shifted' on that day. Figure 18.9 has two displays, the top one shows return to load from a hot start within 15 minutes of the synchonising time required by the Grid Control Centre throughout 1981 and early 1982. The lower display gives greater detail from December 1981. In most cases, hot start synchronising times are within ±5 minutes of that required and full load is achieved in 20 minutes. Figure 18.10 demonstrates an occasion when a unit came off load at 18.43 hours; two hours later the station was informed that it would not be required for load for at least 30 hours. At 06.00 hours the following morning the station was requested to bring the unit on load as soon as possible because a machine elsewhere had become unavailable. The set at Grain was synchronised about 1.5 hours after this request and loaded to 500 MW.

The costs of two-shift operation have been halved as experience has been gained. Figure 18.11 shows two-shift off-load costs based on a six-hour shutdown period. The costs are broken down into fuel oil costs on main and auxiliary boilers, the cost of power consumed by the electrical auxiliary plant, and cost of heat losses in water. The reduction in cost has been achieved by close operational monitoring.

A contributory factor towards improving the load shape of national systems has been the interconnected high voltage transmission network. This transmission interconnection enables generating plant to be pooled on a national basis so that the system demand at time of winter peak is met with minimum installed plant capacity. It also ensures that imbalance between demand and generation, resulting from local variations in weather or plant availability, can be made good at minimum cost from other geographically dispersed areas. Additionally, it enables the demand to be met at all times in the most economic manner having

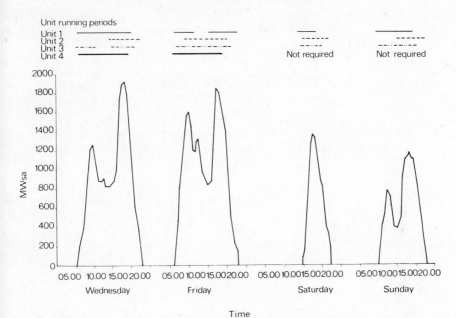

Fig. 18.6 – Typical winter operation, Fawley.

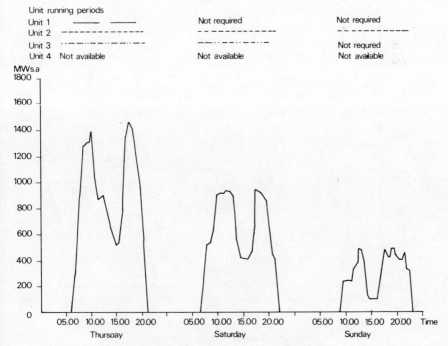

Fig. 18.7 – Typical winter operation, Pembroke.

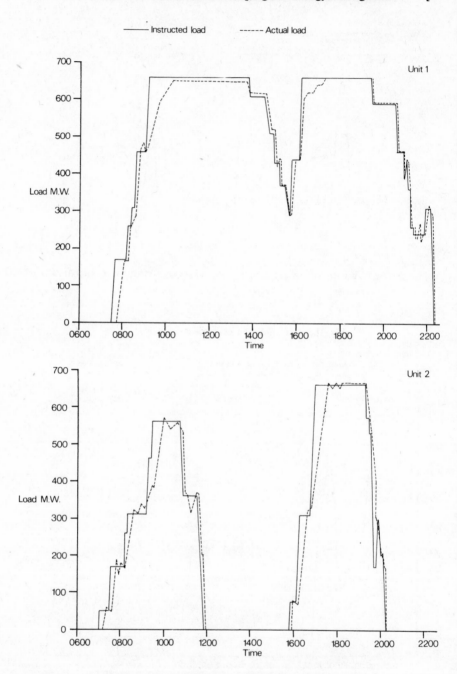

Fig. 18.8 – Station load pattern on 30 December, 1987.

UNIT 2 HOT START RECORD 1981/82

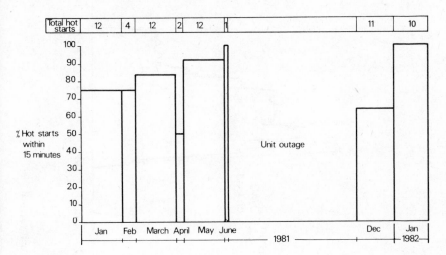

UNIT 2 HOT START RECORD SINCE DEC 1981

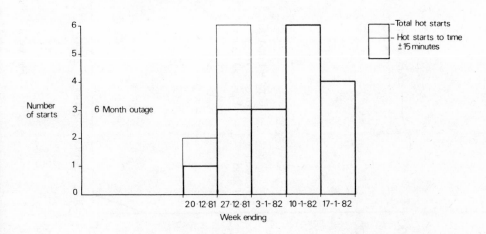

Fig. 18.9 – Hot start records.

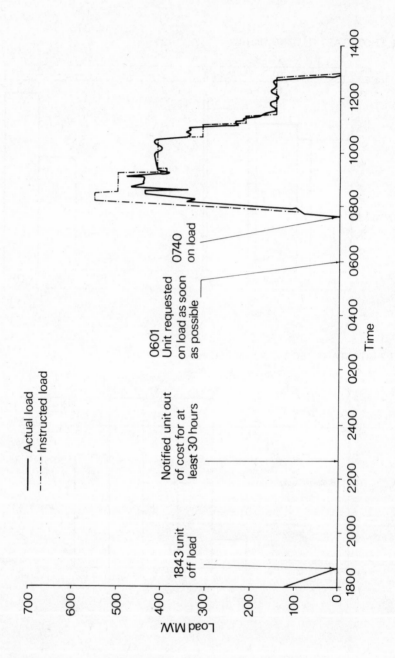

Fig. 18.10 – Short notification unit required.

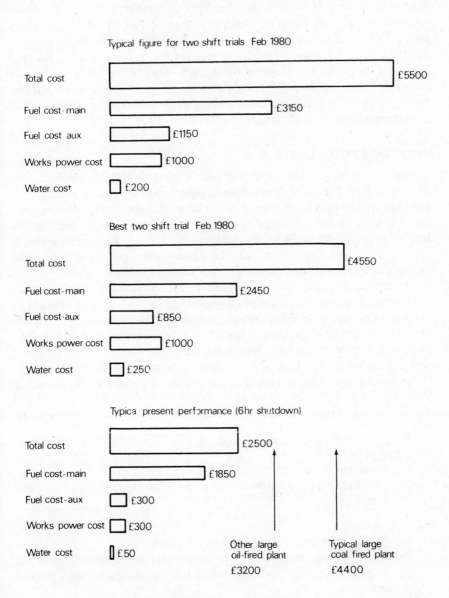

Fig. 18.11 – Improvements in two-shift costs.

due regard to the wide fluctuations which have occurred in recent years in relative prices of coal and oil. The system enables savings from economies of scale by its ability to accommodate larger unit size and more efficient generating plant. A smaller amount of reserve capacity is required for day-to-day cover for contingencies. In the United Kingdom the decision taken in the 1960s to transmit energy by electricity from the coal field as opposed to siting the power stations remote from the coal field and transporting coal by rail has proved fully justified in economic terms.

## IMMEDIATE FUTURE

The immediate future developments in load shaping are likely to continue to be those described but to be increasingly complemented by energy management including remote interactive control of consumer demand. Energy management is concerned with the transfer of energy, from one part of the demand curve to another in a way acceptable to the consumer. In contrast, load management is concerned with reducing load at peaks. Further use of storage devices, 'renewable' sources and agreements with auto-generation are also envisaged.

In addition the new 1800 MW pumped storage station in the United Kingdom at Dinorwic in North Wales will be commissioned in 1982. It is an example of more plant of this type being installed worldwide. The principle is extremely simple. There are two reservoirs separated by several hundred metres in height. The plant contributes to load shape improvement by pumping from the lower reservoir during the low load periods, when electrical energy is cheap, and also by generating electricity at high load times when the water is run through a water turbine from the upper reservoir. Figure 18.12 illustrates some pumped storage schemes. Main plant and control facilities are provided to enable operation in any one or a combination of the following regimes:

(1) Pumping
  (a) Overnight demand trough in-filling.
  (b) Overnight rapid reduction in pumping load to compensate for a sudden loss of generation.

(2) Generation
  (a) Peak lopping and/or order of merit operation.
  (b) Rapid increase in generation to compensate for an instantaneous loss of generation and/or to meet random fluctuations in generation/demand mismatch.

The operational use of pumped storage plant on any particular day is dictated by the relative economic/security requirements.

A later figure gives an indication of how the operation of pumped storage schemes is incorporated in the overall strategy of improving the pattern of thermal generation.

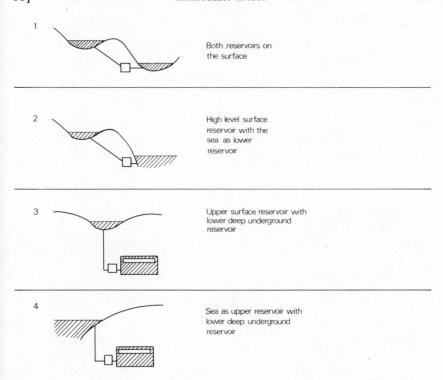

Fig. 18.12 – Methods of pumped storage development.

One of the proposed additional international transmission connections is the 2000 MW high voltage direct current cross-channel underwater link between England and France. Because of the time difference in the occurrence of the annual winter peak demands of the respective utilities, the installation of the interconnection saves in planting capacity requirements of both countries. Additionally, support during emergencies is provided and also it is envisaged that Electricité de France will import 1000–1500 MW from the CEGB during winter nights which will provide additional load for our efficient coal-fired plant. The exchanged energy will be balanced on an adjusted barter basis during daytime hours on spring, summer and autumn days when the CEGB will import from France, thus eliminating the use of marginally expensive oil-fired generation. The other advantage of the new link is illustrated in Fig. 18.13. This shows the CEGB/EdeF daily demand curves for a day in June 1981 and it can be seen that there is a diversity between the two systems and opportunity trading can take place when a peak and a trough coincide, for example, at 12.00 and 17.00 BST on this particular day.

Similar advantages can be further exploited by entering into contracts with

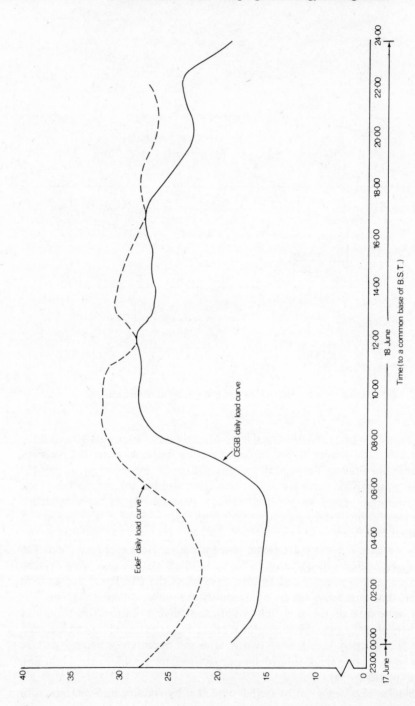

Fig. 18.13 – Comparison of CEGB and E de F daily load curve for Thursday, 18 June, 1981.

industrial consumers/exporters of electricity when the processes require additional electrical energy or when a surplus arises, especially where combined heat and power installations are being utilised. Combined heat and power schemes are much more efficient than conventional power stations because of the elimination of latent heat losses which are incurred in converting the steam from the turbine exhausts back to water for the boilers.

## FUTURE PROBLEMS AND DEVELOPMENTS

### Storage methods

Energy storage will have a major role to play if the renewable energy sources (wind, wave, solar, geothermal and tidal) are developed on a large scale. Since the output from these sources is variable and, with the exception of geothermal and tidal energy, subject to large unpredictable changes in weather, fluctuations in output must be accommodated on the system by regulating the output of conventional plant or by providing storage. Terrestrially generated solar electricity is an extreme example. To obtain significant benefit from solar cells, long-term (seasonal) storage would be required to match the peak output in summer with the peak demand in winter. Storage to perform such a duty would be under-utilised and expensive.

### Pumped storage

Turning to the types of storage plant, I have already mentioned pumped storage which is the only method in widespread use today. In the world, there are some 45 GW operating and a further 10 GW under construction.

In the context of the theme of this book, that is the long term, pumped storage is viewed as an expensive method of daily load shape improvement because of the 25% conversion loss incurred during the daily cycle of pumping water into the upper reservoir and then running it down later for generation. However, direct energy management of consumer demand does not incur such losses and will become the preferred alternative.

### Gas turbine compressed air storage

This was first patented by Stal Laval as long ago as 1949, but it was only recently that the world's first system was installed and commissioned at Huntorf, near Bremen, in West Germany. A gas turbine compressed-air storage scheme makes use of gas turbine components but inserts a time delay between the compression period and the power production period, so that the power may be used at times when the electricity supply system has its peak demand or when it requires stored output for similar reasons to those described for pumped storage. Figure 18.14 shows schematically a gas turbine compressed-air storage arrangement.

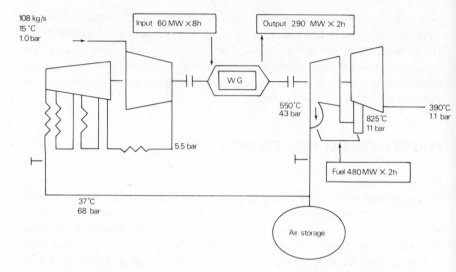

Huntorf uses a salt cavern to store the large quantity of air required. Proposals have been made for untilising mined caverns in hard rock. A major disadvantage of a gas turbine compressed-air storage system is its demand for a premium-grade fossil fuel and therefore it is not as attractive as direct energy management of consumer demand.

*Flywheel storage*

The technology is not new and recently flywheels have provided the short period pulses of power needed, for example, in fusion research. At the Max Planck Institute for Plasma Physics at Garching, the flywheel is of steel, weighing 223 tonnes, and it can supply up to 150 MW for 10 seconds decreasing in speed from 1650 to 1275 revolutions per minute. Flywheels should be constructed of materials with high specific strength. The largest amount of energy that it is thought technically feasible to store in a flywheel is 10 MWh and this is about the smallest amount that would be of interest even on a distribution network. Unless special techniques can be devised to increase the safe working stress of the flywheel materials, it is unlikely that they can be made economic for large-scale electricity storage applications.

*Battery storage*

Battery storage is a topic on its own. Rechargeable electrochemical batteries (usually lead–acid) have a long history of application in electricity systems mainly for emergency duty. Recently, the attractive features of electric vehicle propulsion with regard to the environment and energy conservation have given new impetus to the development of cheaper batteries with high energy densities. Batteries can, in principle, be used for peak-lopping and daily-smoothing the system demand curve. They are suitable for siting close to the consumer and are, therefore, able to ease the maximum load on the transmission and distribution networks. Cost estimates show that the currently available lead–acid and nickel–iron batteries are somewhat above the break-even cost for a load-shaping role. The sodium–sulphur and zinc–chloride batteries, which are both at an advanced stage of development, may show appreciable savings against the break-even figure.

*Magnetic energy storage*

I now come to the most speculative but perhaps the most intriguing scheme, since it is the only one which stores electrical (or magnetic) energy directly. It is based on a large superconducting coil carrying a circulating (DC) current giving rise to a magnetic field.

To keep the specific cost down to a reasonable level it is necessary to store very large amounts of energy in a single installation. For example, a store of 10 GWh is typical of that being studied, and for this a very large coil is needed, at least 100 m in diameter. The design of large coils is dominated by the enormous electromagnetic forces tending to burst them and, for solenoids, to crush them axially. A magnetic field of 5 tesla might be created and this gives stresses equivalent to a pressure of 100 atm. The cost of a self-supporting structure to contain these forces would render the store uneconomic, so all proposals for large superconducting magnetic energy stores envisage placing the windings in annular tunnels cut in bedrock. If the installation is deep enough the weight of the overburden ensures that the forces in the rock remain predominantly compressive. There are a number of design studies, that at the University of Wisconsin being particularly noteworthy. Design problems include coping with stray magnetic fields at the surface, and ensuring that the vast energy stored in the magnetic field (equivalent to 10 000 t of TNT) cannot be released in an uncontrolled way.

## Renewable Resources

The generation of electricity from renewable energy resources, mentioned earlier, is being considered worldwide. It is important to consider renewables not only in their own right as substitutes for fuels but also with regard to their influence on load-shaping requirements.

*Wind energy*

Wind energy in the short to medium term is one of the front runners. The technology, while far from being fully developed has·reached the pilot scheme stage and proposals for the first installation on land are proceeding.

Figure 18.15 illustrates the results of a mathematical model of clusters each of 1000 wind generators at four geographically dispersed sites. The output from these sites has been summated and is shown in Fig. 18.16 in the upper curve together with its integration into the system demand illustrated in the lower curve. The problems involved in incorporating the wind power output into the utility network and of matching generation and demand are being studied.

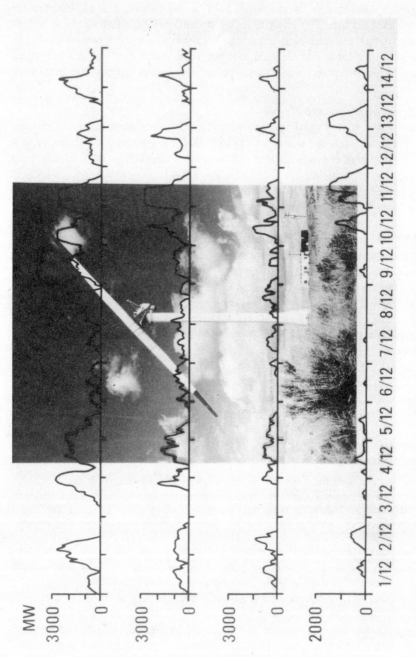

Fig. 18.15 – Wind power output from different sites.

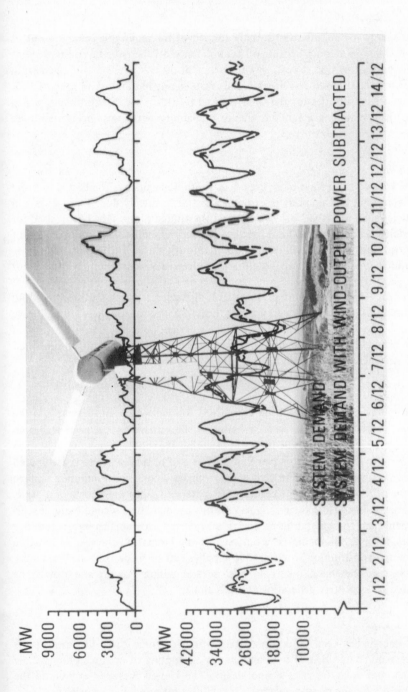

MW
9000
6000
3000
0

MW
42000
34000
26000
18000
10000

—— SYSTEM DEMAND
---- SYSTEM DEMAND WITH WIND-OUTPUT POWER SUBTRACTED

1/12 2/12 3/12 4/12 5/12 6/12 7/12 8/12 9/12 10/12 11/12 12/12 13/12 14/12

Fig. 18.16 – Total wind power output and integration with system demands.

*Wave energy*

Wave energy is now the most heavily funded of the renewable energy resources in the United Kingdom because of its abundance off the Atlantic coasts. Present estimates suggest that a peak power output of 30 GW could be obtained from the waves. The resource is sizeable and wave energy has important potential for the electricity utilities for the time covered by this book. The problems which come from matching a random energy availability with consumer demand are obvious.

*Geothermal energy*

There are two main ways of trying to use geothermal energy. The first is to make use of existing hot aquifiers or springs, and this is already done fairly widely in countries such as Italy, New Zealand and the United States. The heat is used for electricity generation and local heating. The second method is the abstraction of heat from hot rocks by circulating water through them. This second technique has much greater long-term potential for electricity generation. Work on hot rock geothermal energy has been going on at Los Alamos in New Mexico and results are encouraging. At present, it is difficult to assess the resources and economics of geothermal energy in the United Kingdom. Unlike wind and wave power which depend on the vagaries of the weather, geothermal energy from hot rocks could maintain a continuous or flexible electrical output and will aid demand/generation matching.

*Tidal energy*

Tidal energy schemes have been considered for some time, one example is the 10 X 24 MW installation near St Malo on the estuary of the river Rance in Northern France. The CEGB has been studying the feasibility of a tidal barrage scheme in the estuary of the river Severn. The cost of tidal schemes is dominated by the massive investment in civil engineering. A simple ebb generation scheme has the diadvantage that electricity is generated according to moon time. However, it is possible to re-time the output by building a second basin next to the main dam for generation on the flood tide and providing an integral storage facility involving pumping between each of the basins and the sea. Although a substantial contribution to electricity supplies can be made from such schemes, the cheapest scheme is likely to cost several billion pounds and cannot be justified on electricity generating grounds alone.

*Solar energy*

Solar energy has been used for water heating for some time. In various parts of the world large R & D programmes are attempting to produce photovoltaic systems and solar furnaces economically. The United States is examining the possibility of using massive solar power satellites for providing electricity.

## ENERGY MANAGEMENT

The utilisation and integration of pumped storage schemes, trading with neighbouring interconnected utilies and with auto-generators, are well established practices. Because of the speculative nature of the various other energy storage schemes and renewable energy resources, the problems associated with integrating the output of these facilities into the utilities' system demand curves require detailed study. The purpose of these studies is to accommodate these outputs in the most economic way, that is, minimising system fixed and running costs.

The principal objective is to match electricity supply and demand in a way that improves the effective utilisation of the generation capacity and encourages the best use of electricity by all consumer sectors, coincident with improving the value of the service to the consumer. From this principal objective can be derived the aim of an improved use of non-renewable resources of fuel, coupled with the increased use of renewable sources of energy as part of a viable conservation programme. If these objectives are to be achieved it is essential that the strategies and managerial implications of an Energy Management System should be included as a significant element in the corporate planning of an Electricity Supply Utility.

As mentioned previously, the utility tariffs have been successful in improving daily load shapes. However, the tariff price mechanism on its own is a relatively blunt instrument and it is very difficult to predict consumer reaction to any change in the structure. This complicates assessment of the effectiveness of any proposed innovation in pricing policy brought about by new methods of load shape improvement.

Cognitive strategies which are directed at changing an individual's beliefs or attitudes in the expectation of a consequent change in behaviour are more hopeful. An example is the provision of hot water. In practice, people do not know when electrical energy is being supplied for hot water which is heated by a thermostatically-controlled heater and do not concern themselves with it — as long as they have reasonably constant hot water — but may, if it is switched remotely, resent interference with personal liberty. So if we can take this a step further and persuade people that the utility is trying to help and not interfere, they will behave more reasonably and let the utility decide when to supply the energy, still meeting the same criterion of constant hot water. In so doing a considerable amount of load regulation becomes possible to the consumer's ultimate benefit. The best way forward is to establish broadly-based long-term aims which command acceptance by the majority of a community and then encourage innovation in the field of energy technology and make the results of such work available and understandable both to the community at large and to the individual. This creates a self-evolving mechanism by involving the consumer in studying his short-term energy usage and utilising the more efficient technology to effect change which is in line with the agreed long-term aims. It is in pursuance of this line of thinking that studies are being carried out on the possibility of the electrical demand being tailored to generation – in contrast to generation following the instantaneous consumer demand. This is accomplished by creating a closer working relationship between the consumer and the utility and by displaying more real-time information on both the units and the price of energy being consumed, thus achieving a feed-back mechanism for any conservation efforts the consumer wishes to invoke. By this means, energy conservation, both for the utility, by improving load shape, and for the individual, arising out of his wise use of energy on lower tariffs, is achieved.

When the consumer is working interactively in real-time with the supply utility, the repercussions on the management system of the utility are numerous and significant. This consideration leads to the next objective that, at all stages of planning, design and implementation of any Energy Management System, it is essential to gain consumer acceptance. The provision of an improved service is one necessary element in gaining this acceptance. Provision of personal energy control facilities is another important aspect. From the above considerations, stems a series of subsidiary objectives which relate to specific aspects of energy management. The first must be to obtain a reliable and economic terminal for installation at the consumer's premises. The second is to develop a communications system, because there must be a high level of functional interaction between the industry and consumers, and hence demand and supply balance, the desirability for reliable two-way communication arises. In turn, this results in the objective that there must be a higher level of information security and trust in the system, and there should be total utility and consumer confidence that this is so.

Some specific outcomes of the achievement of the above stated objectives of energy management are, therefore:

(1)  smoothing of the daily generation load curve;
(2)  improved facility for billing and collection of electricity accounts;
(3)  improved efficiency of local distribution of electricity;
(4)  potentially enhanced value of supply to the consumer.

In the longer term, therefore, Energy Management Systems are envisaged as including interactive load control with consumer participation in load/demand control. The utility and consumer act indepently but cooperate in order to establish a state of mutually advantageous equilibrium. The advance in microelectronics has enabled three methods of interactive load control to be considered: (1) mainsborne, (2) radio teleswitching and (3) a two-way communication microprocessor controlled system which also provides credit management facilities. These methods are at the embryo stage in the United Kingdom and pilot schemes are to be implemented in 1982. A number of countries have experience of methods (1) and (2).

**Energy Management Systems**

There are three main principles underlying the interactive load control philosophy:

(1) Exploitation of advances in communication-computation technology.
(2) Change the customer attitude — utility relationship into a cooperative relationship where customers and utilities are working together.
(3) Maximise the degree of customers' 'free will', for example, the utility does not cross the 'meter line'.

Interactive load control concepts being studied are;

(1) Spot pricing.
(2) Microshedding where the utility seeks a drop in customer load and the customer decides which components of load will be reduced, that is, customer chooses what will be shed and utility determines when; the incentive or rate can be negotiated from every few minutes up to annually.
(3) Dynamic control where a frequency sensor is added to customer device, for example storage heater, to enable the load to be controlled to provide rapid reserve response facilities in order to compensate for demand/generation mismatch without interfering with customer's actual life-style or usage.
(4) Energy Management Unit (EMU) or 'smart meter'. Equipment installed by the utility on the consumers' premises to meter the electrical energy supplied under the published tariffs, to control supplies on separate circuits in a pre-programmed manner and to display to the consumer agreed data. The unit to be capable of operating in a 'stand-alone' mode or in conjunction with the utility through one or more defined communications channels. The EMU needs a 'non-volatile' memory, that is, it does not lose its contents when the mains electricity supply fails.

All the methods of interactive load control need some form of communications. These communications might be provided either by off-line means involving visits to the installation or by radio, power line carrier or telephone.

It is likely that the choice of communications channel will be dictated by technical and economic considerations. Major factors affecting the choice will be population density, type of distribution system, telephone penetration, radio interference and the current state of development in the respective technologies. Thus, it is likely that a variety of communications methods will be in use and that they may be upgraded as the economics and technical developments change. In particular, more off-line and one-way communication might be expected in the early years with a progressive move towards full two-way communications.

The present form of off-line communication involving visual reading of meter registers and manual resetting of timeswitches and maximum demand indicators, is likely to be unsuitable as the move takes place towards more registers per consumer and more precise and detailed timeswitching. The use of one-way communication in a flexible EMS needs to be evaluated as a system is adopted and refined. Clearly, it will have a back-up role to play, particularly in emergency conditions.

The use of two-way communication requires policy decisions in respect of the communications network needed to deal with the information flow to and from consumer premises.

Load management requires the ability to connect and disconnect consumers' circuits. The number of circuits, and the way in which they are switched, is determined largely by the type of tariff involved.

For non-domestic consumers, it is probable that the Energy Management Unit will not actually carry out the switching operations, but will provide output signals which may be used to operate separate switches.

In addition to the ability to switch circuits, features may be required by the consumer where a maximum demand or subscribed demand tariff is applied. In particular, it would be advantageous if the Energy Management Unit could provide signals on the power circuits as a pre-determined demand level is approached. These signals could be used to switch off the consumer's non-essential appliances in a manner programmed by him so as to stay within his subscribed demand. An alternative approach, more applicable to non-domestic consumers, would be for the EMUs to provide an output facility to interface with the consumer's own computerised energy control system.

The introduction of energy management systems using two-way communication with the consumer opens up new possibilities for the way in which the utilities operate.

Comprehensive two-way systems will involve data communication between most utility locations and consumers, with links to a number of external bodies if required. The EMU on the consumer's premises would be linked by the chosen communications channel to a data concentrator located at a telephone exchange or substation. Groups of data concentrators would be under the control of small computers located at a processing node which could also, if required, be connected to a data communications network thus providing access to the other users of the system, for example other utilities and banks.

In the United Kingdom, communication with the CEGB will provide the facility to transmit tariff messages to the distribution Area Board at the appropriate time and integrate load management within the system operation, planning and control function.

In deciding on the most economic medium for use as a particular link in the overall network, the best solution is likely to arise where that medium is already available or can be shared with other uses. For example, the link between the consumer and the concentrator is most likely to be provided by a medium which makes use of existing facilities rather than by provision of new dedicated communications channels. Thus power line signalling makes use of existing power conductors, telephone would use telephony circuits in some non-interfering way, and radio makes use of existing broadcasting facilities. For other communication links in the network the best arrangements are likely to be those which can be shared with other ESI telecommunication users.

The technical aspects of different types of communications media are well documented. Selection of the communications medium for energy management will depend on the functional specification of the EMS it is supporting.

As mentioned previously, trials to be carried out are intended to be used as a basis for assessing the suitability of the various communications options to support a comprehensive Energy Management System.

*Mainsborne*

These systems have the advantage of providing communications wherever there is a power line and are doubly attractive to supply utilities since the utility has total control over this type of communications channel.

Detracting from these advantages however, are a series of problems that arise from the fact that the power system is not designed for telecommunications and is fundamentally a very poor communications medium. The power system has a low signal bandwidth compared with conventional telecommunications systems. Consequently, the use of high data rates on power line carrier is constrained by the available bandwidth and by the high levels of system noise. A wide range of power line signalling systems has been developed in an effort to overcome the hostile electrical environment and optimise the communications path in terms of cost and information transfer rates. The general characteristics of power line signalling systems can be summarised as follows.

(1) Waveform distortion: Systems in the low frequency range are those that operate by distorting the supply-voltage waveform in order to produce coded signals. The main advantages of this technique are the comparative simplicity and low cost of the transmitting equipment, the flexibility of installations and the good signal propagation characteristics.

(2) Ripple control: At the next system level in the higher frequency range are ripple control systems. Ripple systems have good signal propagation characteristics but, because of the complex transmission and signal injection equipment that is required, need large-scale implementation to be cost-viable.

(3) Power line carrier (PLC): Moving into the very high frequency operating range, above 100 kHz, PLC systems are available which permit the use of higher data transmission rates. These systems take advantage of the reduced effect of electrical noise with increasing carrier frequency. However, these high carrier frequencies also suffer from a much higher level of signal attenuation on the power system and, therefore carrier frequencies in the lower range of 5-20 kHz are used in an attempt to optimise the signal characteristics against the power system parameters.

*Non-power line signalling systems*

A major advantage of all non-power line systems is that physical path used for communications is optimised for purpose of telecommunications and not for the purpose of supplying power. Generally, this means a quiet, reliable communications channel with high information capacities and low power requirements. The independence of the communications channel is also advantageous, because the presence of a power system fault will not degrade the communication channel. There is, however, the disadvantage of lack of exclusive control over a shared system. Non-power line systems can be classified as follows.

(1)  Radio: Radio communication has been used in a variety of roles for load control. Mobile radio telephone systems and commercial radio broadcast systems have been adapted for these load control roles, that is, with the load, management communications 'piggy-backing' on the existing telecommunications channel. Radio system coverage is generally considered to be more easily achieved than any other non-power line system.

(2)  Telephone: The early problems of telephone-based systems used for energy management have largely been overcome. Line and data concentrating equipment have been designed which work in parallel with the telephone exchange equipment — the 'idle-line technique', rather than having to access the exchange equipment using the conventional dial-up technique.

*Hybrid communications*

Another strategy is to use hybrid communications network, that is, networks with more than one mode of communication in the overall scheme. Instead of attempting to optimise one common mode of communication to support energy management totally, it is possible to use one type of communication channel to its optimum limits and then interface it with another channel to complete the communications network. Offsetting the advantages to be obtained by this approach are the obvious additional costs associated with the communications interfaces. Nevertheless, the design and application flexibility offered by the hybrid approach can be advantageous.

In many circumstances the shared use of non-PLC and commercial radio broadcast systems may be advantageous, since the capital costs to the power utility can be reduced or eliminated and the operating costs can be shared amongst other users. Counteracting this advantage is the uncertainty of the long-term operating costs of shared communications services, such as telephone-based systems.

*Spot pricing*

One of the most interesting possibilities which could arise from interactive load control is the concept of 'spot pricing'. This concept has a growing number of advocates, particularly centred at the Massachusetts Institute of Technology. After much theoretical work, they claim that it is now feasible to quote either the short-run or the long-run marginal cost per unit of electricity for short intervals of time ahead, and leave the customers to make a choice of the amount of energy they wish to consume during that period. Several claims are made for this approach:

(1) that the load curve for minimum total delivered cost of energy naturally evolves out of the consumer's response to correct price signals, and therefore strategies aimed specifically at demand control through charging for power (kW) are irrelevant;

(2) that by using 'spot' pricing, the economical integration of complementary energy, generated from combined heat and power (CHP) schemes and intermittent sources such as barrage, wind and solar generation, is enhanced. This serves both the consumer financially and the cause of energy conservation.

A number of individual industries have been studied in the United States with a view to the application of this technique. The studies were aimed mainly at determining what percentage of industries would utilise 'spot' prices and the time interval over which 'spot' prices should be quoted. The fact that all factories, even those producing identical products, have differing energy-saving potential, came through clearly, as did the variability of the response dynamics, so the chosen time interval over which the 'spot' prices are quoted has been somewhat arbitrarily chosen in the order of a few minutes.

The concept has many attractive features and the technology to carry it out is available, but a considerable amount of detailed thought, followed by field trials with selected consumers, would have to be carried out to determine its general social acceptance and limitations.

Whilst the application of 'spot' pricing to industrial consumers appears to be most attractive, it does depend on the industrialists having a real-time control on their energy use, which very few have at the moment.

Initially, it is thought, few consumers would avail themselves of this facility, but within a short time the dynamic microprocessor/home computer industry, will likely see this development as a ready market for cheap programmable control equipment. The traditional interface problem of sensor/actuator is minimal and signalling can be conducted over the consumer's wiring. So once the concept is launched, it could provide a very rapidly expanding market.

*Credit and load management system (CALMS)*
This is a third method being considered in the United Kingdom. The unit (CALMU) is installed on the consumer's premises and applies electronic techniques to control three outgoing circuits. The first circuit gives a continuous supply for items such as lights, television, etc., and commands the highest tariff; the second is an interruptible circuit under the control of the utility and supplies thermal lag loads such as water heating, where interruption of supply does not inconvenience or alter the life-style of the consumer and attracts a lower tariff. Thirdly, there is a time-of-day supply with the lowest tariff of all, for supplies such as under-floor and block-storage heating, giving the consumer the benefit of cheap night units.

All these tariffs are optional. It is considered that, by informing the customer of the benefits, he will utilise this extended range of tariffs in his own short-term interest of pursuing lower energy accounts. With the continuous display of units and accounts, consumers can experiment with alternative energy strategies which, whilst enabling them to enjoy the life-style they desire, achieve a minimum cost and, further, the utilities can help by making known successful strategies pursued by others. Thus energy and resource conservation can be achieved voluntarily without legislative help or restraint.

From the utility's point of view, a regime of demand being tailored to generation makes it possible to obtain significant economies in the way in which generating plant is utilised to meet the resulting flattened electrical demand curve.

This mode not only reduces the fuel consumed due to higher operating efficiencies, but also lowers operational costs; there is also the saving resulting from the elimination of spinning reserve; additional long-term advantages result from the reduction in total plant capacity requirements. All of these economies and many others are returned to the consumer in the form of lower tariffs.

The concept in more detail comprises a credit and load management unit (CALMU) which will replace the conventional electrical meters and time-switches on the customers' premises and a two-way communication system with the utilities' central computers. The system can also embrace gas and water utilities' consumer metering and accounting if desired.

The CALMU to be installed in each customer's premises consists of two units:

(1) a mains unit, which handles the customer's electricity supply and contains current and voltage transformers and solid-state switches controlling three outgoing circuits, all operating at mains voltage;

(2) a touch panel containing the display, input keyboard and all the solid-state electronics including microprocessor, which communicates by cable with the mains unit and by a range of alternative methods to central computers. This is illustrated in Fig. 18.17.

The CALMU measures and records electronically within statutory tolerances the following:

(1) electricity consumption (kWh);
(2) demand in kilowatts/kVA.

The microprocessor utilises these values, together with other data, some of which may be centrally communicated, to provide a range of facilities. The credit and accountancy facilities include:

(1) remote meter reading of a multi-rate meter;
(2) computation and display of the running account;
(3) simultaneous instant tariff changes;

Fig. 18.17 – Touch panel of CALMU.

(4) a flexible accounting structure, which allows consumption and demand to be recorded separately for certain circuits on customer's premises or for particular periods during the day.

It can, if required, receive pulses from the gas and water meters and offers similar facilities to those outlined above for the electricity supply. Because CALMS operates with two-way communications and has its own microprocessor it is able to perform a wide range of functions. These include:

(1) flexible time-of-day switching;
(2) utility-interruptible circuit;
(3) pre-set current limits for fixed maximum demand use;
(4) broadcast current limits for emergency use;
(5) selective immunity from emergency load shedding;
(6) earth leakage protection.

A smaller range of ancillary safety and operating functions can be performed on the gas and water supplies.

The benefits CALMS will provide for the customer are two-fold: those which are perceived directly; and those which are indirectly received by improved operation of the utilities. On electricity supply, the customer benefits are:

(1) information about the account, metered units per circuit, maximum demand, switching times for both load and tariff accounting;
(2) frequent and smaller bills more geared to the customer's income cycle;
(3) improved choice of tariffs;
(4) no appointments for meter readings;
(5) protection facilities, overload, earth leakage and low frequency;
(6) automatic load control by the customer of individual appliances which may be programmed so that load remains below a maximum level;
(7) improved voltage control;
(8) more rapid supply restoration following faults;
(9) minimal disturbance during load shedding, with continuous supplies at a controlled low level instead of an area disconnection;
(10) remote payment facility from the home.

## OPERATIONAL USE OF ENERGY MANAGEMENT

The operational use of the various facilities available for load shape improvement with the aim of achieving a practical optimum daily thermal steam plant generation pattern would lead to a constant night generation level and a higher constant daytime 'Plateau' with the nuclear and 'in-merit' fossil fuel fired plant, in both cases, operating at full load. The existing and postulated shapes are illustrated in the next three diagrams for different seasons of the year and days of the week in 1986/7. The annual daily load factor had been estimated to improve from the existing 82% to 91%. Figures 18.18, 18.19, and 18.20 show (see key on each figure):

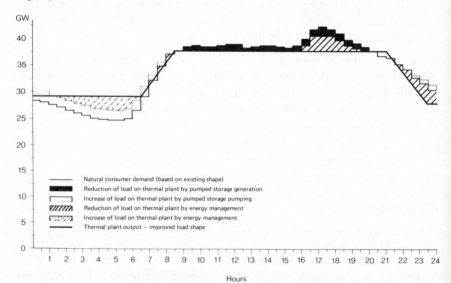

Fig. 18.18 – 1985–86 winter average, weekday.

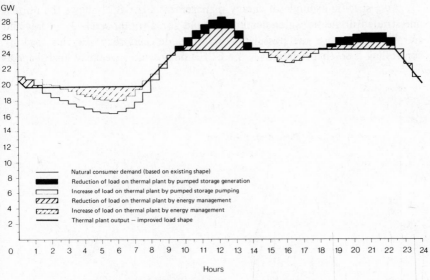

Fig. 18.19 — 1985–86 spring/autumn average, Sunday.

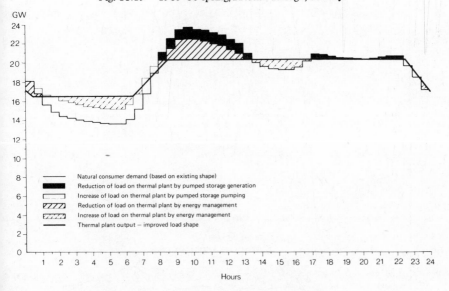

Fig. 18.20 — 1985–86 summer average, Saturday.

(1)  the estimated natural consumer demand for 1986/7;
(2)  the beneficial effect of pumped storage plant for pumping and for generating;
(3)  the scope for infilling of troughs and lopping of peaks by other means, mainly energy management; and
(4)  the practical optimum thermal steam plant generation pattern.

As a specific example of energy management Fig. 18.21 shows the normal unrestricted domestic water heating demand for a winter weekday in England and Wales and also the postulated modified demand shape of this load by controlling it so that it improves the CEGB daily winter weekday load shape as

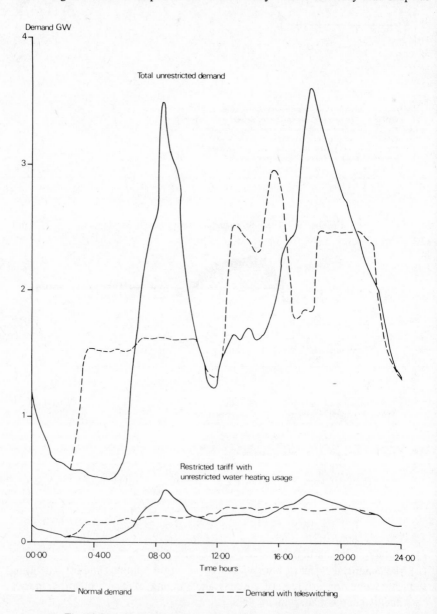

Fig. 18.21 – Unrestricted water heating demand, winter weekday.

illustrated in Fig. 18.22. This is an example of how consumer energy could be transferred from the day 'plateau' to the lowest part of the overnight trough and how peak energy can be transferred to the 'plateau'.

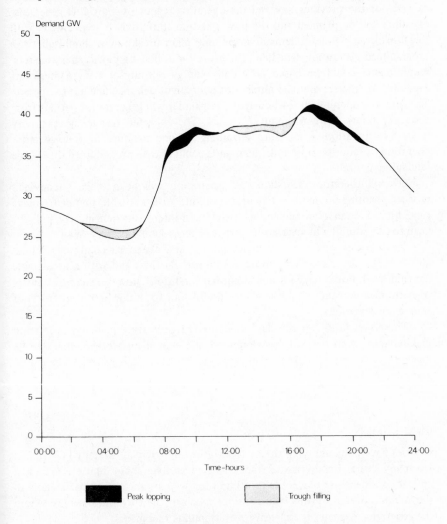

Fig. 18.22 – System demand, winter weekday.

The unrestricted domestic water heating demand is created by some 10 million consumers. Similar results could possibly be obtained with greater facility by astute management of energy supplied to industrial and large commercial consumers. Any thermal inertia demand created by space and water heating and large freezer storage requirements could be utilised to give improved CEGB system demand shapes by controlling a much smaller number of consumers than the 10 million domestic consumers.

With a predictable daily demand shape the scheduling of generation can be carried out the previous day and the operating regime of individual generating machines can be planned and the power station staff advised accordingly. With the postulated two-level demand shape only plant breakdowns should alter the programmed generation schedule. Therefore the following day's generating sets' requirements could be based on a daily energy output for a given megawatt capability of the set and the number of hours that each machine has to operate. The sets will operate for 24 hours, the remainder will generate for periods from about 17 hours down to about 12 hours. The expected frequency sensitivity mode of operation — pressure governing, sliding pressure or free governor operation — required from individual units could also be programmed during the scheduling period.

Overall, therefore, scheduling of generation will be a much streamlined process resulting in more efficient operation. With available pumped storage capability, French interconnection trading and interactive demand control, the need for despatching of generation output changes should be very small.

The control of frequency under normal and abnormal conditions should also be aided by having the above mentioned facilities and with a predictable demand level. For emergency and abnormal conditions, however, remote control of consumer demand would be a very useful tool for immediate restoration of frequency deviations.

Therefore, both 'on the day' and forward generating plant operation and maintenance strategies will be simplified and should enable economies to be made by working to tighter plant margins, by optimisation of overhaul programmes, etc.

## CONCLUSION

Utilities have been applying themselves to the improvement of daily load shapes for many years. Energy management to achieve load shape improvement is a crucial consideration to electricity utilities in the years ahead, and particularly in the period covered by this volume, since much of the flexibility required from the generating capacity is expensive with spiralling fuel prices.

Therefore, it becomes essential to re-think the relationship between utility and consumer to determine the best overall mutual interest and so adopt the least cost method of operation in the future.

The traditions of the electricity utilities worldwide are steeped in the urgent priorities of meeting consumer demand in full at virtually any cost and then requiring the consumer to pay. This is becoming unsatisfactory and will be even more so with the escalating cost of fossil fuel during the period ahead. This will create pressure to persuade the consumers to regulate their activities towards minimising overall costs and therefore significantly reduce fossil fuel consumption by enabling utilities to regulate the use of electrical energy.

In order to achieve the principal objective there will be a need to change attitudes, method of operation and overall management to a completely new approach. The changes will be gradual over a prolonged period.

# 19

# Renewable sources of carbon chemicals through biotechnology

S. J. Pirt, Microbiology Department, Queen Elizabeth College, London.

## INTRODUCTION

Biotechnology (industrial application of microbes, cells or enzymes) is a means for recycling reduced organic matter in a form suitable for chemical feedstocks or an energy source. At the moment biotechnology is not regarded as a means for the primary production of reduced organic matter but primary production of organic matter from carbon dioxide through biotechnology could be rapidly developed.

A classification of feasible biotechnology routes to heavy organic chemicals is given in Table 19.1. The first route is the well-known anaerobic degradation of

**Table 19.1**

Biotechnology routes to carbon chemical feedstocks and energy sources

| Substrate | Major products |
|---|---|
| *Route 1: Anoxic degradation* | |
| Carbohydrate | Ethanol, acetone, butanol, 2:3 butanediol |
| All C, H, O compounds | Methane |
| *Route 2: Anoxic synthesis* | |
| CO | Acetate |
| $CH_3OH$ | Butyrate |
| $CO_2 + H_2$ | Acetate |
| *Route 3: Aerobic synthesis* | |
| $CH_4$ or $CH_3OH$ | Biomass, poly $\beta$-hydroxybutyrate |
| *Route 4: Microbial photosynthesis* | |
| $CO_2$ | Biomass, starch, fat, glycerol, hydrocarbons |

organic matter to ethanol or methane. The second route is the unfamiliar one of anoxic (anaerobic) synthesis from $C_1$ compounds which can be performed by bacteria referred to later. The third route of aerobic synthesis is under development as a novel means of exploiting methane or its derivative, methanol. The fourth route offers the possibility of primary production of organic matter from $CO_2$.

The exploitation of fermentation as a route to heavy organic chemicals is critically dependent on the nature of the substrate or feedstock for the fermentation. Waste organic matter in the form of agricultural residues or urban and industrial waste is the main substrate considered. But the amounts of such residues or wastes are quite insignificant when compared with the consumption of the fossil fuels [19.1]. Another problem is that the organic wastes are diffuse and consequently costly to collect. Also the most abundant waste is lignocellulose which is intractable to enzymic attack without expensive pretreatment. The production of 'energy crops' by conventional agriculture to provide substrate mainly as carbohydrate has been adopted, notably in Brazil, but analysis of the potential of conventional agriculture [19.2] shows that it is in a very similar position to the fossil fuel industry, that is, it is stretched to capacity and now requires supplementation in order to meet the increasing world demand for food and fibre. These substrate problems add point to route 4 in Table 19.1, that is, the biotechnology for primary production of organic matter from $CO_2$ and solar energy. The problems and potential of these various biotechnology routes to heavy organic chemicals are illustrated further by reference to specific examples.

## ETHANOL

Most of the commercial technology for ethanol production is based on primitive batch fermentation. Several lines of development are being followed. Ethanol is an inhibitory product. One aim of the development is to select organisms with an increased resistance to ethanol inhibition, which would permit increase in the fermentation rate with a given ethanol concentration. The bacterial species *Zymomonas mobilis* has been proposed as a more ethanol-tolerant organism [19.3]. However, the substitution of the bacteria for a yeast appears to necessitate a reduction of some 5°C in the fermentation temperature which may offset the advantage.

Since sugar is not an abundant substrate, attention has been turned to the possibility of using cellulose as the substrate for ethanol production, obviously this would increase the substrate availability by orders of magnitude. However, most cellulose is in the form of lignocellulose which must be subject to expensive pretreatment to delignify and pulverize it to make enzymic attack on the cellulose proceed at an adequate rate. The cellulose first has to be saccharified either by the mould, *Tricoderma reesii* [19.4] or the bacterial species *Clostridium thermocellum* [19.5]. However, the rate and degree of degradation of the cellulose are relatively low and the enzymic inactivation rate is high.

The introduction of continuous fermentation technique into ethanol production seems to be slow. Continuous fermentation under vacuum [19.6], which distils off the ethanol as it is produced, offers the advantages of removal of ethanol inhibition and concentration of the biomass with proportional increase in the fermentation rate. Also vacuum fermentation makes use of the heat of fermentation to distil off the alcohol. An ultimate objective is to obtain thermophilic† species of organism so as to elevate the temperature of the fermentation and facilitate vacuum fermentation.

## METHANE GENERATION

In principle, all C, H, O compounds can be completely degraded in anoxic fermentations to $CH_4 + CO_2$. This requires the combined action of bacteria in consort. This route should become increasingly important as a means of recycling any waste carbon compounds in the reduced form and conserving most of the free energy of the substrates [19.7]. However, this degradation process is little understood and the microbial agents are obscure. The mechanism appears to be conversion of the carbon substrate into acetate, $CO_2$ and $H_2$, which can be converted to methane. Fatty acids are formed as intermediates and the process can be stopped at the fatty acid stage by lowering the pH and removing the hydrogen, but this seems to have attracted little attention. The main problems in the methane fermentation are the slowness of the process and its instability. The solution of these problems requires more fundamental research into the mechanism and kinetics of the process.

## ANOXIC FERMENTATIVE SYNTHESIS

Some newly discovered bacteria obtain their energy by reduction of $C_1$ compounds with formation of acetate or butyrate. These biosynthetic routes are:

$$4\,CO + 2\,H_2O \quad \rightarrow \; CH_3COOH + 2\,CO_2 \tag{19.1}$$

$$10\,CH_3OH + 2\,CO_2 \rightarrow 3\,CH_3CH_2CH_2COOH + 8\,H_2O \tag{19.2}$$

$$2\,CO_2 + 4\,H_2 \quad \rightarrow \; CH_3COOH + 2\,H_2O \tag{19.3}$$

These reactions can proceed stoichiometrically under anaerobic conditions with *Butyribacterium methylotrophicum* [19.8] as the agent of reaction (19.1) or (19.2) and *Acetogenium kivui* [19.9] as the agent of reaction (19.3). The lack of side reactions and the mild conditions are advantages reactions (19.1) and (19.3) possess over chemical carboxylation of methanol. Reaction (19.3) occurs at 65°C which is advantageous in cooling the fermentation. The economics of these fermentative processes are still obscure through lack of kinetic data. However, the existence of these microbial processes suggests that fermentations can be

† Adapted to high temperatures.

based on abundant cheap substrate derived from coal. A possible advantage of these synthetic fermentations over the chemical syntheses is that the microbial processes could more effectively utilise low-grade synthesis gas, produced, perhaps by underground coal gasification.

## QUANTITATIVE PERFORMANCE OF A FERMENTATION PROCESS

If fermentation routes are to become of major importance as sources of chemical feedstocks the kinetics, material and energy balances of fermentations have to be given a reliable quantitative basis. Much progress to this end has been made in the last thirty years, especially through the development of continuous fermentation processes [19.10]. The basic equation for the rate of conversion of an energy source for a fermentation is

$$qx = [(\mu/Y_G) + m]x \tag{19.4}$$

where $q$ = specific rate of utilisation of the substrate (g substrate (g biomass)$^{-1}$); $\mu$ = specific growth rate (h$^{-1}$); $Y_G$ = maximum growth yield (g biomass (g substrate)$^{-1}$); $m$ = maintenance energy coefficient (g substrate (g biomass)$^{-1}$ h$^{-1}$); $x$ = biomass concentration (g L$^{-1}$). The value of $q$ is a function of the substrate concentration ($s$) and the product concentration ($p$). The value of $q$ depends on ($p$) because of product inhibition which often occurs in anoxic fermentations. Also the product inhibition limits the maximum product concentration attainable For an anoxic fermentation the usual maximum value of ($\mu + mY_G$) is of the order 0.4 h$^{-1}$, and of $Y_G$, 0.2. Hence the maximum substrate utilisation rate in an anoxic fermentation with a soluble carbon substrate is of the order $2x$ $gL$ $h^{-1}$. Since biomass concentrations up to about 40 g dry weight L$^{-1}$ may be achieved, provided there is concentration of the biomass by recycle or feedback, maximum substrate conversion rates of the order 80 g L$^{-1}$ h$^{-1}$ may be expected. However, overcoming the product inhibition is a basic problem. If the product is volatile it can be evaporated from the culture. This has been achieved in the ethanol fermentation by 'vacuum fermentation'.

The maximum product concentration which can be reached in anoxic fermentations is about 8–15% for neutral products such as ethanol and 2–8% for organic acids such as acetic acid. The efficient and cost-effective separation of these products from dilute solutions presents a challenge to chemical engineers. The maximum yield of product in anoxic fermentations is about 85% of the theoretical. The rest of the substrate is used for biomass synthesis.

## MICROBIAL PHOTOSYNTHESIS

Thorough analysis [19.1] [19.2] shows that conventional agriculture and forestry have virtually no spare capacity after meeting the demand for food and fibre.

For example. Pimentel *et al* [19.1] show that if in the USA the cultivation of crops were optimised and the whole of the agricultural and forestry residues were turned into fuels (ethanol or methane), the amount would be equivalent to only 1% of the country's gasoline consumption. The limit on the agricultural output is not solar irradiation, but the other constraints on the system. These constraints are: the need for arable land, lack of control over climate, poor nutrient and water conservation, $CO_2$ limitation, discontinuity of crop growth, environmental damage (to water and soil), and a low energy ratio, that is, energy stored/energy consumed. Conventional agriculture is in a similar position to the fossil fuel supplies, the productivity is lagging behind demand. It follows that if photosynthetic biomass production is to become a major alternative source of reduced carbon compounds some novel photosynthetic technology must be developed to supplement conventional agriculture. Such novel photosynthetic technology to overcome the constraints on conventional agriculture is offered by biotechnology [19.11]. This novel technology must be based on purpose-built photobioreactors which can be installed on any open space over land or sea and permit microbial (algal or bacterial) photosynthesis to be optimised at all times. Thus the output of biomass $ha^{-1}$ can be increased 20-fold or more over the best obtainable by conventional agriculture and there is virtually no limit to the area available for photosynthesis [19.11].

Attempts at micro-algae cultivation in the past have been based on open ponds and channels [19.12]. These suffer from the disadvantages that they are confined to flat land, and prevent control over the gas phase, and evaporation; also mixing is poorly controlled. These limitations are overcome by use of a tubular loop photobioreactor fabricated of glass or plastic [19.11] [19.13]. The current performance characteristics of such a tubular photobioreactor are summarised in Table 19.2. An output of 2 g biomass dry matter $m^{-2}$ $h^{-1}$ has been achieved. Translated to outdoors this means an output of about 70 t dry matter $ha^{-1}$ $y^{-1}$. A solar irradiation of 90 $Wm^{-2}$ would be sufficient to achieve such a productivity, hence it would be achievable in Britain and exceed the output

### Table 19.2

A summary of the performance characteristics of a tubular loop photobioreactor [19.13]

---

Utilises solar irradiation up to 90 $Wm^{-2}$
Maximum energy storate in biomass, 16 $Wm^{-2}$
Efficiency corresponds to about 70 t biomass dry matter $ha^{-1}$ $y^{-1}$
Energy ratio (energy stored/energy dissipated in process) = 2.5
Estimated cost of biomass production in slurry form, $700 $t^{-1}$
(over land about 75% of the cost would be capital and 25% recurrent cost).

---

of good conventional agriculture by a factor exceeding 10. Development of the process has the potential to increase the output of the process by more than two-fold. The photobioreactor can be used to produce a variety of products including protein, starch, fat, glycerol and hydrocarbons. Also the photobioreactor could be used for nitrogen fixation, and the oxygen released by photosynthesis may be recovered in concentrated form mixed only with a little $CO_2$. The photobioreactor is a means of recycling $CO_2$, which could be scaled up without limit. This could become an essential function of the reactor if it becomes necessary to limit the increase in the $CO_2$ content of the atmosphere in order to prevent the $CO_2$ 'greenhouse effect'. As an example of the capability of a photobioreactor with the performance summarised in Table 19.2 consider its coupling with the output of $CO_2$ from an ammonia synthesis plant which produces 1000 t $d^{-1}$ of ammonia. This would produce also about 1000 t $d^{-1}$ $CO_2$, which would suffice to produce about 200 kt $y^{-1}$ biomass dry matter and 300 kt $O_2$ $y^{-1}$ from 3000 ha photobioreactors.

## CONCLUSION

Critical analysis shows that conventional agriculture and forestry can make nothing more than a marginal contribution to chemical feedstocks and energy sources. Agricultural technology is essentially neolithic in concept and development of it is showing reducing returns. The inherent limitations on photosynthesis imposed by conventional agriculture can be overcome only by novel technology based on purpose-built photobioreactors for cultivation of micro-algae. Such photosynthetic technology based on solar energy and $CO_2$ recycling offers the possibility of unlimited production of organic matter.

## REFERENCES

[19.1]  Pimentel, D., Moran, M. A., Fast, S., Weber, G., Bukantis, R., Balliett, L., Boveng, P., Cleveland, C., Hindman, S. and Young, M. (1982) *Science* **212** 1110-15.

[19.2]  Pimentel, D., Dritschilo, W., Krummel, J. and Kutzman, J. (1975) *Science* **190** 754-61.

[19.3]  Rogers, P. L., Lee, K. J., Skotnicki, M. L. and Tribe, D. E. (1980) *Advances in Biotechnology 2*, Moo-Young, M. and Robinson, C. W. (eds) Pergamon, 189-94.

[19.4]  Vikari, L., Nybergh, P. and Linko, M. (1980) *Advances in Biotechnology 2*, Moo-Young, M. and Robinson, W. W. (eds) Pergamon, 137-42.

[19.5]  Angerinos, G. C., Fang, H. Y., Biocic, I. and Wang, D. I. C. (1980) *Advances in Biotechnology 2*, Moo-Young, M. and Robinson, C. W. (eds) Pergamon, 119-30.

[19.6]  Cysewski, G. R. and Wilke, C. R. (1977) *Biotech. Bioeng.* **19** 1125-43.

[19.7] Pirt, S. J. (1978) *J. Appl. Chem. Biotechnol.* **28** 232–6.
[19.8] Zeikus, J. G. (1981) *Trends in the biology of fermentations for fuels and chemicals* Plenum Press.
[19.9] Leigh, J. A., Mayer, F. and Wolfe, R. S. (1981) *Arch. Microbiol.* **129** 275–80.
[19.10] Pirt, S. J. (1975) *Principles of microbe and cell cultivation* Blackwell Scientific Publications.
[19.11] Pirt, S. J. (1982) *J. Chem. Tech. Biotechnol.* **32** 198–202.
[19.12] Kawaguchi, K. [1980] *Algae biomass,* Shelef, G. and Soeder, C. J. (eds) Elsevier, Amsterdam, 25–33.
[19.13] Pirt, S. J., Balyuzi, H. H., Bazin, M. J., Lee, Y.-K., Pirt, M. W. and Walach, M. R. *A photobioreactor to store solar energy by microbial photosynthesis,* Microbiology Department, Queen Elizabeth College.

# 20

# The prospects for renewable energy sources and improved efficiency of energy use

**Dr. Derek Pooley,** Director of Non-nuclear Energy Research, UKAEA

## INTRODUCTION

The bulk of this book is concerned with energy supplies from the four great commercial fuels, oil, gas, coal and uranium. These provide the lion's share of world energy at present and, despite occasional difficulties, will continue to do so for several decades at least. However, renewable energy sources, usually in the shape of hydroelectricity (highly processed solar energy) and to a lesser extent geothermal steam (heat from large scale natural radioactivity), are already important in some countries. Hydroelectric power in particular has been exploited fairly extensively; it provided 40% of the electricity generated in the world in 1925 and, even in 1978, 25% of worldwide generating capacity was hydroelectric. Moreover, few energy analysts doubt that the overall role played by renewables in world energy supply will grow in the years ahead. A more difficult judgement to make is which particular renewable technologies will be successful, where their contributions will be made and how large those contributions will be.

Figure 20.1 shows the origins of renewable energy and how the energy provided might be delivered to the final users. Not all sources are renewable in the same sense nor on the same timescale; for example direct solar radiation, wind, waves and tides are renewed daily, rainfall renews hydroelectricity essentially annually, biomass is renewed over at least several seasons; and geothermal heat over tens or hundreds of years. The important common feature is that in every case the physical resource is large or very large and can, in principle, be used for very long periods of time without depletion.

Because of this, renewable energy sources have been seen by many in recent years as the ideal replacements for oil and gas as these become scarcer. As energy income and apparently benign they have been particularly contrasted with coal and nuclear power which have often been castigated as capital intensive, polluting and dangerous. Partly as a result of their prominent place in the energy debate, substantial funds have been used in the assessment and development of renewables and we now know much more about them than we did a decade ago.

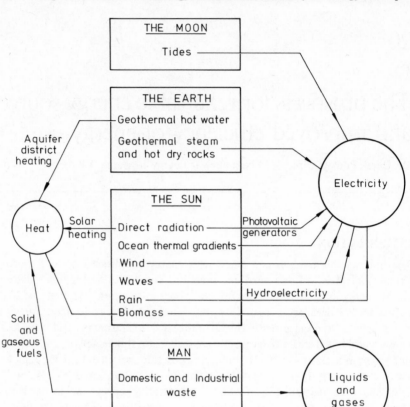

Fig. 20.1 – The origins of renewable energy sources and the form in which the energy might be delivered.

Figure 20.2 shows that £60 million has so far been allocated to renewable energy R & D by the British Department of Energy, primarily to enable that department to make a firm and sensible assessment of the role they might play in energy supply in the United Kingdom. Even within the United Kingdom more has been allocated because the Science and Engineering Research Council, the Central Electricity Generating Board and many industrial companies have also funded research on renewables, but Britain has spent proportionately less heavily than the United States and several European countries. The worldwide funding of renewable energy research has therefore been enormous. How do their prospects look now after so much work?

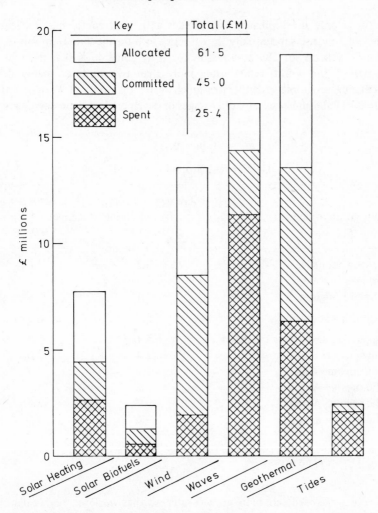

Fig. 20.2 – The size and distribution of renewable energy funding by the United Kingdom Department of Energy at February 1982.

## THE GENERAL PROSPECTS FOR RENEWABLES

Most renewables are not truly competitive against conventional fuels at present prices for the latter, but their prospects will improve with time if conventional energy prices rise in real terms, as most have during the 1970s. Their value is primarily the cost of the conventional fuels they displace and their cost effectiveness depends, more than anything else, on conventional fuel prices. Of course, oil prices are currently falling and real energy costs have fallen over most of this century, so only the foolhardy will predict with total confidence that energy prices will now rise substantially in real terms into the medium-term future, but they may well do so. The best reason for their doing so is that most of the countries of the world, representing more than two-thirds of mankind, are experiencing economic growth rates much faster than ours in Western Europe (Table 20.1) and most are at earlier stages of development where energy growth

**Table 20.1**

GDP growth around the world

| Area or group of countries | Average percentage GDP growth per year | |
| --- | --- | --- |
| | achieved during 1970–80 | expected during 1980–90 |
| Developed countries | 3.3 | 3.6 |
| Western Europe | 2.8 | 3.3 |
| United Kingdom | 2.1 | |
| Oil exporting countries | 5.2 | 6.5 |
| Oil importing developing countries | 5.1 | 5.4 |
| Low-income Africa | 2.4 | 3.0 |
| Low-income Asia | 3.2 | 4.4 |
| Mid-income Asia | 8.2 | 8.1 |
| Mid-income America | 6.0 | 5.6 |
| Mid-income Europe | 4.6 | 4.6 |

*Source:* World Bank *World Development Report 1981;* data for United Kingdom from *Annual Abstract of Statistics 1981*

rates actually exceed economic growth rates, so that world energy demand will continue to rise if these economies continue to grow. Without any doubt economic growth is badly needed in these countries to provide all the benefits of good health, education, etc., which are made possible by a high level of economic activity, but the energy required to sustain the growth will have to be derived from more and more costly sources as the lowest-cost oil and gas fields are depleted. Price rises in real terms therefore seem inevitable to many energy analysts and it is prudent to explore the value/cost relationship for renewable energy sources in a variety of energy price futures. Indeed if renewable energy sources have much practical relevance to the chemical industry it will probably be indirectly, through the extent to which they might mitigate the real energy price rises which would otherwise occur as world demand grows and current oil and gas fields are depleted.

## LIQUIDS AND GASES

It is true that renewable energy sources could also help the chemical industry more directly; liquids and gases from biomass could provide a totally renewable feedstock for the industry. In Britain we believe that the thermal processing of wood to produce synthesis gas will not be economic for some time although it is probably the most likely way in which gases or liquids (methanol) can be derived on a large scale from biomass, in temperate climates like ours. It is, of course, possible to produce ethanol by fermentation and this could also be used as a chemical feedstock. However, at present in the United Kingdom the materials which could be used in this process — sugar beet and cereals, or even agricultural and municipal wastes — have higher values in other markets, either as foodstuffs or a starting point for other fuel supply routes. Figure 20.3 shows the different routes available for thermal processing and Fig. 20.4 highlights the probable economics of steam gasification, which is currently thought to be the most attractive route.

If oil prices rise over the next twenty years at the sort of rate they have risen during the 1970s the production of methanol from wood will be clearly economic, given what we now know about the likely cost of thermal processing equipment and assuming that a rate of return on capital of 5% more than the rate of inflation is required. If thermal processing becomes economic it by no means follows that wood energy crops are the best use for the land nor that thermal processing is the best route for the exploitation of those crops if grown, even if the technology proves reliable. The high price of food means that only marginal land can even be considered for fuel wood growth and even there the farmer might be better advised to rear sheep or grow timber for the construction and furniture industries. The contribution of land costs to the overall production cost estimate in Fig. 20.4 is $26 per dry ton of wood and is based on an estimate of the likely increase in timber prices and hence the opportunity costs of the land, over the life of the energy plantation. Part of the United Kingdom programme on biofuels is looking in considerable detail at this land utilisation aspect and the indications are that in Britain, where we have quite a lot of marginal land, the conversion of wood from 6% of the total land area could contribute 5.8 Mt† of methanol each year, that is 4.4 Mtce or about one-third of the United Kingdom non-energy use of petroleum.

Quite apart from the land use aspects, it seems to me extremely unlikely that wood waste and fuel wood will ever be converted into gases or liquids for chemical synthesis while natural liquid and gas fuels still supply a part of the heat market. It will be better to use the wood in the heat market directly and so reduce the demand for natural gases and liquids. I believe this is an important point because our need for heat in North Western Europe is very large (Fig. 20.5), and a lot of it is currently supplied as oil and gas. There are few direct

† $10^6$ tonnes.

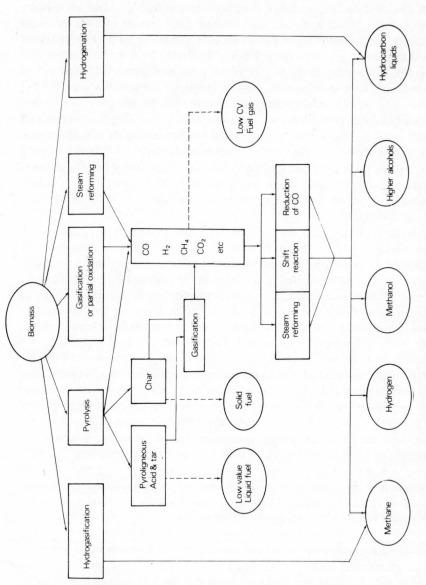

Fig. 20.3 – Possible routes for the thermal processing of wood.

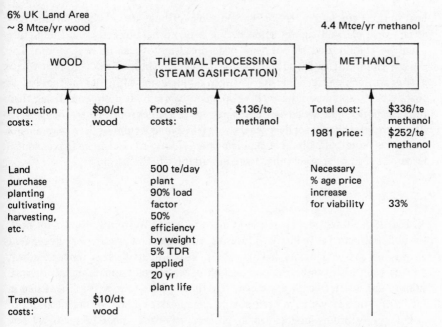

Fig. 20.4 – The economics of thermal processing of wood by steam gasification.

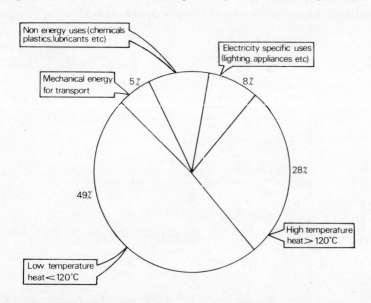

Assuming 20% average efficiency of converting liquid fuel to mechanical energy

Fig. 20.5 – The ultimate use for energy in the United Kingdom. (*Source:* Energy paper No. 39.)

heat options among the renewables and some of those are not economically very attractive. Coal will supply some of the required heat directly and the picture could be complicated by the very widespread use of electric heat pumps. The general picture of fuel wood and wood waste being burned for heat, near where it is grown or produced and in relatively simple combustors, leaving natural or coal-derived gases and liquids to be supplied centrally, is more convincing than the transport of the wood to a complex central gasifier followed by the national supply of gas or liquid for heat generation. Whether any biomass-derived methanol would be more valuable as a supplementary transport fuel than as a chemical feedstock is another aspect that I will not attempt to discuss now.

## HEAT

As Fig. 20.5 shows our major energy need in North Western Europe is for heat, for buildings and for industrial processes, and right now we are very dependent on oil and gas to generate it (Fig. 20.6). For Britain the best renewable heat sources seem likely to be the combustion of waste and biomass and geothermal aquifers for district heating — effectively three of the four possibilities listed in Fig. 20.1. Already we have been able to encourage a number of organisations to install waste-burning boilers which promise pay-back times of relatively few years. For example an independent school at West Dean, Sussex has installed two 250 kW boilers burning wood chips supplied from their own extensive

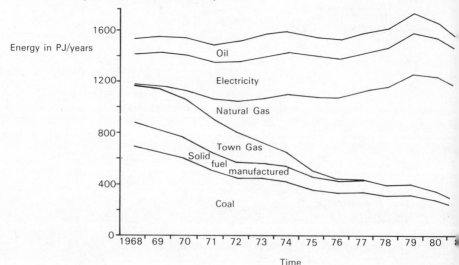

Note: 100 Peta joules = 4 M tonnes coal equivalent.

Fig. 20.6 – Fuel supplied to the domestic sector in the United Kingdom.

forestry holdings of over 1000 acres. Figure 20.7 shows hydraulic shafts pulling wood onto a horixontal screw auger, which matches with a vertical auger feeding the furnaces. This is an example of the use of green wood chips as solid fuel. Projects like these are included as demonstration projects in the Energy Conservation Demonstration Projects Scheme (ECDPS) which is managed for the United Kingdom Department of Energy by their Energy Technology Support Unit (ETSU) at Harwell and they have the aim of showing other organisations how effective and profitable energy conservation projects can be, by careful monitoring of the savings which result, and promotion of the information. Ultimately 10-15 Mtce of heat could be provided in the United Kingdom by burning biomass, a small but useful fraction of the current United Kingdom heat requirement, which is equivalent to a primary energy of roughly 200 Mtce per year. This estimate ignores air pollution constraints.

Fig. 20.7 – A demonstration wood-chip boiler at West Dean, Sussex in the United Kingdom.

The conventional geothermal aquifer heating scheme is shown in Fig. 20.8, consisting of two holes for withdrawal and return of the highly saline water from the aquifer. However, if the aquifer is near the sea the water can be discharged there, saving the cost of a second hole. Geothermal aquifer heating schemes involving only a single hole (singlets) look very attractive if the prices of competitive fuels, in this case coal, are assumed to rise during the 1980s and 1990s. Having to drill a second hole to reinject the extracted water makes the economic advantage of geothermal heating much less certain, perhaps insufficiently certain

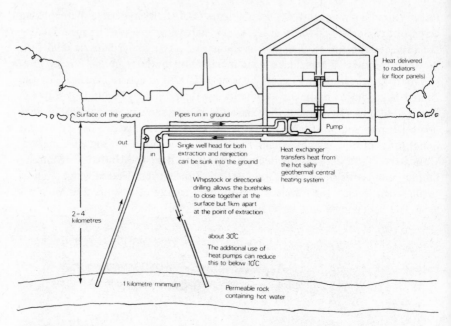

Heat delivered
to radiators
(or floor panels)

Surface of the ground       Pipes run in ground

Pump

out

in        Single well head for both
          extraction and reinjection
          can be sunk into the ground

          Whipstock or directional
          drilling allows the boreholes
          to close together at the
          surface but 1km apart
          at the point of extraction

2–4
kilometres

Heat exchanger
transfers heat from
the hot salty
geothermal central
heating system

about 30°C

The additional use of
heat pumps can reduce
this to below 10°C

1 kilometre minimum

Permeable rock
containing hot water

Fig. 20.8 – The conventional scheme for obtaining heat from geothermal aquifers.

to overcome the disadvantage of capital intensity which is inevitable in a geo-thermal scheme. The use of geothermal hot water is therefore rather constrained. An aquifer at the right depth and temperature, in rock of adequate permeability, is the first requirement and a heat load of the right size and type on the surface is the second. Thirdly the site should preferably be near the sea if the costs of a reinjection hole are to be avoided. For these reasons we currently see the potential in the United Kingdom for this kind of heating as about 1–3 Mtce only.

Solar heat now looks a less attractive way of supplying heat than was thought in the mid-seventies. For Britain this is perhaps not very surprising given our cloudy northern climate, which produces a large seasonal mismatch between the heat demand (predominantly in the winter) and the quite respectable amount of solar energy we receive during the summer. For us, and presumably for other countries in Northern Europe, the most cost-effective way to use solar energy is through buildings designed specifically to capture sunlight without special collecting and distribution systems – passive solar heating. We have a number of experimental buildings; for example, those at Milton Keynes which have the living rooms fitted with large windows facing south (Fig. 20.9) and the utility rooms with small windows facing north (Fig. 20.10). This kind of design is not significantly more expensive than a conventional house and can reduce the heat load in a well-insulated house by 20–30%. Passive solar design can usually only be used in new houses and then only if the houses can be oriented nearly south

Fig. 20.9 — A passive solar house at Milton Keynes, the south facing wall.

Fig. 20.10 — A passive solar house at Milton Keynes, the north facing wall.

and are free from overshading. We do not yet know how acceptable these designs will be to the house buying public, nor how difficult it will be to fit them into an average modern housing estate, nor whether there will be socio-technical problems such as the widespread use by the house owners, for reasons of privacy, of net curtains which reduce the solar gain. Active solar heating, for space and water or for water alone, does not look economically attractive in the United Kingdom, even if fuel prices double in real terms, unless the total installed cost of the collector, pipework, etc. falls by a factor of two in real terms. Achieving this will be a formidable task and may not be possible.

Although heat-producing renewables are disappointing, there are, of course, many excellent ways to mitigate the problems of future heat supply. Below are mentioned some of the energy conservation methods being encouraged in the ECDPS. The general approaches are to reduce heat demand by:

(1) reducing both the amount of heat which needs to be supplied, via thermal insulation in particular, and also the amount which is unnecessarily supplied when it is not needed, by improved heat control systems;
(2) recovering and re-using waste heat whenever process losses are inevitable in an air or fluid stream which has to be exhausted. Several of the ECDPS projects involve simple heat recovery, such as the one in a malt kiln shown in Fig. 20.11;
(3) using heat pumps to reclaim heat which has been degraded to lower temperature than the process demands, but is still substantially above ambient. Heat pumps are particularly effective in drying processes where they can often recover the latent heat of evaporation of the water. A good example is again in a malt kiln (Fig. 20.12) where this time the gas burner has been replaced by a gas-engine driven heat pump.

Evidence that large reductions in the use of fuel for heat are now occurring is provided by the data in Table 20.2. These approaches will likely have a much bigger impact on heat supply than renewable heat technologies: indeed the ECDPS claims that the projects already in hand, if replicated to a reasonable extent, will save in the order of 5–7 Mtce per year.

## ELECTRICITY

Hydroelectric power makes an important contribution to electricity supply around the world; not surprisingly more can be exploited if electricity prices rise. Within OECD countries it is thought that only about half of the total hydroelectric resource of 520 GW is exploited and in Western Europe the hydroelectric capacity in use in 1970 was about 47 GW, roughly one-third of the estimated total resource of 160 GW. Unfortunately, the United Kingdom does not have many good hydroelectric sites and has an installed capacity of only 1.3 GW. In spite of the dearth of hydroelectricity, Britain is very generously

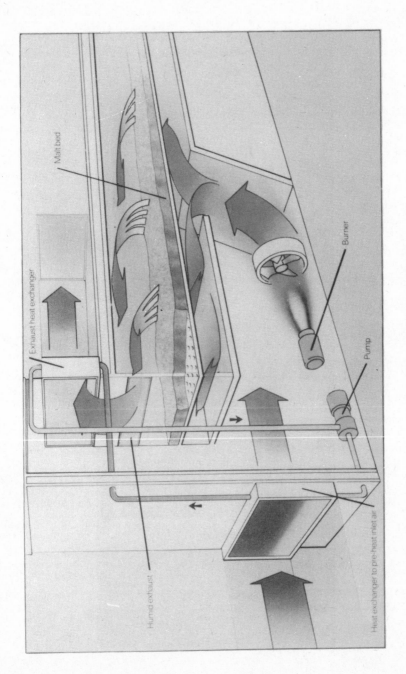

Fig. 20.11 – Heat recovery in a malt kiln.

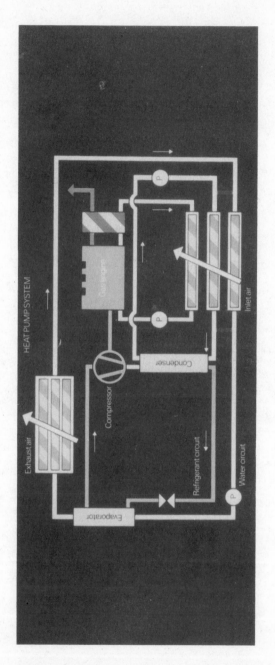

Fig. 20.12 – The use of a gas-engine drive heat pump in a malt kiln.

## Table 20.2

Recent trends in the Use of some fuels in the UK

| Use of fuel | Consumption in Mtce in 1979 | 1981 | % change |
|---|---|---|---|
| *Transport fuel* | | | |
| Gasoline | 28.6 | 28.6 | 0 |
| DERV | 9.3 | 8.5 | −9 |
| Aviation fuel | 7.1 | 6.9 | −3 |
| Total | 45.0 | 44.0 | −2 |
| *Electricity supply* | | | |
| Nuclear | 12.3 | 12.2 | −1 |
| Coal | 88.8 | 87.2 | −2 |
| Oil | 18.1 | 8.7 | −52 |
| Total | 119.2 | 108.1 | −9 |
| *Heat supply* | | | |
| Natural gas | 17.3 | 17.2 | −1 |
| Coal in Industry | 9.2 | 7.0 | −24 |
| Domestic coal | 8.9 | 7.0 | −21 |
| Gas oil | 21.1 | 17.7 | −16 |
| Fuel oil (not including power station use) | 23.9 | 15.3 | −36 |
| Total | 80.4 | 64.2 | −20 |

*Notes:*
(1) Data from Department of Energy Bulletin *Energy trends* May 1982.
(2) Oil and gas have been converted from Mt oil and G therms to Mt coal equivalent by multiplying by 1.5 and 3.6 respectively.
(3) Parts of the heat supply market, for example, the steel industry have been omitted.

endowed with possibilities for electricity generation, both conventional and renewable. We have large coal reserves and substantial experience of nuclear power, the two energy sources from which we generate most of the 260 TWh per year we currently use. Among renewable electricity sources we have the best tidal power sites in Europe from which we could generate 40 TWh at costs in the range 3-6 p/kWh. Roughly one-third of the total tidal resource is associated with the proposed Severn inner barrage for which estimated costs are around 3 p/kWh.

The coastal parts of Britain and North Western Europe are quite windy and on- and off-shore wind generators are real possibilities, offering Britain as much as 200 TWh/year in total. We face the windy north Atlantic and therefore have a sizeable wave energy resource comparable with that for wind. We are having very good success with research on the use of geothermal hot rocks as a heat source for small thermal power plants and the present resource estimate there is of the order of 20-30 TWh/year. We do not see solar photovoltaic generators, solar power towers or ocean thermal gradients as relevant to North Western Europe, but could still, if we had to double our electricity production without using more coal or nuclear stations — but at a substantial price.

It is worth considering those factors which influence the cost of electricity from renewable sources and those which influence the value of the product to the electricity supply system. The comparaison between the two will determine cost effectiveness but neither cost nor value are quite as straightforward for the electricity producing renewables as first appears.

The capital cost of the generator is quite clearly a crucial factor and the renewables face their biggest difficulty in the relatively low power density with which they have to work. The average linear energy density in the waves off our western coasts is of the order of 50 kW m$^{-1}$ and the cross-section of the generator would be several hundred square metres, so that the energy density is only 200-300 W m$^{-3}$. A 100 m diameter wind generator would be rated typically at 5 MW, also 50 kW m$^{-1}$ in linear terms but perhaps 5 kW m$^{-3}$ in volume terms. The designers of wind and wave generators have been very clever in seeking to reduce the weight and hence the cost of their designs. The Boeing 'Mod 2' aerogenerators (Fig. 20.13) are an example of what can be done; they have a light tower which has a fundamental vibration frequency lower than the rotor operating frequency, that is, the system is vibrationally supercritical. They also have a low compliance, hinged rotor mounting and a teetering hub. Boeing claim that in sizeable batch production the Mod 2 generators could cost as little as £1.5 million, and in a United Kingdom lowland site they might generate 7 GWh/year.

However, as Table 20.3 indicates there are other important factors in determining costs whichare likely to be crucial for wind and wave generators, namely the operation and maintenance charges ($m$) and the possibly limited lifetime ($l$) of the highly stressed and heavily vibrated structures. Figure 20.14

Fig. 20.13 – The Boeing 'Mod 2' aerogenerator.

### Table 20.3

A simple expression for the cost of electricity supplied from a wind or wave generator

---

*Renewable electricity generation*

Cost of electricity =

$$\frac{c}{d}\left[m + \frac{r}{1 - (1+r)^{-l}} \times \frac{(1+r)^b - 1}{br}\right]$$

where

$c$ = generator and transmission system capital cost

$d$ = quantity of electricity delivered annually

$m$ = cost of maintenance, as a fraction of the capital cost per year

$r$ = required real rate of return (usually 0.05 per year, that is 5%)

$l$ = generator lifetime

$b$ = time to build and connect the generator

---

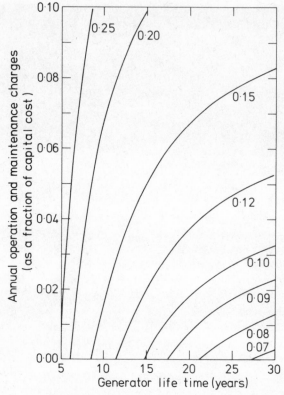

COST OF ENERGY FROM WIND GENERATORS
Given as a fraction of capital cost divided
by average annual output: Assumes 2 year
build time, 5% real rate of return on capital

Fig. 20.14 – Contours of annual rate of charge for renewable electricity generators
as a function of generator lifetime and operation and maintenance charges.

shows the contours of the factor inside the brackets of Table 20.3 plotted as a function of $m$ and $l$. Experience with aircraft suggests that $m$ could be anywhere between 0.5% and 10% and we currently do not know how long wind and wave generators will survive the pounding they will take. Using $m = 0.07$/year and $l = 10$ years implies an energy cost about 2½ times as high as when $m = 0.01$/year and $l = 25$ years, so these parameters are really just as critical as capital cost and output. For the optimistic numbers given above, the cost of electricity would be only 1.7 p/kWh.

The value to an electricity supply system of a statistically fluctuating source like a wind or wave generator is often misunderstood, in the sense that it is often assumed that the variable nature means that a specific back-up supply or a storage system will be needed. That this is not so can perhaps best be seen by noting that a statistically fluctuating renewable generator is formally no different from an extra fluctuating demand, negative in sign but presenting the supply system with qualitatively the same tasks as does an ordinary fluctuating demand. A sudden drop in the wind would be the same as the end of the Coronation Street programme on British television. Nor is it true that no capacity credit can be claimed for variable electricity-producing renewables; in fact it is fairly easy to show that adding a small amount of a variable generator allows the system safely to meet a simultaneous maximum demand which is larger by the mean output of the new generator. This capacity credit does fall fairly rapidly as relatively large amounts of fluctuating renewable are fed into the supply system but initially it is significant.

It is not even true that the renewables and nuclear power would be competitors in most national supply systems at present. Both would be used when available to reduce oil and coal burn in conventional thermal stations and the main component of their value would be the value of the coal or oil which might otherwise be burned. If the electricity supply systems became predominantly nuclear or contained very large renewable contributions (although this would probably be difficult to handle for operational reasons), then the two would compete and reduce each other's value to the supply system, but we are a very long way from that situation at present. In the United Kingdom current fuel costs in coal-burning power stations are about 1.6 p/kWh so that, to a first approximation, the value of wind-generated electricity and the most optimistic estimates of its costs are comparable. There are many other complicating factors such as the extra cost of substantial load following in coal-burning power stations and the doubtful environmental acceptability of on-shore wind generators but the overall message is clear, if, for on-shore wind generators, the cost, maintenance and reliability targets can be achieved, then it will be a serious contender for some bulk electricity generation.


## CONCLUSIONS

Renewable energy sources probably do have a part to play in the energy future of North Western Europe, almost certainly through the use of industrial and perhaps domestic waste; probably through the supply of geothermal heat and the passive solar design of buildings; possibly through electricity generation from winds and tides. Their use will help to limit the rise in energy prices which will have to occur if the renewables are to find widespread application in North Western Europe, but I cannot see their making a contribution big enough to affect any of the main energy markets substantially before well into the twenty-first century.

# 21

# Biotechnology: The energy balance

**P. P. King,** General Secretary, Society of Chemical Industry

Biotechnology has been proposed as a possible solution to the energy and feed-stocks problems which will arise for the chemical industry as the availability of cheap oil declines. Without entering into an argument as to when the decline in oil supply will take place, one can, in advance, make some general statements about what biotechnology will be able to contribute when its time comes. It is my contention that the help we can expect from biotechnology in this context will be limited because of the underlying characteristics of biological processes. Without exception they are much less intense than those we currently employ in the petrochemical industry. The 'process streams' are aqueous and dilute; the range of reaction temperatures is strictly limited. As an example, we will look at the production of single cell protein by growing *Methylophilus methylotrophus* on methanol [21.1]. Figure 21.1 shows the organism, a simple aerobic bacterium which grows in water with methanol as its sole carbon and energy source, plus a few inorganic salts. Figure 21.2 is a familiar photograph of ICI's $80 million 'Pruteen' Plant at Billingham from which some points about the practical use of a microorganism to synthesize protein are immediately obvious. Those engaged on the project were proud of designing, building, erecting and commissioning the largest fermenter in the world with a volume of 2000 m³. It is placed in the centre of the plant and is so large because of the dilute nature of the culture which is continuously grown in it. Microbiologists speak with pride of cell densities of 30 or 40 or 50 grams per litre; the chemical engineer groans when he hears of a product stream with a concentration of only 3 or 4 or 5%. Because the product stream from the fermenter is so dilute, the capital cost of the downstream equipment for dewatering and drying is correspondingly high – actually a good deal higher than the cost of the fermenter itself. Apart from the capital cost of product recovery there are the running costs – by no means negligible since considerable quantities of energy are inevitably consumed.

To the right of the plant is a large cooling tower which illustrates another problem in biotechnology. The temperature in the fermenter is around 40°C.

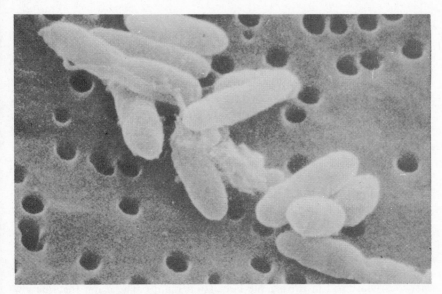

Fig. 21.1 – *Methylophilus methylotrophus.*

Fig. 21.2 – ICI 'Pruteen' Plant.

The growth of cells in the fermenter is really no more than a series of chemical reactions which, as one would expect from elementary thermodynamics, are less than 100% efficient. This inefficiency reveals itself as an exothermic heat of reaction. We should not blame the bugs; all living creatures give out waste heat. The problem is that heat at 40°C is of no practical value. It cannot economically be turned into work and must be dissipated in something like a cooling tower. If only fermentations could be run at 200–300°C, the waste heat could be used to raise steam and generate electricity. Unfortunately, no one has yet found a living creature capable of working at these temperatures although, of course, industrial catalysts frequently do. This means that if a conventional petro-chemical process liberates heat it can be recovered in a variety of ways and the penalty of thermodynamic imperfection can be mitigated. In the 'Pruteen' process a significant fraction of the methanol fed to the fermenter appears as carbon dioxide in the exhaust gas. If the heat of composition of that methanol was released at the adiabatic flame temperature of approximately 2000°C it would have real value; at around 40°C it actually incurs cost. The cooling system required to dissipate low-grade heat is expensive in capital terms and electrical energy is consumed in driving the circulation pumps. The claim that the low temperatures of biological processes are in some way advantageous in energy terms is misplaced. If any process is exothermic, a low reaction temperature is a disadvantage.

One way of looking at the low intensity of biological processes is to consider adenosine triphosphate (ATP) as an energy store. ATP is the universal energy carrier of all living creatures. When hydrolysed to adenosine diphosphate (ADP) and inorganic phosphate, it liberates approximately 40 KJ/g mol which is used to drive a huge variety of biological functions ranging from muscle contraction to the rear lights of fireflies. ATP is regenerated from ADP and inorganic phosphate in an equally wide variety of chemical reactions including photo-synthesis and the oxidation of sugars. Given its importance, it is surprising that so few people can quote off-hand its molecular weight, which is what determines the energy density of ATP.

If we had a motor car that in some hypothetical way could be charged with ATP to provide the motive power, with the ADP being subsequently regenerated at a suitable charging station, we would be deeply dissatisfied with the product. It would be even more cumbersome than the battery-operated milk floats which deliver milk door-to-door in England. As Table 21.1 shows, the energy density of ATP is only just over half that of the conventional lead–acid battery, which in turn is feeble when compared with petrol. The biotechnological equivalent of the motor car would be a very slow donkey.

The second characteristic of biological processes is that they are usually inefficient in energy terms. This is exemplified by nitrogen fixation in *Rhizobia* living symbiotically in the root nodules of leguminous plants. A number of studies have shown that when biological fixation of nitrogen occurs, 6 g of carbon

### Table 21.1

ATP as an energy store

| | kJ/kg |
|---|---|
| ATP $\longrightarrow$ ADP + P$_i$ (mol. wt. 525) | $\sim 80$ |
| Lead–Acid battery | $\sim 150$ |
| Petrol | $\sim 45\,000$ |

### Table 21.2

Energy requirements of nitrogen fixation

| | MJ/kg |
|---|---|
| NH$_3$ from air and water | $\Delta$H $\sim 23$ |
| NH$_3$ via Haber–Bosch, actual | $\sim 30$ |
| Fertiliser nitrogen as NH$_4$NO$_3$ on field, total | $\sim 60$ |
| Rhizobial N energy consumption (6 g C respired per 1g N fixed) | $\sim 235$ |

from carbohydrate are respired for every 1 g of nitrogen fixed. Thermodynamically it is not in the same league as the Haber–Bosch process (Table 21.2). The $\Delta$H of formation of ammonia from nitrogen in the atmosphere and hydrogen in water is 23 MJ/kg. That is the energy that would be released if ammonia was burned as a fuel to nitrogen and water. With the latest designs, the actual consumption in a modern ammonia plant is around 30 MJ/kg – an energy efficiency of better than 75%. It is perhaps worth pointing out that in every known case of biological nitrogen fixation, ammonia is the product first made, so we are comparing like with like. By the time we have converted Haber–Bosch ammonia to a typical solid fertiliser such as ammonium nitrate and allowed for all the extra energy in building the plants, packaging, transport and spreading the product on the land, the total energy consumed is still only 60 MJ/kg nitrogen. Compared with this, *Rhizobia* energy consumption is at least 235 MJ/kg nitrogen. Not surprisingly, crops with lodgers that fix nitrogen in their roots, suffer accordingly.

In one fascinating piece of experimental work by G. J. A. Ryle *et al.* [21.2], this has been elegantly demonstrated. He took three legumes, soyabean, cowpea and white clover, and grew each in two sets of conditions. In the first case he carefully ensured that the young growing plants were kept in aseptic conditions so that they could not become infected with *Rhizobia*. He fed them with nitrate nitrogen and recorded the yield as measured by shoot weight. In the second case, he innoculated the soil surrounding the young plants with *Rhizobia,* allowed them to grow normally without nitrate nitrogen and again recorded shoot weight. The yield depression suffered by legumes fixing their own nitrogen as compared with those which were germ-free and nitrate-fed, was formidable as Table 21.3 shows. One should be careful in interpreting these results because under field conditions legumes can and do take up some of their nitrogen as nitrate from the soil particularly during the early stages of growth. Ryle's figures are for a set of well-defined but artificial conditions. Having said that, they are nevertheless a solemn warning to those biotechnologists who seek a nitrogen-

**Table 21.3**

Yield depression in legumes fixing nitrogen

|              | Yield depression % |
| ------------ | :----------------: |
| Soyabean     | 70                 |
| Cowpea       | 45                 |
| White clover | 60                 |

Fig. 21.3 – Ammonia plant.

fixing wheat. I hope for their sakes that they never find it. With the depressed yields they will record, they will never sell the seed.

Some people may argue that this is being unfair; that synthetic ammonia made in a plant such as that shown in Fig. 21.3 depends on fossil energy whereas biological nitrogen fixation effectively uses solar energy. However, even if the supply of fossil energy was severely limited, it would still be advantageous in energy terms to synthesise ammonia and apply it as a fertiliser to cereal crops. Figure 21.4 shows the response of wheat to fertiliser nitrogen. It is an average of

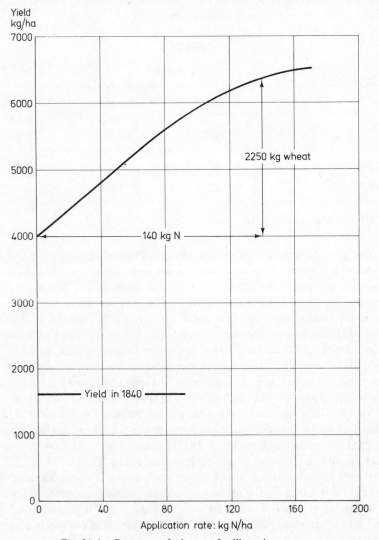

Fig. 21.4 – Response of wheat to fertiliser nitrogen.

a number of trials over the years on not particularly good soil, as is seen from the absolute magnitude of the yields. For an application rate of 140 kg of nitrogen per hectare we obtain, on average, an increased yield of 2250 kg of wheat. If we look at this on an energy accounting basis the conclusion is clear as Table 21.4 shows. The energy content of the fertiliser nitrogen at 140 kg/ha and 60 MJ/kg is 8.4 GJ/ha; the energy content of the extra wheat at 2250 kg/ha and 15 MJ/kg is approximately 34 GJ/ha. The energy-output to energy-input ratio is 4.0. This means that if we were desperately short of oil and natural gas it would pay us, in energy terms, to use hydrocarbons to make ammonia, fertilise the wheat and burn the extra grain in a power station!

Table 21.4

Energy account

|  | Rate kg/ha | Energy per unit MJ/kg | Energy content GJ/ha |
|---|---|---|---|
| Fertiliser nitrogen | 140 | 60 | 8.4 |
| Extra wheat | 2250 | 15 | 34 |

Output/Input ratio = 4.0

The idea of burning food to raise electricity disturbs some people and as an alternative it has been suggested that in some unspecified way we might use the straw, not the grain, for energy, that is, use straw as the feedstock for an ammonia plant. It is an intriguing thought. So far as I know there is no established process for making ammonia from straw but one would presumably start with partial oxidation to produce carbon monoxide. For the purposes of argument I would consider straw to be a bulky, low-grade version of coal. As Table 21.5 shows, the overall agronomic and thermodynamic parameters can be met. In order to set tolerable limits to transport costs I have postulated a 10 miles radius circle planted, year in year out, wall-to-wall with wheat. From the straw yield and its calorific value we have a total energy available of 3 million GJ/year. Taking 140 kg/ha as a reasonable nitrogen application rate in Europe, we arrive at a total ammonia requirement of 14 000 tonnes a year. This means that the energy available to make each tonne of ammonia is 215 GJ.

A modern Haber–Bosch plant running on natural gas, would consume about 30 GJ/te. Coal-based ammonia plants used to consume about 70 GJ/te. The 100 GJ/te of ammonia from straw is only a guess but it suggests that there would be enough energy available even if the 10 mile circle was partly planted in break crops.

### Table 21.5

Ammonia from straw

| | |
|---|---|
| Area of 10 mile radius circle | 80 000 ha |
| Straw yield | 2.5 tonne/ha |
| Straw production | 200 000 tonne/year |
| Calorific value of straw | 15 GJ/tonne |
| Total energy available | 3 million GL/year |
| Fertiliser nitrogen application rate | 140 kg/ha |
| Total nitrogen requirement | 11 000 tonnes/year |
| Total nitrogen, as ammonia | 14 000 tonnes/year |
| Energy available per tonne ammonia | 215 GJ |
| Energy required per tonne ammonia $\approx$ | 100 GJ |

The problem with making ammonia from straw (even if it was technically feasible) arises from the economics of the operation. The plant would be a very small one, making approximately 40 tonnes a day, compared with 1200 tonnes a day typical of a modern gas-based plant. As Table 21.6 shows, this leads to very high finance and fixed costs per tonne of ammonia. The capital cost of the plant has been estimated by applying the well-known two-thirds power rule of the chemical engineer starting from a full-scale plant. The estimate clearly, therefore, contains a wide range of error. An allowance has been made for the costs of installing refrigerated storage for, say, 10 000 tonnes of liquid ammonia which would subsequently be dissolved in water for application to the land. This is necessary because the ammonia plant would have to run continuously throughout the year, whereas wheat's demand for fertiliser nitrogen is highly seasonal. I have made no charge for the straw as such. When one compares the total cost of $900/tonne with current commercial prices of around $400/tonne

### Table 21.6

Cost of ammonia from straw

| | | |
|---|---|---|
| Scale, tonne/day | 40 | |
| Capital Cost, $ million | 60 | |
| Finance Cost†, $/te NH$_3$ | | 600 |
| Fixed Costs, $/te NH$_3$ | | 200 |
| Services, catalysts, straw-handling, etc. | | 100 |
| Total | | $900/tonne |

† 2 years construction, 20-year life, 10% DCF return.

delivered to the farm and applied in aqueous solution, one can see the proposal is killed by the small scale of the operation. If alternatively, one contemplated a 400 000 tonne/year ammonia plant fed with straw, it would consume approximately 3 million tonnes of straw a year – a mind-boggling quantity which would have a volume of around 15 million $m^3$! This illustrates a general difficulty which would face any large-scale biotechnological venture which depended on an agricultural waste product.

At the moment, biotechnological processes tend to incur high feedstock costs. The production of polyhydroxybutyrate (PHB) using *Alcaligenes eutrophus* is entirely feasible [21.3]. Figure 21.5 shows the formulae of the monomer and the polymer; $n$ typically lies between 5000 and 10 000.

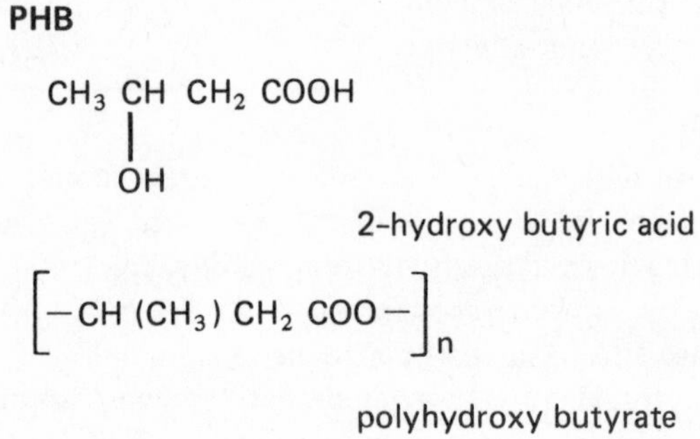

**PHB**

$$CH_3 \; CH \; CH_2 \; COOH$$
$$|$$
$$OH$$

2-hydroxy butyric acid

$$\left[ -CH\,(CH_3)\; CH_2 \; COO- \right]_n$$

polyhydroxy butyrate

Fig. 21.5 – Polyhydroxybutyrate.

Figure 21.6 is a transmission electron micrograph of cells of *A. eutrophus* which have been persuaded to produce 80% of their cell weight as PHB; the white granules of the polymer can be clearly seen. PHB has a number of interesting properties which are shown in Fig. 21.7 where it is compared with propylene homopolymer. PHB is a perfectly respectable thermoplastic which can be processed in conventional equipment. Unfortunately, it is quite uneconomic because of the high cost of the fermentation substrate, glucose, compared with that of naphtha, the starting point for conventional thermoplastics. This is disappointing when we see how the price of crude oil has moved relative to that of maize (Fig. 21.8). Relatively speaking, oil has become ten times as expensive as maize over the last ten years but that is still not enough to bring about a switch from oil.

One way of looking at the problem is in terms of the costs of energy. Naphtha, the oil fraction which is the raw material for most of the petrochemical industry, costs about $7/GJ. The cheapest glucose, in syrup, made from maize in the Midwest of the United States costs around \$20/GJ. Oil will have to become very much more expensive in real terms before it is displaced as a chemical feedstock.

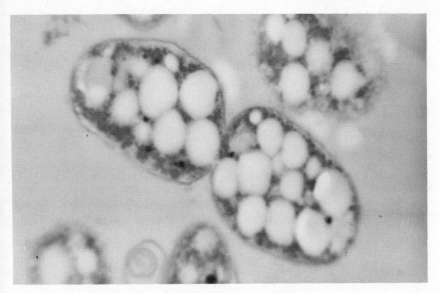

Fig. 21.6 – *Alcaligenes eutrophus.*

**PHB compared with Propylene Homopolymer**

| Property | Units | PHB | PP |
|---|---|---|---|
| Crystalline melting point | °C | 175 | 176 |
| Crystallinity | % | 80 | 70 |
| Molecular weight | Daltons | $5 \times 10^5$ | $2 \times 10^5$ |
| Glass transition temperature | °C | 15 | −10 |
| Density | g.cm$^{-3}$ | 1.250 | 0.905 |
| Flexural modulus | GPa | 4.0 | 1.7 |
| Tensile strength | MPa | 40 | 38 |
| Extension to break | % | 6 | 400 |
| UV resistance | | good | poor |
| Solvent resistance | | poor | good |

Fig. 21.7 – PHB compared with propylene homopolymer.

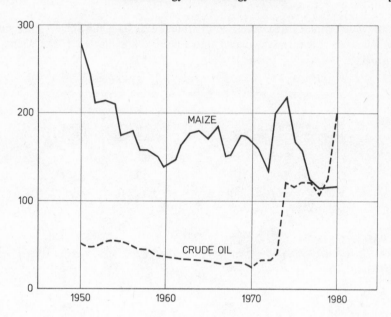

Fig. 21.8 – Price of crude oil and maize.

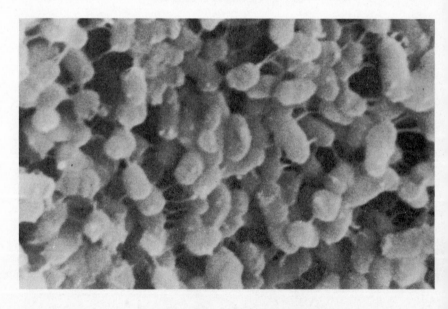

Fig. 21.9 – Immobilised *Arthrobacter*.

Biotechnology faces a final difficulty that is common to all novel technology — simply that it is novel. A good example of this is the fate in Europe of a product composed of immobilised *Arthrobacter* containing glucose isomerase, an enzyme which readily converts glucose to fructose. Figure 21.9 shows whole dead cells of *Arthrobacter* held in an open structure by strands of flocculating agent. This product would have permitted the low-cost production of 'isoglucose', an ideal sweetner for the confectionery and soft drinks industries. It was blocked politically in Europe although it is making great strides in the United States.

All this will have sounded deeply pessimistic but the question being addressed concerned the role of biotechnology in meeting Europe's future needs for energy and feedstocks — not biotechnology as such. I believe that biotechnology could well have a great future, but it will succeed according to the strength of demand for its products in the market place. They will need to be complex, ingenious and desirable. That is the way biotechnology will live up to its promise, not because of its contribution to the broader problems of energy and feedstock supply.

# REFERENCES

[21.1] Smith, S. R. L. (1980) Phil. Trans. R. Soc. Lond. **B290** 342-54.
[21.2] Ryle, G. J. A. *et al.* (1979) *J. Exptl Bot.* **30**.
[21.3] King, P. P. (1982) *J. Chem. Tech. & Biotech.* **32** 2-8.

# Index